Introduction to Intelligent Systems, Control, and Machine Learning using MATLAB

Dive into the foundations of intelligent systems, machine learning, and control with this hands-on, project-based introductory textbook, featuring:

- Precise, clear introductions to core topics in fuzzy logic, neural networks, optimization, deep learning, and machine learning, avoiding the use of complex mathematical proofs, and supported by over 70 examples.
- Modular chapters built around a consistent learning framework, enabling tailored course offerings in intelligent systems, controls, and machine learning.
- Over 180 open-ended review questions to support self-review and class discussion, and over 120 end-of-chapter problems to cement student understanding.
- Over 20 hands-on Arduino assignments connecting theory to practice, supported by downloadable MATLAB and Simulink code.
- Comprehensive appendices reviewing the fundamentals of modern control, and practical information for implementing hands-on assignments using MATLAB, Simulink, and Arduino.

Accompanied by solutions for instructors, this is the ideal guide for senior undergraduate and graduate engineering students, and professional engineers, looking for an engaging and practical introduction to the field.

Marco P. Schoen is Professor of Mechanical Engineering at Idaho State University, where he is the Director of the Measurement and Control Engineering Research Center. He is a Member of ASME, a Senior Member of IEEE and AIAA, and the recipient of the 2019 ISU Distinguished Teacher Award.

Introduction to Intelligent Systems, Control, and Machine Learning using MATLAB

Marco P. Schoen

Idaho State University

CAMBRIDGE
UNIVERSITY PRESS

CAMBRIDGE
UNIVERSITY PRESS

Shaftesbury Road, Cambridge CB2 8EA, United Kingdom

One Liberty Plaza, 20th Floor, New York, NY 10006, USA

477 Williamstown Road, Port Melbourne, VIC 3207, Australia

314–321, 3rd Floor, Plot 3, Splendor Forum, Jasola District Centre, New Delhi – 110025, India

103 Penang Road, #05–06/07, Visioncrest Commercial, Singapore 238467

Cambridge University Press is part of Cambridge University Press & Assessment, a department of the University of Cambridge.

We share the University's mission to contribute to society through the pursuit of education, learning and research at the highest international levels of excellence.

www.cambridge.org
Information on this title: www.cambridge.org/highereducation/isbn/9781316518250

DOI: 10.1017/9781009004992

First published 2024

Printed in the United Kingdom by CPI Group Ltd, Croydon CR0 4YY

A catalogue record for this publication is available from the British Library.

Library of Congress Cataloging-in-Publication Data
Names: Schoen, Marco P., author.
Title: Introduction to intelligent systems, control, and machine
learning using MATLAB / Marco P. Schoen.
Description: Cambridge ; New York, NY : Cambridge University Press, 2024. |
Includes bibliographical references and index.
Identifiers: LCCN 2023023920 (print) | LCCN 2023023921 (ebook) |
ISBN 9781316518250 (hardback) | ISBN 9781009004992 (ebook)
Subjects: LCSH: Intelligent control systems – Mathematical models. | Machine
learning – Mathematical models. | MATLAB.
Classification: LCC TJ217.5 .S36 2024 (print) | LCC TJ217.5 (ebook) |
DDC 510.285/536–dc23/eng/20230729
LC record available at https://lccn.loc.gov/2023023920
LC ebook record available at https://lccn.loc.gov/2023023921

ISBN 978-1-316-51825-0 Hardback

This book is dedicated to
my mother and father,
Georgette and Friedrich Schoen,
who always inspired and supported me
&
my family,
Renata, Mathias, and Natalie Schoen,
who made this book project possible

Contents

Preface

This book is for engineering students and practicing engineers interested in intelligent systems, machine learning, and their application to controls and modeling. It offers insight into the underlying mathematics and presents solutions in terms of programming using MATLAB and Simulink. As the fields of machine learning and controls are vast and continuously expanding, the book provides for the understanding of the key concepts along with practice material to allow for learning and mastery of many of these data-driven methods and algorithms. The individual chapters are constructed such that the depth of the material is modularly built with the progression of the chapter, allowing the reader or course instructor to include or leave out certain topics toward the end of a chapter if necessary. In addition, courses in intelligent systems and controls can be constructed by assembling different chapters from the book to provide emphasis and nuances that allow for tailoring the course to the audience. Examples throughout each chapter are presented with solutions as well as MATLAB and Simulink code and programs. The reader is also challenged by a set of review questions and end-of-chapter problems.

At first glance we may not be aware of the presence of automatic control when engaging with everyday activities such as driving a car, cooking dinner, or surfing the internet. The comportment of controls is only possible through its capability of perceiving and reacting to changes in the world around our activities. Despite its inconspicuous presence, control has been a major factor in improving efficiency, reliability, and safety in our daily lives. For much of its history, the development of automatic control algorithms has relied on rather precise mathematical formulations of the environment it is supposed to control. The derivation of these descriptions often depends on a first-principles approach: utilizing the underlying physics along with the mathematical relations that characterize these processes and expressing these relations as compact dynamical models. Much of the corresponding analysis, in particular the investigations into the resulting stability of the controlled system, has been based on these derived mathematical models. A competing approach to building controllers based on derived dynamic models is the "model-free" approach. *Model-free* refers to the fact that we neglect to consider some or all of the underlying physical principles when developing the characterization of the process or system we control. There are a number of model-free approaches, both statistical and deterministic. However, neural networks and machine learning approaches have garnered much popularity in recent years as alternative model-free modeling tools. The popularity is partially due to some of the recent advancements made in deep learning, along with the availability of increased computational resources and computational capabilities. The progress in the development of these intelligent methods manifests itself in a plethora of available computer code, often using the Python programming language.

IT departments, academia, and data and computer scientists alike are not only served by this infrastructure, but actively built on it to allow the steady advancement in intelligent systems to permeate into a broad field of applications. Nonetheless, we have yet to see these model-free data-driven intelligent systems conquer a number of applications that are under the supervision of regulatory committees, such as specific flight, medical, and food safety applications. Proven and reliable methods such as PID and model predictive controllers are still the standard in many of these industries. Engineers are trained to develop, design, tune, and implement these standard controller architectures using common tools and skills acquired during their educational journey. Although Python programming has made its way into some of the current engineering curricula, control and control design is still very much done using MATLAB and Simulink. One of the reasons for the adherence to these programs is the very rich set of control design and analysis tools, including machine learning and deep learning apps, that they provide.

Is This Book for Me?

This book is primarily for engineering students in all disciplines at the advanced undergraduate or graduate level who are taking courses on intelligent control systems, machine learning, and advanced control systems. It is also suitable for information science, business, and psychology majors, among others. Each topic included in this book is clearly derived, explained, and presented in a gentle way while refraining from the use of mathematical proofs. Prior exposure to MATLAB and Simulink is a plus, but is not considered necessary. Special appendices provide tutorials on both the program environments of MATLAB and Simulink and their general use beyond control-related applications. For readers lacking a formal controls background, Appendix A details a tutorial on modern control systems, including a brief introduction and review of the major topics that a traditional controls course would cover. The use of MATLAB is emphasized in this tutorial in preparation for the primary chapters in this book.

Each topic in each chapter provides examples how to implement the concepts presented using MATLAB and Simulink. There is no emphasis on using or generating an efficient and compact MATLAB code due to the aim to remain generally readable to novice users of these programming environments. Many routines and programs in MATLAB presented in the book do not require more than a few toolboxes. However, the later chapters benefit from having access to

MATLAB apps such as the Deep Network Designer, Reinforcement Learning Designer, Classification Learner, Fuzzy Logic Designer, and Regression Learner App.

How Is the Book Structured?

The book introduces fuzzy logic-based methods and algorithms first, followed by neural networks and their use in modeling. Both topics are presented in a way that consideration of controls can be included or left out. As optimization is an integral part of many of the data-driven learning methods, heuristic as well as iterative optimization methods are detailed in separate chapters prior to their use in the respective discussions of machine learning algorithms. The later chapters emphasize deep learning concepts as well as pure machine learning algorithms. A project-based and hands-on class, or even laboratory section accompanying a traditional class, can be offered by making use of the project information as well as the microcontroller programming material described in the appendices.

How Can I Use This Book for Teaching a Class in Intelligent Systems and Controls?

The organization of this book is kept deliberately so that one can construct different types of courses. Hence, the book can be used to offer courses serving diverse interests and a range of student demographics, including graduate and undergraduate courses, as well as different disciplines in engineering and other science programs. In the following, some sample topic outlines are provided with different course emphases.

An undergraduate course in controls and intelligent systems may include:

- Appendix A – overview of modern control systems using MATLAB;
- Chapter 1 – introduction to intelligent systems, controls, and machine learning;
- Chapter 2 – principles of fuzzy logic;
- Chapter 3 – fuzzy inference systems, leaving out Section 3.4;
- Chapter 4 – optimization;
- Chapter 6 – fundamentals of neural networks;
- Chapter 7 – control design using neuro-fuzzy systems;
- Chapter 9 – introduction to machine learning concepts.

A graduate course in controls and intelligent systems can be offered by selecting the following chapters as part of the list of topics for the course:

- Chapter 2 – principles of fuzzy logic;
- Chapter 3 – fuzzy inference systems;
- Chapter 5 – intelligent optimization algorithms;
- Chapter 6 – fundamentals of neural networks;
- Chapter 7 – control design using neuro-fuzzy systems;
- Chapter 8 – deep learning;
- Chapter 9 – introduction to machine learning and reinforcement learning.

An undergraduate course that is more focused on intelligent systems than controls can be constructed by addressing the topics of the following chapters:

- Chapter 1 – introduction to intelligent systems, controls, and machine learning;
- Chapter 2 – principles of fuzzy logic;
- Chapter 3 – fuzzy inference systems, leaving out Section 3.5;
- Chapter 4 – optimization;
- Chapter 5 – intelligent optimization algorithms;
- Chapter 6 – fundamentals of neural networks;
- Chapter 8 – deep learning;
- Chapter 9 – introduction to machine learning concepts.

In addition to the various combinations of chapters to construct a course for a specific target audience, the book also includes a project for implementing some of the controls topics on an embedded system using simple components for physical demonstration and testing. The appendices provide for the necessary details on the project realization, including specifics on Arduino microcontrollers and how to program these devices using Simulink. Implementing a laboratory section for hands-on work that accompanies the lecture material is possible with the given project descriptions and implementation instructions.

What Are the Features of This Book?

Having taught intelligent control systems for a number of years, I have found that current resources that cover the range of topics entailed in the present book are usually distributed across many different books and separate tutorials, as well as online articles. Many methods and algorithms are also available online as Python code. One of the goals of this book is to combine systematically the various intelligent systems topics into one volume. In addition, an emphasis is placed on providing MATLAB-based solutions throughout the different chapters.

Having seen the benefits of incorporating hands-on projects into courses, this book allows for practicing the theoretical material on assignments that are based on a simple physical setup. These projects only involve simple physical systems and employ readily available embedded system platforms. Hands-on or project-based components in a class or laboratory setting not only allow for a "reinforcement" learning environment, but may also provide for inspiration to expand on the material covered and to include other domain-specific projects and topics. Some of the other features of this book are:

- A consistent presentation of theory and practice: All topics are developed in a precise and clear manner, supported by examples throughout the book, supplemented with end-of-chapter problems, questionnaires, and summaries.
- The book is composed in a modular fashion, allowing for tailored course offerings in intelligent systems, controls, and machine learning.
- Many chapters include assignments that involve the practical implementation of the theory covered using a simple physical setup, a common microcontroller, and Simulink.
- The book provides for a solid introduction to machine learning using MATLAB. Included in this introduction are common classification and regression methods, as well as image processing algorithms for object identification and guidance and control applications.
- A section detailing the concept of reinforcement learning is included in Chapter 9. Although this chapter builds on prior chapters in the book, it presents a clear and easy introduction to reinforcement learning and reinforcement learning control using the appropriate MATLAB app.
- Instructors will have access to a set of PowerPoint slides covering all the images in the book. In addition, instructors can utilize full-length syllabi for some versions detailed above of a course in intelligent control systems and machine learning.
- Instructors will have access to a solutions manual. Students and instructors will have access to a companion webpage that holds a repository of all the MATLAB and Simulink code used throughout the book, as well as many of the appendices.
- For courses that build on a curriculum missing the traditional modern controls course, or students who need a review of modern controls, the book includes an entire appendix with the main concept of modern controls using MATLAB.
- Students and practitioners who are new or need a refresher in the use of MATLAB and/or Simulink have separate appendices covering a range of topics to allow them to become sufficient in the use of MATLAB and MATLAB programming, as well as Simulink programming.

1 What Is an Intelligent System?

1.1 Introduction

As we will work with intelligent systems and controls throughout this book, it might be rather beneficial to attempt answering the question posed in the title of this introductory chapter. Although a clear definition is still under debate, the word *intelligence* has commonly been used to express some association with the capacity to understand, the ability to reason with logic, and the facility to acquire and apply knowledge. From a broader perspective, an intelligent system utilizes these traits to create some level of self-awareness. For such a system, this type of cognition – however limited – is then utilized to interact with the system's environment. The objective of such interactions may involve altering the behavior of itself or its environment, hence performing some type of control task. When we combine intelligence with control, we are referring to a subset of control techniques that implements some of the intelligent system's capabilities to learn about the characteristics of its environment and to use this inference to achieve some predefined outcome.

Often, we find the term artificial intelligence (AI) associated with the capabilities listed above. The term *artificial* arises from specifically relying on man-made competences that can be categorized as intelligent. AI has the objective to mimic human behavior by incorporating human-like capabilities. This type of nonbiological intelligence is expressed through the use of mathematical rules. Some of the conceived AI systems belong to the field of machine learning (ML). ML comprises algorithms that attempt to learn mathematical rules automatically from acquired or observed data. It is found to be useful not only for methods and algorithms, but also in data analysis and analytics, in so-called big data, and in software tools. Another term that is linked with AI and ML is deep learning (DL). DL is generally considered a subfield of ML and to some degree also part of AI. DL particularly constitutes algorithms that are associated with multilayered neural networks.

Intelligent control techniques are especially convenient when dealing with incomplete or inadequate system representation. Their application also extends to situations where we deal with partial specification as well as uncertain environments. As there is missing information on both the target and the approach, learning becomes a key function in achieving the desired control objective. Hence, the learning component in such control systems emulates human intelligence. As we noted, a clear definition for intelligence is still deliberated upon, and so a clear definition of intelligent controls as a field is derived from and affected by this debate. An intelligent control system does leverage some of the advancements and algorithms developed for ML systems, such as neural networks, fuzzy inference systems, evolutionary algorithms, and combinations of any of these methods. In many applications of these methods the control design benefits from modeling a system based on data rather than on deriving governing equations utilizing physical laws, assumptions, and simplifications.

Intelligent systems are heavily dependent on computational resources. Hence, AI's development is rather tightly connected to any advancements achieved in computer technology. The early 1940s, when efforts were exerted to extract intelligence from coded communications during World War II, may serve as the origin of the development of modern computers. After World War II, in 1950, Alan Turing introduced the Turing test, which essentially provides for a means to determine if a computer can achieve the same level of thinking as a human. During the early 1950s computational statistics advanced to a level that allowed the inception of the field of ML. The generally accepted birthday of AI is traced back to 1956, when a small group of researchers attended the Dartmouth summer research project on AI. In particular, mathematics professor John McCarthy had proposed "to proceed on the basis of the conjecture that every aspect of learning or any other feature of intelligence can in principle be so precisely described that a machine can be made to simulate it" [1]. The original inception of AI was followed by a number of scientific contributions in the subsequent years. For example, in 1965, Lotfi Zadeh expanded the infinite-logic field by introducing fuzzy logic [2]. The 1960s also saw major developments in the fields of natural language processing and computer vision. Those contributions were an ideal playground to make major advancements in robotics during the 1970s. For example, the first so-called intelligent robot, the WABOT 1 introduced in 1972, was able to engage in simple communications and perceive its environment by the use of artificial ears and eyes [3]. The original enthusiasm for AI and the ensuing expectations resulted in some disappointment as AI's capabilities were very much curtailed by the limitations of the available hardware. Intelligent systems and, in particular, AI have experienced repeated waves of attention followed by periods of disillusionment, where limited progress was achieved. Such epochs of disappointments and resulting loss of research efforts are termed *AI winters*.

The period from 1975 to 1980 has been termed the first AI winter, while the second AI winter occurred between 1987 and 1993. However, singular achievements that caused public interest in this subject have repeatedly led to research activity and excitement in this field. For example, in 1997 an IBM supercomputer called Deep Blue competed against the world chess champion Garry Kasparov and defeated him in a six-game match [4]. Another such public exposure of advanced capabilities in intelligent systems that caused a revival of interest in AI was in 2015, when a neural network called AlphaGo beat the world's best Go player [5]. With the AlphaGo event the term *deep learning* was made popular beyond the research community. However, in between those times, intelligent systems invaded our homes in the form of smart appliances and device-based assistants. Today we are seeing AI capabilities embedded in all kinds of devices and systems. However, we are also dealing with some of the negative implications and challenges of intelligent systems. For example, we find ourselves debating liability and the legal ramifications of self-driving cars in the event of accidents, personal data and its analysis in consumer behavior, AI's role in bias and discrimination, and even legal personhood issues.

Although we introduce the underlying mathematical foundation of many of these intelligent system algorithms, a prominent topic throughout this book is the application of control systems using intelligent systems. Advancements in auto-mation and in particular control systems have been a result of a shared and ceaseless quest by intelligent systems researchers and controls engineers to design machines that exhibit complete autonomy. Guided by a persistent belief that such levels of autonomy can be achieved by drawing expertise and knowledge from different fields, computer scientists, engineers, and mathematicians have been collaborating and contributing to this goal. Often, inspired by processes and behaviors found in nature, contributions are also made by researchers well outside of these disciplines. This broad alliance of disciplines is also reflected in the application of intelligent controls. We find intelligent controls algorithms nowadays applied to robotics, propulsion, communications, security, transpor-tation, and logistics, to mention a few such applications.

1.2 Concepts in Machine Learning

Commonly, we categorize ML into three distinct groups: supervised learning, unsupervised learning, and reinforcement learning. When referring to supervi-sion we imply having access to data that has labels. An example of labeled data is a repository of images depicting different leaves from different trees. Each picture itself is a data point, and its association with a tree is its label. Data with labels allow us to perform classification or solve regression problems. Classification will permit us to assign the proper label to a new image of a leaf. A regression problem is when we are seeking a model that lets us predict its output value. An example of a prediction problem is the regression model of forecasting house prices based on features such as number of bedrooms, location, age, etc. The labeled data are the data available from the market (i.e., houses sold and their feature information).

Unsupervised learning handles data that does not have any association with labels. As there is no reference information or labels, unsupervised learning is used to perform clustering tasks, association analysis, and dimensionality reduction. Clustering is a function that authorizes us to group data into clusters. There is no label for the individual group or clusters, but the elements of each group share some distinct features. In ML we refer to a feature as a measurable characteristic. Features entail not only the characteristics of a data point, but also can be a descriptive attribute of the data point. In contrast to clustering, the association analysis refers to the process of finding prominent relations between features in a data set. An example of association analysis in practice is the process retailers perform on sales data to determine consumer behavior and preferences, which in turn can be used to drive promotion of certain products. The third possible application of unsupervised learning, the dimensionality reduction process, is a methodical approach to reduce the number of features in a data set. Generally, the more features a data set exhibits, the better an ML algorithm works for making distinctions between its data points. However, dealing with more features requires more computational resources. Hence, dimensionality reduction may help with the efficiency of an ML algorithm by supporting the elimination of features that contribute less than others to the clustering process.

Reinforcement learning (RL) refers to algorithms that employ a rewarding and penalizing mechanism to achieve a specific objective. The learning is subsidized by creating a reward when the algorithm moves in the right direction (i.e., approaches the stated goal). However, penalties are imposed when the opposite occurs. The rewards and penalties are a result of trials in the algorithm which is incrementally attempting to reach a stated goal. There are several types of RL methods: for example, the model-based and the model-free RL algorithms. The distinction is expressed by the type of implemented policy that the algorithm uses when deciding on rewards and penalties. When using model-based RL, the algorithm is interested in the underlying predictive model that defines the behavior of the system. This is achieved by considering the application of some type of intervention or input and then considering the results of the intervention, the next state of the system, and the immediate reward of the intervention. An example of model-based learning is dynamic programming. A model-free RL algorithm does not explicitly reference a model. Instead, it samples from experience or real observations from the environment rather than using predictions of the next state and next reward in order to change its behavior. An example of model-free RL is Monte Carlo control, Q-learning, or actor–critic learning algorithms. We will discuss those algorithms in more detail in Chapter 9.

Classification using supervised learning may entail methods such as the K-nearest neighbor, support vector machines (SVM), naïve Bayes classifier, decision trees, and random forests, to mention some of the more popular algorithms. The K-nearest neighbor algorithm labels an unlabeled data point based on the majority label of its K-nearest neighbors. The number K (i.e., the number of neighbors) has great influence on the resulting class label: a small value of K allows for a greater influence of noise on the results; a larger number of neighbors, and hence a higher K, increases the computational cost in making the classification determination. Additionally, the choice of how we compute the distance to determine the neighbors is also a choice for the ML designer. Common distances utilized are the Hamming distance, the Manhattan distance, and the Minkowski distance. Support vector machines utilize a hyperplane to divide data points into classes. In order to evaluate the hyperplane parameters, and hence determine class association, perpendicular distances of the nearest data points close to the hyperplane are computed and maximized. This optimization problem is essentially a computational approach to maximize the margin between two classes. There are also multiclass SVM algorithms, which we will treat in more detail in Chapter 9. As the name indicates, the naïve Bayes classifier algorithm is a probabilistic classifier. This algorithm is an adaptation of Bayes' theorem to make prediction based on a data point's probability. The naïve reference indicates that the algorithm is indifferent to any dependencies of the different features used (i.e., we assume that all features applied in the algorithm are independent of each other).

Decision trees are simple algorithms using supervision to partition the data into smaller subsets. The partitioning – or branching – is a result of binary outcomes to a set of rules or questions. Essentially, each branch node of the tree embodies a choice among a number of possibilities and each leaf node represents a decision. Ultimately, the branching will cover all possible outcomes of a decision. The resulting structure of the flow chart resembles a tree; hence the name decision trees. An extension to decision trees is the random forests algorithm. A random forest algorithm is essentially a combination of multiple decision trees, where each tree is responsible for one unique class. The classification is determined by a majority voting computation among all the different classes and trees.

Unsupervised learning methods entail algorithms such as K-means, hierarchical clustering, and principal component analysis (PCA), to mention a few. The K-means clustering algorithm refers to a process where we determine partitions of a set of data points into K groups, which are called clusters. Although similar in principle to the K-nearest neighbor algorithm, it does not have access to data labels, and hence uses a similarity property of the data to form clusters. The discriminator utilized for finding similarities is a geometric property that is found by assessing the nearest mean or centroid. This point becomes a characterizing representative of the cluster. Another method for clustering is the hierarchical clustering algorithm. Beginning with considering the entire data set being assigned to individual clusters for each data point, the two nearest clusters are merged into a new cluster, eliminating the original two clusters. This process is repeated until only a single cluster remains. There are different implementations of hierarchical clustering available, such as agglomerative hierarchical clustering and divisive hierarchical clustering. Principal component analysis is widely used in many disciplines, including in unsupervised ML. When we reference principal components we seek the most information of a data set by respecting the variance in the data. Here, the directions of the maximum information are the principal components. The process employs an orthogonal transformation to ensure that the variables of the process become uncorrelated.

1.3   Concepts in Deep Learning

Deep learning algorithms are closely associated with neural networks. The development of neural networks was inspired by observations of how the human brain functions. The human brain is literally composed of billions of neurons. A neuron is a nerve cell intertwined with other neurons and responsible for processing and transmitting chemical and electrical signals. From a biological perspective, dendrites are connections that obtain information from other neurons. Axons are used to transmit or send information, while a synapse is the connection between dendrites and axons. Processing of neural stimulation is handled by receiving multiple signals from different dendrites into the cell body. Once the accumulated signal meets or surpasses a threshold value, the cell body generates a signal that is conducted to other neurons by the corresponding axon. Since artificial neural networks are computer algorithms, we can use an analogy between the biological terms just introduced and the terms used to describe an artificial neural network. In this analogy, a biological neuron becomes an artificial neuron, while preserving the basic mathematical functioning of a biological neuron. A cell, also sometimes referred to as a soma, becomes a node, dendrites are inputs, and synapses become weights – which also function as interconnections. Finally, an axon is simply an output in the artificial neural network terminology.

The neuron in the artificial neural network not only contains the accumulation function but also houses a nonlinear activation function. This activation function supports the threshold operation, comparable to what the biological cell does. In an artificial neural network the inputs are weighted. The values of these weights are determined during the training – or learning – algorithm. Unlike humans, who have an astounding capability to learn using very few examples or data points, neural networks are much more in need of data and examples. For artificial neural networks backpropagation has shown to be one of the most effective learning algorithms. To apply this learning method we first process input data with the untrained network to generate an output, or prediction. This prediction is compared to the label associated with the input, also called ground truth. The comparison is the essence of supervision for training neural networks, as the error is now fed back to the network to adjust the weights and biases of the network. The adjustment mechanism is essentially an optimization algorithm. For this optimization, we minimize an error, which is defined by a loss function. Popular loss functions in use are the mean square error or the cross-entropy loss functions.

A common optimization algorithm employed for minimizing the loss function is the stochastic gradient descent algorithm. When we feed the error back to the network, we compute the gradients and use this information along with a learning rate to adjust the weights and bias terms of the network. Since we use a lot of data and many examples to train such a network, we use specialized vocabulary to quantify learning iterations. For example, one *epoch* implies that we conducted one iteration using the entire data set. A *batch* is a subset of the entire data set. Hence, if we have 100,000 data points and a batch size of 2,000, then an epoch should contain 100,000/2,000 = 50 iterations.

Unlike in real life, for neural networks there is something like "too much learning." This is called *overfitting*, and occurs when we train a network for a data set to a degree that it only works for this single data set and not for other data sets, rendering the network useless. A network that is trained on one data set and can function well on another data set – representing similar characteristics – is a network that is *generalizable*. To monitor the training – or learning – process, we use a second set of data that is usually generated by splitting the original data set into two: a training set and a validation set. At instances during the training we determine the error of the network using the validation data set. At the moment the error based on the validation data set bottoms out, we assume that the network is trained, and any further training – even if the training data set provides lower subsequent errors – results in overfitting.

Up to this point we have described regular artificial neural networks. However, based on the section heading, we should wonder what DL is. The *deep* in deep neural network, and similarly in DL, is a reference to the number of layers such a network is constructed of. Regular networks have an input layer, a hidden layer, and an output layer. For such regular neural networks the described learning process works well. However, if we add more hidden layers the meaning of the error term used in the backpropagation algorithm is lost, and training becomes ineffective. Networks composed of multiple hidden layers are called deep neural networks. A concept called *dropout* – which is a form of regularization – and the use of different activation functions allow these deep neural networks to be trained in a similar fashion as regular neural networks. This inclusion in the learning process is referred to as deep learning.

With a multitude of layers, other operations can now be included into such networks. One such structure attempts to achieve greater autonomy and independence of human interaction by automatically extracting the features of the input data using operations such as convolution and pooling. We provide a thorough review of these methods and computations in Chapter 8. Deep neural networks have found applications and provided solutions to problems that originate in face recognition, image classification, speech recognition, handwriting transcription, and medical diagnoses. Deep learning has been useful in decision-making, operation of equipment, including a range of different vehicles, social media, and – not to leave it out – the gaming industry.

1.4   Concepts in Intelligent Control

Control systems have been facilitating human activities for much of our species' existence. Seminal discoveries and inventions may have spurred new interests and areas of technologies utilizing controls or advancing controls itself. Nowadays, we find controls in every part of our lives, including in business, government, technology, and medicine, to mention a few. However, in this book, we deal with dynamical systems and its control in terms of the dynamical behavior of such systems. To describe the behavior we use mathematics, and hence the usage of control theory is dependent on employing this language. A driver for the application and the advancement of dynamic control system is the desire to reduce the necessary human work by increasing automation. Automation is dependent on controls engineering in order to perform automation tasks. The controller used in automation has the function of embedding some kind of strategic method and computation to achieve a specific outcome. Among control action outcomes, the stability of the system is one of the primary goals, especially if the system without control exhibits unstable behavior. However, the application of closed-loop control may also induce instability of the system. Hence, the study of stability is closely associated with control engineering. Many of the tools developed for these studies are dependent on precise system dynamic descriptions (i.e., mathematical equations characterizing the behavior of the closed-loop system). In some instances the acquisition of such descriptions is rather difficult or even impossible. For such situations, automatic inference of the system description through the control system has been used. These types of readings and extractions have found topical homes in system theories such as system identification, estimation, and adaptive control theory.

System identification is a process that collects input and output data from the operation of the system and uses these data to extract a system description, circumventing the utilization of physical principles to describe the system dynamics. Estimation theory makes ample use of statistics and allows for the estimation of system parameter values in stochastic environments. Adaptive control systems make use of estimation theory and system identification to adapt to changes in the system characteristics. As control systems depend on intervention (i.e., using actuators), the accounting of the exerted energy may be of interest, in particular for optimizing effort and costs. Such considerations lead to the field of optimal control theory.

Other control system disciplines address issues in the robustness of the controller in the presence of system uncertainties. Dealing with exclusively nonlinear system behavior has resulted in theories encapsulated as nonlinear control systems. Specialization of specific control applications has resulted in a wealth of different specific control fields, such as reconfigurable control systems, resilient control systems, digital controls, and multivariable control systems, to mention a few. However, in this book we specifically look at intelligent control systems. The intelligent part in this field of study may function as an aid in evaluating controller parameters, defining and adjusting the controller structure, estimating the controller or system parameters, finding optimality, or even as a tool to define the entire system description.

Intelligent control systems is a broad field. In this book we focus on a few important concepts of intelligent control systems. In particular, we study fuzzy logic and how to utilize fuzzy logic theory to construct controllers. We use intelligent optimization methods such as genetic algorithms and particle swarm optimization to find optimal controller parameters, and we employ ML to infer models that represent the system to be controlled.

Considering a typical fuzzy logic control structure (Figure 1.1), the main components of such a system are the fuzzifier, the fuzzy knowledge base, a fuzzy rule base, an inference engine, and the defuzzification unit. The fuzzifier has the assignment to convert so-called crisp signals into fuzzy quantities. A crisp quantity is a reference to a signal that can have a numeric value – for example, a voltage reading of 3.309 volts is a crisp quantity. The fuzzifier associates this type of number into a degree of membership to a class. The class could be "high," "low," etc. and hence a degree of membership is expressed as some percentage of the voltage to be characterized as high and as low. The fuzzy rule base is responsible for modeling the knowledge of the system using fuzzy sets. For

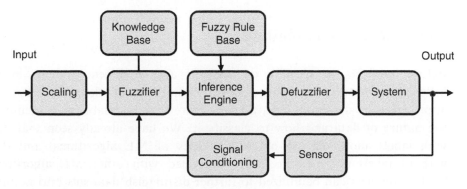

Figure 1.1 Fuzzy logic control system schematic.

example, the fuzzy rule base details the input and output relations using fuzzy relationship operators and membership functions. The inference engine embodies an abstract form of human decision-making by utilizing approximate reasoning, which is used to define the control strategy. Once the control action is decided upon, the decision – which is in the form of a fuzzy quantity – needs to be converted into a crisp output using the defuzzification process.

A system as depicted by Figure 1.1 can be hybridized with other intelligent control systems. For example, the fuzzification using membership functions to associate signals with linguistic variables can be optimized using heuristic methods such as tabu search, genetic algorithms, etc. A similar approach can be taken to find optimal rule sets for the inference engine. If simulations are required before the implementation, the system block is no longer a physical system, but a block representing the system dynamics. Such a block can be inferred using ML – for example, a trained neural network can be utilized to represent the physical system. Additionally, neural networks and fuzzy logic can be married to form neuro-fuzzy inference systems. An extension to such neuro-fuzzy inference systems is an adaptive neuro-fuzzy inference system where the inference system adapts to the changes of the system itself.

Extending the use of neural networks, and in particular deep neural networks, a different type of control system framework can be constructed: the RL controller or the reinforcement DL controllers (RL controllers). Such controllers represent rather powerful approaches to search-and-find-optimal controllers in the presence of nonlinearities, noise, and uncertainties. They can also be adaptive to system changes. RL controllers work sequentially by determining the impact of decisions made and control signals applied, which includes the consideration of future interactions. RL systems entail a reward function, which is used to evaluate and iteratively improve the policy, which determines the control action. A policy in RL is often given by a deep neural network, hence the learning component of the RL controller is responsible for optimizing the policy by adjusting the deep neural network using a cumulative assessment of the reward achieved.

1.5   Data, Signals, and Methods

Before indulging in specifics on controls and intelligent systems in the following chapters, we must recognize that all of these systems and approaches are driven by data. So, it is more than justifiable to contemplate the nature of data and how to classify it. We have already seen that data with labels allow the use of certain types of ML algorithms, and data without labels are open to being processed with other ML algorithms. Such differences can be utilized to further distinguish data sets into nominal versus ordinal data. Nominal data, from the Latin *nomen*, names or labels variables in the absence of any type of quantitative value. Our example for leaves associated to trees is a case of nominal data. Ordinal data has a structure or order that organizes itself by value, position, or scale. The English alphabet is an ordinal data set, as there is an order to the data.

We can also distinguish between structured data and unstructured data. Structured data are often housed in matrices or vectors or relational databases. Unstructured data are usually not configurable in rows and columns, or relational databases. An example of an unstructured data set is a repository of images or videos. Structure in data usually allows for smaller storage requirements. One last categorization option of data we should mention in this brief review: data stemming from stochastic or deterministic signals. In real-life applications we deal with signals that are contaminated by noise and interference. Those random contributions to our signals may originate from the measurement itself, or are part of the process. In simulations we often neglect the nature of this probabilistic quantity and work with deterministic signals. However, there are few instances in controls where we can make such an approximation unless other processes are engaged to allow for treating signals as deterministic. For example, using filters, in particular low-pass filters, for control applications may limit the effect of noise and interference on the operation of a control system.

Another form of filtering signals or data may occur by following the temptation to cherry-pick data. Often, we have preconceived notions on how the system and hence the data should behave or evaluate. However, this notion should not be applied by only using the data that fit this expectation as we may bias the outcomes of the intelligent system. If the data set is incomplete, we also may end up with a tainted outcome, commonly referred to as survivor bias, as the data have survived the applied selection criteria. Another man-made hazard for achieving unbiased results is gerrymandering of data. Gerrymandering is a result of purposely selecting groups of data sets to fit a preconceived outcome. We also caution about overfitting and creating a sampling bias. The latter is achieved by drawing conclusions from a data set that does not represent the population we are attempting to understand.

Finally, some words on how to choose the right approach when applying AI methods. As the following chapters will provide all the insight and usage of the different methods and algorithms, it might be of value to have a general guide for the decision process in order to aid us in finding a suitable algorithm for our application. For this, consider Figure 1.2, which is essentially a flow chart steered by questions and binary responses. It may not cover the entire spectrum of applications and algorithms we present in this book, but it helps us to determine whether supervised learning, unsupervised learning, or reinforcement learning is applicable.

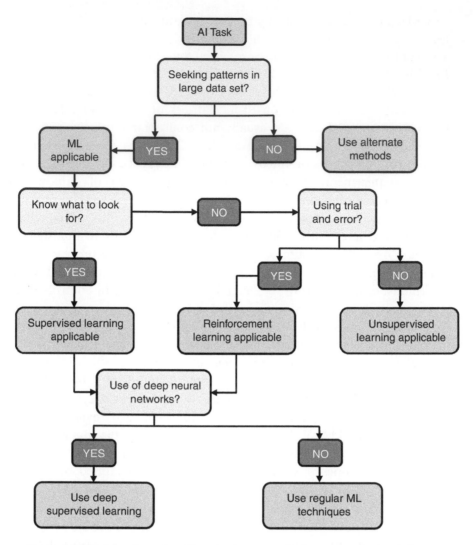

Figure 1.2 Decision flow chart for selecting supervised, unsupervised, reinforcement, and deep learning methods.

1.6   Conclusions

This chapter presented some introductory concepts we find in intelligent systems and control, mostly dealing with AI and ML. We defined what the word *deep* implies in deep learning, and discussed some of the nuances of different ML algorithms. A detailed treatment of each of these concepts is presented in the following chapters, including how to implement these algorithms using MATLAB and, in some instances, Simulink.

SUMMARY

Artificial Intelligence

AI refers to systems that are capable of mimicking human intelligence, mostly using computational resources and applied to a vast application range, including cognition, decision-making, and automation.

Control System

A closed-loop control system observes a system's output, also called plant output, and uses mathematical rules to adjust the input to the system in order to achieve a specific output. An open-loop control system does not engage measurements of the plant's output during the operation of the plant, but predetermines the input to the plant based on prior experience.

Deep Learning

Deep learning is part of ML, where deep neural networks are used to identify features from data with successive levels of complexity.

Fuzzy Logic

Binary logic entails true or false; however, fuzzy logic allows for all shades between true and false, which is called many-valued logic.

Machine Learning

Machine learning uses data to infer automatically information from large sets of data. Machine learning is considered to be a part of AI.

Overfitting

Overfitting occurs when we are training an ML model to the point that it is no longer generalizable (i.e., applicable to new data). During the learning process overfitting includes modeling of the noise and interferences contained in the training data set, which is not part of the system's characteristics.

Supervised Learning

Accessing labeled training data to evaluate the ML model's performance, which is used to adjust the ML model's parameters during the training process, is a form of supervised learning.

Unsupervised Learning

Contrary to supervised learning, unsupervised learning does not have access to a ground truth or labeled data during the training process of an ML model. Rather, it utilizes characteristics of the data to perform the learning task.

REVIEW QUESTIONS

1. What is the difference between ML and DL?
2. What is the meaning and implication of the word "artificial" in AI?
3. Can you apply unsupervised ML algorithms to labeled data?
4. What is an AI winter?
5. How do you define biased data?
6. What are possible causes of overfitting?
7. What is an epoch in supervised learning?
8. How many layers does a regular neural network have?
9. What is the function of a fuzzifier in a fuzzy logic control system?
10. What is the difference between the K-means clustering algorithm and the K-nearest neighbor algorithm?
11. What is the meaning of naïve in the naïve Bayes classifier algorithm?
12. Considering a dendrite of a biological neural system, explain what the analogy is for an artificial neural system.
13. What is the principle behind the SVM algorithm and for what type of data can it be used?
14. What is a loss function and for what purpose is it used in the process of training neural networks?
15. What is the difference between a decision tree algorithm and a random forest algorithm?

PROBLEMS

1.1 Using an online search engine, look up definitions for *intelligent*, *systems*, and *controls*. Develop your own interpretation of what an intelligent control system is, using those definitions.

1.2 Look up reports on Deep Blue for chess games, AlphaGo, and AlphaZero. Note what changes and improvements have occurred to the algorithms and what capabilities they obtained by these changes.

1.3 Research examples where the use of ML may be useful. Provide reasons for the use of ML techniques over traditional methods.

1.4 Look up recent publications on intelligent control systems in technical journals/magazines and comment on the latest trends and where they have been applied.

1.5 Suppose you have a dataset consisting of video clips detailing different cars driving by an intersection. Your intuition is to apply a K-nearest neighbor algorithm to build a model of manufacturer names of cars. Use Figure 1.2 to find out if this ML method is suitable and why, and, if not, why not and what alternative method could be used.

1.6 Consider recording audio signals and storing them digitally in folders, each folder having a name. Define the type of data we are dealing with. Use all possible descriptors introduced in Section 1.5.

1.7 Suppose you have a complex system that can be modeled by an ordinary differential second-order equation with constant coefficients. State why you need an adaptive control scheme (or why not) to impose a particular response of the system.

1.8 Consider training a deep neural network using a large data set. Describe the process that ensures no overfitting will occur during the training process.

1.9 What is the difference between ML and DL? Provide an example.

1.10 List five different examples of regression problems and five different classification problems. Describe the necessary data set and what type of methods could be used to either find a model or find patterns in the data.

REFERENCES

[1] McCarthy JMM, Rochester N, Shannon CE (2006). A proposal for the Dartmouth summer research project. *AI Magazine*, 27(4): 12.

[2] Zadeh LA (1965). Fuzzy sets. *Information and Control*, 8(3): 338–353.

[3] Takanishi A (2019). Historical perspective of humanoid robot research in Asia. In: Goswami A, Vadakkepat P (eds), *Humanoid Robotics: A Reference*. Springer, Dordrecht.

[4] Newborn M (1997). *Kasparov versus Deep Blue: Computer Chess Comes of Age*. Springer, Dordrecht.

[5] Silver D, Huang A, Maddison C (2016). Mastering the game of Go with deep neural networks and tree search. *Nature* 529: 484–489.

2 Fuzzy Sets and Fuzzy Logic

2.1 Introduction

In this chapter we will be laying the foundation for fuzzy logic-based control, addressing topics such as crisp versus fuzzy sets and classical set operators versus fuzzy set operators, and defining membership functions and fuzzy set mathematical operators. We will also look at fuzzy relations and fuzzy compositions. All of these topics will be supplemented with easy-to-follow examples.

First, however, we start with some motivation for studying these topics. Fuzzy logic is based on many-valued logic, which is a form of propositional calculus where there are more than two truth values. In fuzzy logic the truth values will be expressed by variables that can take on any measure between 0 and 1. Here, a number less than 1 and greater than 0 represents a partial truth, while the number 1 is completely true and 0 is completely false [1]. Using this concept for control, one can easily imagine a controller interacting with a system by various degrees of participation, which can be expressed as partial truths. A common household machine where fuzzy logic is used is the washing machine. Consider the various attributes the machine possesses, such as temperature settings of *Cold*, *Warm*, or *Hot*. These attributes can easily be seen as partial truths of the value *Hot*, where *Warm* may represent the number 0.5, *Cold* the number 0, and *Hot* the number 1. More distinction can be created by taking additional intermediate steps between 0 and 1 into consideration. A controller can now easily be fashioned by using those numbers and mapping them to a corresponding valve position for the cold and hot water lines feeding the washing machine. The same concept could be used for an electric stove's cooking temperature settings by engaging the corresponding resistor value.

Fuzzy logic has its origin in the studies of infinite-value logic, which started in the 1920s [2]. However, the term "fuzzy logic" was not introduced until 1965 by Lotfi Zadeh [3], who had noted that the computer logic used in those days could not operate on data that represented subjective or imprecise concepts. This observation led to the creation of fuzzy logic, allowing computing machines to incorporate human reasoning concepts.

There are now many fuzzy logic-based systems operating in everyday life. For example, the city of Sendai in Japan utilizes a 16-station subway system which is controlled by a fuzzy logic-based computer. Other examples include an automatic transmission designed by the Nissan Corporation for its fuzzy anti-skid braking systems, and Hitachi's handwriting- and voice-recognition systems.

Going back to basics: What does a control system do and what is its objective? In a general formulation, the aim of a control system is to produce a desired output of a system by providing a set of inputs. These inputs are computed by the controller. In other words, controls address the change of the dynamic characteristics of a system. Hence, traditional control techniques such as optimal, adaptive, robust, etc. control require a system model in order to calculate what the inputs to the system should be. These models are usually represented by differential equations, which are obtained by using the underlying physics of the system. However, many systems are rather complex, and the physics of the system is difficult or impossible to accurately quantify. In such instances, an alternative approach may be used to infer the mathematical representation of the system: using input–output data gathered from an experiment. This approach of inferring a dynamical model from observed experimental data is the subject of system identification.

On the other hand, fuzzy logic-based control systems generally do not require an explicit equation form of the system dynamics in order to design the controller and hence compute the controller output/system input. This last fact will be rather a theme of this book, where we will encounter a number of algorithms and approaches that allow us to neglect precise mathematical formulations of the underlying dynamics of the system to be controlled. To illustrate this approach, consider the temperature control of a room. Suppose the temperature is controlled by a heating/cooling system (actuator) and its output is measured by a temperature probe (sensor). The temperature of the room may be influenced by the outside temperature (winter, summer, etc.) and the type of insulation used for the windows and walls. Another factor affecting the temperature in the room may be a door to an adjacent room, the number of people in the room, etc. To use traditional controls, the dynamics of each of these influence factors need to be assessed and expressed by equations, thus characterizing the room temperature dynamics. A simple block diagram representation of the system for room temperature control – including the controller – is given in Figure 2.1.

The block entitled "Controller algorithm" includes the information of the entire system dynamics, and functions as a computer that calculates what the settings on the heating or cooling system should be. It computes these settings based on the error temperature of the room, which is the desired temperature (given by the temperature settings on the thermostat) and the actual temperature at the given time, measured by a temperature probe. An alternative approach for this control system would be a fuzzy logic-based control, where one neglects the precise system dynamics (i.e., the governing equations representing the temperature dynamics of the room with respect to the influence parameters). Instead, the

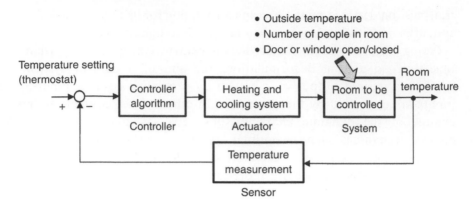

Figure 2.1 Block diagram for temperature control of a room.

controller has a set of simple rules that state, for example: If the room is cooler than the set temperature, the heater is turned on at half power; if the room is much cooler than the set temperature, the heater is turned on at full power; if the room is about the set temperature, the heater is turned off, etc. These are simple "IF–THEN" rules and do not include any complex mathematical models to determine how much the heater has to contribute in order to maintain the temperature in the room.

There are a number of benefits to using fuzzy logic-based control systems, such as reduced memory requirements, no heavy number-crunching demands for formula-based solutions, no need for intricate mathematical models, and only a practical understanding of the overall system behavior being necessary. However, there are also some drawbacks to using fuzzy logic-based control systems: traditional control systems, and in particular linear control systems, have a vast armory of analysis and design tools which bring insight and additional understanding of the systems to be controlled. This type of analysis is much less developed for fuzzy logic-based control systems.

This chapter on fuzzy sets and fuzzy logic includes a review on classical set theory, which is then used to establish fuzzy sets, where we also introduce fuzzy types and fuzzy set operations. We will proceed to introduce fuzzy relations by first reviewing crisp relations theory and then follow up by discussing set-theoretic operations. To arrive at the "IF–THEN" rules described above, we will introduce linguistic variables and fuzzy reasoning. All of this material serves as the background information for Chapter 3, where we will introduce a number of fuzzy inference systems, such as the Mamdani fuzzy inference system and the Sugeno fuzzy inference system, which all utilize the material covered here. Both chapters entail the use of MATLAB; hence, some of the examples introduced will make use of MATLAB scripts and commands.

2.2   A Review on Classical Sets: Crisp Sets

Table 2.1 lists some of the more common set notations and their meanings. For this table, lower case symbols are elements and upper case symbols are sets.

In this case a set is a collection of elements. Operations of these sets and variables can be depicted easily with diagrams, as shown in Figures 2.2–2.5. The first operation listed here is the UNION operation. The UNION operation combines all elements among the sets operated on by the UNION operator. A graphical representation is shown in Figure 2.2.

In Figure 2.2 the outer frame represents the border that comprises the universe (i.e., all elements of all sets). Usually this outer limit comprising all possible elements and all sets is called the *universe of discourse*. The two shaded rectangles are the sets A and B, respectively. Although they overlap, the union of A and B (i.e., $A \cup B$) is now all the area that is shaded. Note, the area that is overlapping by both sets is not counted twice for the UNION operator. The overlapping implies that the elements in that area are members

Table 2.1 Crisp sets: overview of notation and meaning

Notation	Explanation
$x \in X$	x belongs to X
$x \in A$	x belongs to A
$x \notin B$	x does not belong to B
$A \subset B$	A is fully contained in B (IF $x \in A$ THEN $x \in B$)
$x \subseteq B$	A is contained in or is equivalent to B
$(A \leftrightarrow B)$	A is equivalent to B

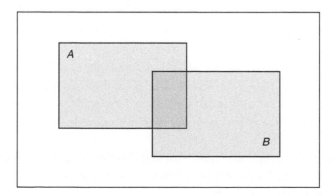

Figure 2.2 UNION operation for two sets: set A and set B.

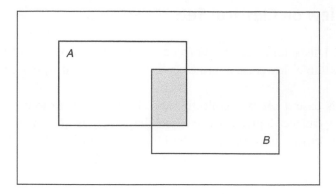

Figure 2.3 INTERSECTION operation for two sets: set A and set B.

of both sets (i.e., of set A and set B). Mathematically properly expressed, this can be stated as: $A \cup B = \{x | x \in A \text{ or } x \in B\}$. Note, in this formulation the term OR is used to indicate the dual use of the same elements who have membership in set A or set B.

Figure 2.3 presents the INTERSECTION operator. Here, the resulting set from the INTERSECTION operator is given by the area that contains the elements that are common to both sets (i.e., $A \cap B$ = shaded area in Figure 2.3).

In contrast to the UNION operator, the INTERSECTION operator can be described by the use of the word AND. Mathematically, Figure 2.3 can be written as

$$A \cap B = \{x | x \in A \text{ and } x \in B\}.$$

A common set operator used to describe the elements that are not part of a particular set is the COMPLEMENT operator. In Figure 2.4 the shaded area represents the complement of set A. One can express this operation as: $\overline{A} = \{x | x \notin A, x \in X\}$.

There exists a number of other operators which we will not review here (as this will be our focus in the next section); however, they can be easily derived and visualized in the same manner as above. For example, the DIFFERENCE operator is defined by: $A | B = \{x | x \in A \text{ and } x \notin B\}$. Applying this to the given sets A and B as used in the previous figures, the difference between A and B can be depicted graphically as shown in Figure 2.5. Note, the difference between A and B is not equal to the difference between B and A.

Having defined operators, we can also briefly review the basic properties of classical sets. The four main properties – commutativity, associativity, distributivity, and idempotency – are:

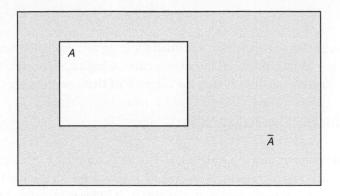

Figure 2.4 The complement of set A is given by $\overline{A}$, the shaded area.

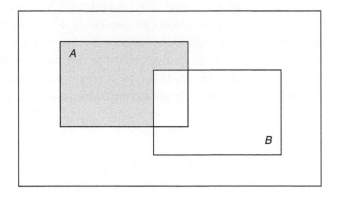

Figure 2.5 The difference between set A and set B.

$$
\text{Commutativity}: \quad
\begin{aligned}
A \cup B &= B \cup A \\
A \cap B &= B \cap A,
\end{aligned}
\tag{2.1}
$$

$$
\text{Associativity}: \quad
\begin{aligned}
A \cup (B \cup C) &= (A \cup B) \cup C \\
A \cap (B \cup C) &= (A \cap B) \cap C,
\end{aligned}
\tag{2.2}
$$

$$
\text{Distributivity}: \quad
\begin{aligned}
A \cup (B \cap C) &= (A \cup B) \cap (A \cup C) \\
A \cap (B \cup C) &= (A \cap B) \cup (A \cap C),
\end{aligned}
\tag{2.3}
$$

$$
\text{Idempotency}: \quad
\begin{aligned}
A \cup A &= A \\
A \cap A &= A.
\end{aligned}
\tag{2.4}
$$

We can also bring back the overall set, the universe of discourse X, and introduce the null set ($\emptyset$) and define some identities:

$$
A \cap \emptyset = \emptyset,
\tag{2.5}
$$

$$A \cup X = X. \tag{2.6}$$

Although these properties are defined for crisp sets, we will primarily focus on fuzzy sets – as introduced in the next section. Challenge yourself to draw membership diagrams such as Figure 2.4 for each of these properties.

2.3   Introduction to Fuzzy Sets

In the previous section the statement "*x is an element of A*" would make perfect sense: either this statement is true, and x is indeed an element of A, or the statement is false, and x is not an element of A. This is visualized by the distinct border lines between sets in each of the diagrams of Section 2.2. We used sharp lines of infinitely small width, indicating the boundaries and distinction between different sets. We call these sets "crisp sets." A statement such as "*x is somewhat an element of A*" or "*x is to some degree an element of A*" would not be possible with the definitions and properties introduced in Section 2.2. The notation "*somewhat*" or "*to some degree*" implies that the absolute terms of "*true*" or "*false*" no longer apply. Also, the sharp lines categorizing each set cannot be used in depicting such vagueness. As a simple example, take the local weather characterization on a given day: suppose there is some cloud cover mixed with some sunshine and blue sky. The question now is: Do we declare that it is cloudy or shall we announce it is a sunny day? Suppose we cannot make that decision without assessing some numerical value. Hence, we compute the percentage of cloud cover (i.e., area of clouds vs. total area). Suppose this simple measure yields 29% cloud coverage and we have arbitrarily established that it takes 30% cloud coverage to declare a "cloudy day." Will this suffice to make a determination of a sunny day? Comparing this with the crisp set theory, we can see that we need something less determinant, less black and white. We need something mathematical that allows for stating the gray shades or vagueness of an attribute. This is where fuzzy sets come into play. A fuzzy set is a mathematical tool that allows for assigning a grade of membership to an individual element. Going back to our weather example: a fuzzy set could be constructed for the attribute *sunny* by assigning a degree of belonging of 1 to cloud cover of 0%, 0.75 to cloud cover of 25%, 0.50 to cloud cover of 50%, and 0 to cloud cover of 100%.

In the next section we will detail the basic fuzzy set types, how to note them, by what means to program them in MATLAB, and in what manner to use them.

2.4   Basic Fuzzy Sets Types

A convenient way to quantify the degree of membership or, as stated above, the vagueness of an element belonging to a set is the use of membership functions. The membership function takes on the value of an element and computes how much this element belongs to a certain set. The output of a membership function is usually scaled with the unit interval [0,1]. Considering the universal set X (the universe of discourse), then a membership function μ_A creates a set A by feeding each element of X to μ_A and recording the corresponding output as the new fuzzy set A:

$$\mu_A : X \rightarrow [0, 1]. \tag{2.7}$$

There are other forms of notation found in the literature for Equation (2.7), such as

$$A : X \rightarrow [0, 1].$$

However, they have the same meaning and functionality. We will be using the notation shown in Equation (2.7).

It is easier to visualize the creation of these new fuzzy sets by using graphs of the membership functions. Consider Figure 2.6, where a triangular membership function μ_A is defined.

By utilizing Equation (2.7), we can take an element from X (i.e., $x = 2.5$) and evaluate its degree of membership to the set A using its membership function. According to Figure 2.6, for $x = 2.5$ we will obtain a degree of membership of 1.0. We can evaluate any value of x and determine its degree of membership to A using the function depicted in Figure 2.6. For example, $x = 5$ will yield a degree of membership of 0, or $x = 2$ implies $\mu_A = 0.5$, etc.

Using several fuzzy sets we can start representing linguistic concepts such as low, medium, and high. This approach is used to define the states of a variable (i.e., a fuzzy variable). To visualize this, consider the five membership functions for a variable describing the pressure inside a pressure tank, as shown in Figure 2.7. Accordingly, we can have the fuzzy variable take on linguistic values of *Very Low*, *Low*, *Medium*, *High*, and *Very High*. Upon close inspection of Figure 2.7, you will notice that the pressure variable can take on two different values – for example, x can be *Low* as well as *Medium*. To quantify *Low* and *Medium* we can

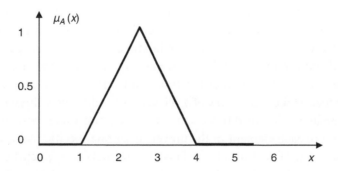

Figure 2.6 Triangular membership function.

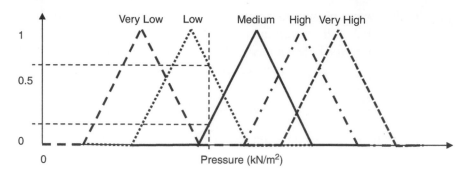

Figure 2.7 Pressure variable expressed as a fuzzy variable.

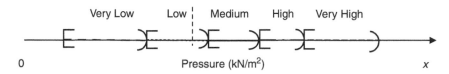

Figure 2.8 Pressure variable expressed as a crisp variable.

assess the degree of membership to each of these membership functions. In Figure 2.7 the thin dashed line indicates the value of x. Reading the corresponding values for the two-membership function, we see that x has 0.6 degree of membership to *Low* and 0.2 degree of membership to *Medium*.

We can go back to the classical set theory for this example and compare how these fuzzy sets relate to crisp sets. This is shown in Figure 2.8, where x is shown as a traditional (crisp) variable.

Note that from our example the current pressure indicated by the thin vertical dashed line would be categorized as *Low*. There is no indication if x is closer to *Very Low* or to *Medium*.

Equation (2.7) defines an ordinary fuzzy set μ_A for a universal set X. Note that the determination of the degree of membership is a simple look-up process with a defined relationship between x and μ_A. If some fuzziness or vagueness needs to be introduced to this relationship, an additional area has to be included in the design of a fuzzy membership function. Consider the Bell membership function shown in Figure 2.9. In contrast to the membership function shown in Figure 2.6, the shape is curved and rounded, and there is some gray area attached. The addition of the gray area allows this membership function to describe an interval-valued fuzzy set. Suppose x takes on a value of a. Then $\mu_A(a)$ yields two values: $\mu_A(a) = [\alpha_1, \alpha_2]$. Hence, an interval-valued fuzzy set is a fuzzy set whose membership function does not assign to each element of the universal set one specific real number, rather a closed interval of real numbers. This closed interval is given by the lower and upper bounds of α_1 and α_2. Interval fuzzy sets are used when ordinary fuzzy sets

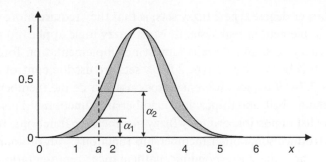

Figure 2.9 Bell membership function as an interval-valued fuzzy set.

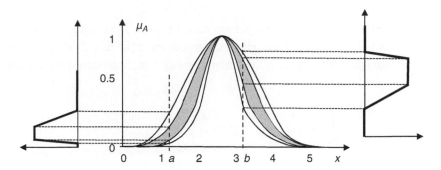

Figure 2.10 Example of a type 2 fuzzy set.

are considered too precise and some additional degree of vagueness needs to be included.

This concept can be extended further by introducing a membership function for the gray area of the interval fuzzy set. Consider Figure 2.10, where x takes on a value of a and b.

Using the arbitrarily selected values of $x = a$ and $x = b$, and choosing a trapezoidal form for the fuzzy sets of the gray areas, we can construct two fuzzy sets, each applicable to a and b, respectively, as shown in Figure 2.10. The membership function μ_A is called a type 2 membership function and results in type 2 fuzzy sets. The membership function of a type 2 fuzzy set can be written as

$$\mu_A: X \to \mathcal{F}[0, 1], \tag{2.8}$$

where $\mathcal{F}$ symbolizes that the set is also called a fuzzy power set. Usually, the power set is scaled to be within the interval [0,1]. A power set is basically a set of all ordinary fuzzy sets that can be defined with the universal set, in this case [0,1].

In this introductory text to fuzzy sets, fuzzy logic, and ultimately fuzzy logic-based control, we will only be dealing with ordinary fuzzy sets. However, it is noteworthy to mention that interval-valued fuzzy sets and type 2 fuzzy sets have some neat modeling capabilities. The disadvantage of interval-valued fuzzy sets, and to a larger degree type 2 fuzzy sets, is that they require more computational power to implement in real systems. Most fuzzy logic applications are based on ordinary fuzzy sets, and there is almost no implementation found – as of the writing of this book – where type 2 fuzzy sets are used in control applications.

Figures 2.7–2.9 depict different shapes used to define membership functions (i.e., triangular, bell, and trapezoidal membership functions). This is by no means a complete list of possible ordinary fuzzy membership functions. In the following we will introduce some other membership functions – all ordinary membership functions – and the corresponding mathematical representation, as well as a MATLAB implementation.

Example 2.1

Consider the problem of classifying a person based on age into categories of *Young*, *Middle-Aged*, and *Old*. Let us use straight lines to construct the membership functions, and define an age span of 0–90 years. The membership functions could be designed as follows:

$$\mu_{Young} = \begin{cases} 1 & \text{if } x \le 20 \\ (35 - x)/15 & \text{if } 20 < x < 35 \\ 0 & \text{if } x \ge 35 \end{cases}$$

$$\mu_{MiddleAged} = \begin{cases} 0 & \text{if either } x \le 20 \text{ or } x \ge 60 \\ (x - 20)/15 & \text{if } 20 < x < 35 \\ (60 - x)/15 & \text{if } 45 < x \le 60 \\ 1 & \text{if } 35 \le x \le 45 \end{cases}.$$

$$\mu_{Old} = \begin{cases} 0 & \text{if } x \le 45 \\ (x - 45)/15 & \text{if } 45 < x < 60 \\ 1 & \text{if } x \ge 60 \end{cases} \tag{2.9}$$

A graphical representation of Equation (2.9) is given in Figure 2.11. Note that we used trapezoidal membership functions to express the age groups.

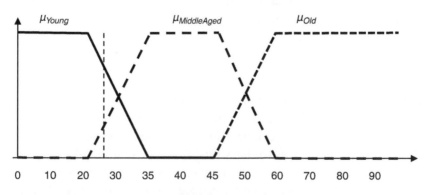

Figure 2.11 Membership functions to categorize age groups.

In this example, a 22-year-old senior student is considered *Young* as well as *Middle-Aged*. However, and much to the relief of many senior students, the degree of membership to the group *Middle-Aged* is much smaller than the degree of membership to the group *Young*. Using Equation (2.9) we can determine the degree of membership for each of these groups in a precise manner: *Young*: 0.87; *Middle-Aged*: 0.13. Having simple representations for the membership functions allows for simple implementations of them in MATLAB; for example, Equation (2.9) can be written in MATLAB code as follows:

```
% mfyoung.m

% trapezoidal membership function for "Young"
% dom...degree of membership
clear;
year=zeros(1000,1);dom=year;
for i = 1 : 1001
    year(i,1) = (i-1)*0.1;
    if year(i,1) >=0 && year(i,1) <= 20
        dom(i,1) = 1.0;
    elseif year(i,1)<35 && year(i,1)>20
        dom(i,1) = (35-year(i,1))/15;
    else
        dom(i,1) = 0;
    end;
end
plot(year,dom);
title('Membership function Young'); xlabel('Years'); ylabel('Degree of
Membership');
```

Note that, in the MATLAB code in Example 2.1, we used two variables: `year` and `dom`. As each of these variables are computed repeatedly in a for-loop the variables are initialized to zero at the beginning of the code to avoid resizing the corresponding arrays in each iteration of the for-loop. For-loops are not exactly state-of-the-art; a more effective and efficient implementation would be to vectorize this code, which is left for the reader to implement.

Considering the resulting graph of the membership functions in Figure 2.11, we may also find the transition rather sudden – that is, the degree of membership for a 21.9-year-old to a 22.1-year-old, symbolized by a sharp corner at year 22, followed by a straight line going to zero by age 35. The general description of fuzzy logic promises a gentler transition. Incorporating such gentleness could be done by introducing different membership functions that have slow transition expressed by curves. Rather than writing your own membership functions in MATLAB code, MATLAB provides a built-in function. Some of these built-in functions and the depiction of the corresponding membership function graphs are given in Table 2.2.

In any of the MATLAB functions introduced in Table 2.2, the parameters *a*, *b*, *c*, and in some instances *d* specify the actual shape of the membership functions. By

Table 2.2 Selected MATLAB fuzzy logic membership functions

Function name	MATLAB syntax	Equation
Trapezoidal MF	trapmf(x, [a b c d])	$f(x; a, b, c, d) = \max\left(\min\left(\frac{x-a}{b-a}, 1, \frac{d-x}{d-c} \right), 0 \right)$
Gaussian MF	gaussmf(x, [sig c])	$f(x; sig, c) = \exp\left(\frac{-(x-c)^2}{2sig^2} \right)$
Sigmoidal MF	sigmf(x, [a c])	$f(x; a, c) = \dfrac{1}{1 + \exp\left(-a(x-c)\right)}$
Triangular MF	trimf(x, [a b c])	$f(x; a, b, c) = \max\left(\min\left(\frac{x-a}{b-a}, \frac{c-x}{c-b} \right), 0 \right)$
Z-shaped MF	zmf(x, [a b])	$f(x; a, b) = \begin{cases} 1, & x \le a \\ 1 - 2\left(\dfrac{x-a}{b-a}\right)^2 & a \le x \le \dfrac{a+b}{2} \\ 2\left(\dfrac{x-b}{b-a}\right)^2 & \dfrac{a+b}{2} \le x \le b \\ 0 & x \ge b \end{cases}$
Generalized bell-shaped MF	gbellmf(x, [a b c])	$f(x; a, b, c) = \dfrac{1}{1 + \left\lvert \dfrac{x-c}{a} \right\rvert^{2b}}$

choosing different values of these parameters, one can tailor the membership function to characterize the desired properties of the linguistic variables. There are also some requirements regarding the relationship for the parameters: for example, the trapezoidal membership function `trapmf(.)` allows only parameters that have the property $a \leq b$ and $c \leq d$. If $b \geq c$, MATLAB will interpret the trapezoidal membership function as a triangular membership function. Examples of the different shapes of the introduced membership functions in Table 2.2 are shown in Figure 2.12.

There are many more membership functions that MATLAB has incorporated in its Fuzzy Logic Toolbox, which are easily explored with the `help` command in MATLAB. In order to use these membership function in the context of fuzzy logic, fuzzy reasoning, and ultimately in fuzzy inference systems, we need to discuss how to do mathematical operations with such membership functions and sets, which is the topic of the next section.

2.5   Fuzzy Set Mathematical Operations

Let's start with some basic definitions. We will make use of the universal fuzzy set X and define a subset A.

Definition 2.1
The *support* of A is the crisp subset of X consisting of all elements with membership grade greater than zero:

$$\text{support}(A) = \{x | \mu_A(x) > 0 \text{ and } x \in X\}. \tag{2.10}$$

Definition 2.2
The *core* of A is the crisp subset of X consisting of all elements with membership grade equal to 1.0:

$$\text{core}(A) = \{x | \mu_A(x) = 1 \text{ and } x \in X\}. \tag{2.11}$$

Definition 2.3
The α-cut of a fuzzy set A is the crisp set that contains all the elements of the universal set X whose membership grades in A are greater than or equal to the value specified by α:

$$^{\alpha}A = \{x | \mu_A(x) \geq \alpha \text{ and } x \in X\}. \tag{2.12}$$

Definition 2.4
The strong α-cut of a fuzzy set A is the crisp set that contains all the elements of the universal set X whose membership grades in A are greater than the value specified by α:

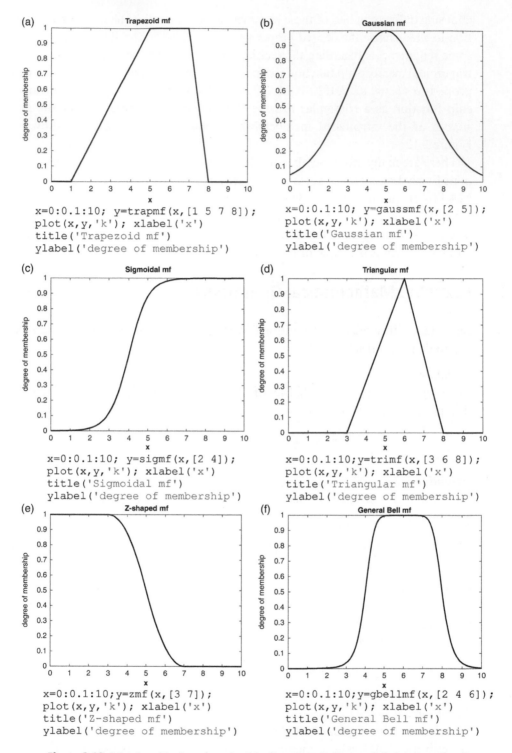

Figure 2.12 Membership function graphical representations and their corresponding MATLAB code: (a) trapezoidal membership function, (b) Gaussian membership function, (c) sigmoidal membership function, (d) triangular membership function, (e) Z-shaped membership function, (f) general bell membership function.

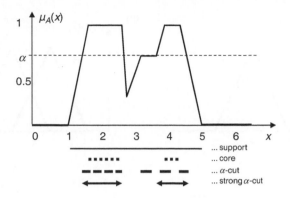

Figure 2.13 Graphical representation of Definitions 2.1–2.4.

$$^{\alpha+}A = \{x|\mu_A(x) > \alpha \text{ and } x \in X\}. \tag{2.13}$$

The definitions introduced above can be easily visualized using some generic membership function. Consider Figure 2.13. Here, the support is given by the solid line below the figure, showing a set that contains any number of x that has a membership grade greater than 0 (i.e., $0 \le x \le 5$). Looking at the two dotted lines, we see that these lines span the range of values where x has full membership (i.e., $\mu(x) = 1$). For the α-cut and the strong α-cut, we see from Figure 2.13 that elements which attain a membership value of α belong to the α-cut, but to belong to the strong α-cut their membership grade needs to be greater than α.

Just like with crisp sets, fuzzy sets can be operated on using the standard set operators such as AND, OR, NOT, etc. Figure 2.14 shows graphically the implication of such operations on fuzzy sets A and B.

Recall from the review material of crisp sets, INTERSECTION means AND while UNION means OR, which are expressed by the symbols $\cap$ and $\cup$, respectively. Let's expand on this and define the standard complement of a fuzzy set A as follows.

Definition 2.5
The standard complement of a fuzzy set A with respect to the universal set X is defined for all $x \in X$ as

$$\overline{A}(x) = 1 - A(x). \tag{2.14}$$

To explore this concept more closely, let's revisit the fuzzy sets for categorizing age, as shown in Figure 2.11. In this figure, one can find points where the degree of membership of one set is equal to the degree of membership of another fuzzy set – that is, having an age of 27.5 years will give a degree of membership of the fuzzy set *Young* equal to 0.5 as well as for the fuzzy set *Middle-Aged*. Instead of comparing two different fuzzy sets, we can compare a fuzzy set with its

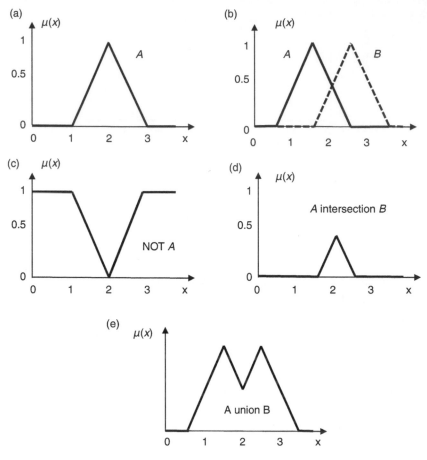

Figure 2.14 Graphical representation of set operations on fuzzy sets: (a) set A; (b) set A with solid line and set B with dashed line; (c) fuzzy set NOT A; (d) fuzzy set A INTERSECTION B; (e) fuzzy set A UNION B.

complement. Then, if $\overline{A}(x) = A(x)$ for some x, these points x are called equilibrium points.

Suppose we have two fuzzy sets, A and B. We can perform the operations of UNION and INTERSECTION as shown in Figure 2.15.

Note that these two operations can now be written as

$$A(x) \cap B(x) = \min[A(x), B(x)], \tag{2.15}$$

$$A(x) \cup B(x) = \max[A(x), B(x)], \tag{2.16}$$

where MIN and MAX donate the minimum operator and the maximum operator, respectively.

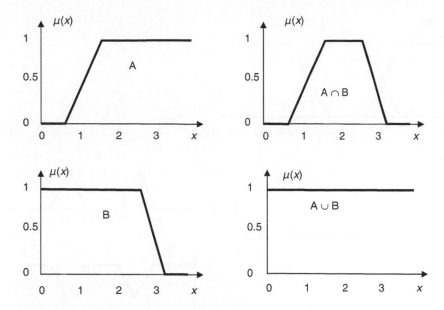

Figure 2.15 Min and MAX operator on fuzzy sets.

Definition 2.6

A fuzzy subset of X is called *normal* if there exists at least one element $x \in X$ such that $\mu_A(x) = 1$. A fuzzy subset that is not normal is called *subnormal*.

An example of Definition 2.6 is any crisp set, with the exception of the null set. A crisp set has a membership degree of either 1 or 0, and since we exclude the null set, crisp sets are always normal. Having defined normal and subnormal, we can also introduce the term *height of a set*. This term is applicable to subnormal sets, where the degree of belonging never reaches 1.0. The height of a subnormal fuzzy set is the maximum value of $\mu_A(x)$:

$$height(A) = \max_x(\mu_A(x)). \tag{2.17}$$

Consider the three fuzzy sets A, B, and C shown in Figure 2.16(a). Fuzzy sets A, B, and C all are normal – that is, they each have at least one element that yields a degree of membership of 1. Considering Figure 2.16(b) and (c), we see that simple fuzzy operations such as the AND operator do not necessarily yield a fuzzy set that is normal. In fact, we can also note that the resulting fuzzy sets shown in Figure 2.16(b) and (c) are convex; however, the fuzzy sets $D \cup E$ and $\overline{D \cup E}$ as shown in Figure 2.16(d) and (e) are not convex.

Generally, we can state that normality and convexity may be lost when standard operators such as AND, COMPLEMENT, and OR are applied to fuzzy sets. For readers interested in this topic, you may find that the two fuzzy operations AND and OR on fuzzy sets will satisfy all the properties (such as associativity,

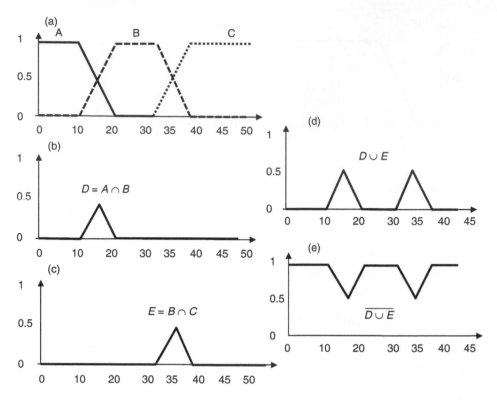

Figure 2.16 Fuzzy set operations: (a) three fuzzy sets A, B, and C; (b) intersection of fuzzy set A with B; (c) intersection of fuzzy set B with C; (d) union of fuzzy sets D and E; (e) complement of union of fuzzy sets D and E.

distributivity, idempotency, absorption, etc.) of the crisp set operators with the exception of the Law of Contradiction and the Law of Excluded Middle.

Until now we have used continuous number theory. However, much of today's world is digital, and hence we may want to talk about discrete fuzzy sets. Although this is not quite digital, it gives us the basic background to deal with implementation issues on digital devices such as microcontrollers. Suppose we have a fuzzy set A with n elements of x, each having a specified degree of membership. To record this type of set we will use the following notation:

$$A = \mu_A(x_1)/x_1 + \mu_A(x_2)/x_2 + \cdots + \mu_A(x_n)/x_n. \tag{2.18}$$

Here, $\mu_A(x_i)/x_i$ is called a singleton, which is a pair composed of the grade of membership and the element that belongs to a finite universe of discourse A:

$$A = \{x_1, x_2, \ldots, x_n\}. \tag{2.19}$$

Example 2.2

As an example of this type of fuzzy set, consider preparing a dinner for a preschooler. Your choices of food available are chicken nuggets, mac and cheese, broccoli, and apple sauce. Suppose there is no other food item available and nutrition is not your concern for this specific meal. Which items would you choose to prepare and serve to the preschooler? You may know the preschooler and hence you have some idea which food item she likes more than another. In this example, the food choices constitute the finite universe of discourse, that is

$$A = \{chicken\ nuggets, mac\ and\ cheese, broccoli, apple\ sauce\}.$$

For simplicity, we abbreviate and use

$$A = \{CN, MC, B, AS\}.$$

The knowledge of which item your preschooler prefers more than the other can now be noted as a fuzzy set with the degree of membership expressing likeability of the food item. Hence, we could define the discrete fuzzy set B as

$$B = \{(CN, 0.8), (MC, 0.7), (B, 0.1), (AS, 0.4)\}.$$

Using the notation introduced by Equation (2.18), we can write

$$B = 0.8/CN + 0.7/MC + 0.1/B + 0.4/AS.$$

The plus sign in this notation is not an arithmetic operation, it simply implies that the next element also belongs to the set.

Let's expand on discrete fuzzy sets and introduce new terms.

Definition 2.7

A fuzzy set A is considered *equal* to a fuzzy set B if and only if

$$\mu_A(x) = \mu_B(x)\ \forall x \in X. \tag{2.20}$$

Definition 2.8

A fuzzy set A is *included* in a fuzzy set B if and only if

$$\mu_A(x) \leq \mu_B(x)\ \forall x \in X. \tag{2.21}$$

Definition 2.9

The *cardinality* of a fuzzy set A is the sum of the values of the membership function of A, $\mu_A(x)$:

$$\text{card}_A = \sum_{i=1}^{n} \mu_A(x_i)\ \ \text{for } i = 1, 2, \ldots, n. \tag{2.22}$$

Definition 2.10

A fuzzy set A is *empty* if and only if

$$\mu_A(x) = 0 \ \forall x \in X. \tag{2.23}$$

Example 2.3

Let's illustrate these definitions using some examples. Consider the following fuzzy sets A, B, and C:

$$A = 0.3/1 + 0.4/2 + 0.65/3 + 0.25/4,$$

$$B = 0.29/1 + 0.35/2 + 0.5/3 + 0/4,$$

$$C = 0.3/1 + 0.4/2 + 0.65/3 + 0.25/4.$$

In this example, we notice that all three fuzzy sets have the same members (i.e., 1, 2, 3, and 4). Also, fuzzy set A and C have the same degree of membership for each of their members, and since they have the same members, fuzzy set A and C are equal. Comparing fuzzy sets A and B we notice that each member of fuzzy set B has a lower membership value than fuzzy set A for each of its corresponding members. Hence, fuzzy set B is included in fuzzy set A. Using Definition 2.9, we can compute the cardinality of each fuzzy set – that is, card$_A$ = 1.6, card$_B$ = 1.14, and since fuzzy set C is equal to fuzzy set B, we have card$_C$ = card$_A$. One additional note on this example, in particular with the observation that fuzzy set B is included in fuzzy set A: when this occurs we can also state that fuzzy set B is a subset of A: $B \subseteq A$.

Example 2.4

Suppose there is an additional fuzzy set, defined by

$$D = 0/1 + 0/2 + 0/3 + 0/4.$$

We see that fuzzy set D has the same members as the fuzzy sets in Example 2.3 (i.e., 1, 2, 3, and 4). However, the degree of membership for each and every element in fuzzy set D is equal to zero. According to Definition 2.10 such sets are called *empty* fuzzy sets.

Example 2.5

We can also use the definitions introduced for continuous fuzzy sets, such as α-cut, for discrete sets. Consider α to be equal to 0.3; then we have

$$^{\alpha}A = \{1, 2, 3\},$$

$$^{\alpha+}A = \{2, 3\}.$$

Mathematical operations to fuzzy sets are not limited to the ones so far introduced. We can also perform "traditional" operations such as multiplication, power, etc. For example, the multiplication of a set by a scalar λ is computed by

$$\lambda A = \{\lambda \mu_A(x), \forall x \in A\}. \tag{2.24}$$

To compute the power of a set we can follow the same procedure as introduced by Equation (2.24) – that is, we compute the power of each element of the set:

$$A^\beta = \left\{\mu_A(x)^\beta, \forall x \in A\right\}. \tag{2.25}$$

Example 2.6
Suppose

$$A = 0.5/a + 0.3/b + 0.2/c + 1/d$$

and $\lambda = 0.5, \beta = 2$. Then,

$$\lambda A = 0.25/a + 0.15/b + 0.1/c + 0.5/d$$

and

$$A^\beta = 0.25/a + 0.09/b + 0.04/c + 1/d.$$

Having discussed some of the basic mathematical operations on fuzzy sets, we will next introduce the concept of relations for fuzzy sets.

2.6 Fuzzy Relations

Let's start with two crisp sets A and B. For such crisp sets there are only two degrees of relationships between the elements of the two sets: The elements are either *completely related* or *not at all related*. For crisp sets, a relation indicates the presence of association, interaction, or interconnectedness between each element of the two crisp sets. On the other hand, it also tells us the lack of such association, interaction, and interconnectedness. For crisp sets the relation is a binary outcome, often stated as true or false.

Example 2.7
Consider a set A representing the members of a student club A and a set B representing the members of another student club B on campus. Students are free to choose and become members of either club or both clubs or neither club. A graphical representation could be made using Table 2.3.

Students are either members of a club or not, symbolized by an X if membership is current and a 0 if not.

Table 2.3 Crisp relationship between two sets *A* and *B*

Club	Member				
	Natalie	Matthew	Isabella	Jason	Jennifer
Club A	0	X	X	0	0
Club B	X	0	X	0	X

If we expand this concept of relations to fuzzy sets, we note that an element in a fuzzy set may only be partially belonging to this set. This means that fuzzy relations map a degree of belonging into the relationship properties. How do we compute this property into a relationship? For this we need to discuss what a Cartesian product is. Consider the crisp sets $A_1, A_2, \ldots, A_n$, then the set of n-tuples $a_1, a_2, \ldots, a_n$ where $a_1 \in A_1$, $a_2 \in A_2$, $a_3 \in A_3, \ldots, a_n \in A_n$ is called the Cartesian product of $A_1, A_2, \ldots, A_n$. The Cartesian product is usually denoted by

$$A_1 \times A_2 \times \cdots \times A_n. \tag{2.26}$$

Example 2.8

Here are some examples on how to compute the Cartesian product of crisp sets. Consider two crisp sets $A = \{5, 8, 4, 9\}$ and $B = \{r, u\}$.

We can compute the following Cartesian products:

$A \times B$: $A \times B = \{(5, r), (5, u), (8, r), (8, u), (4, r), (4, u), (9, r), (9, u)\}$

$B \times A$: $B \times A = \{(r, 5), (r, 8), (r, 4), (r, 9), (u, 5), (u, 8), (u, 4), (u, 9)\}$

$A \times A$: $A \times A = \{(r, r), (r, u), (u, r), (u, u)\}$.

Going back to fuzzy relations, we can define a fuzzy relation as fuzzy subsets of $A \times B$, which is a mapping from $A \rightarrow B$. Fuzzy relations map each element of one set A to those of another set B through the Cartesian product of the two sets.

Definition 2.11

A *fuzzy relation R* is a mapping from the Cartesian space $A \times B$ to the interval [0 1]. The mapping strength is given by the membership function of the relation for the ordered pairs:

$$R = \left\{ \big((a, b), \mu_R(a, b)\big) \big| (a, b) \in A \times B \right\}. \tag{2.27}$$

Definition 2.12

Suppose A is a fuzzy set on the universe X and B is a fuzzy set on the universe Y, then the *Cartesian product* between the two fuzzy sets A and B will produce a fuzzy relation R that is contained within the Cartesian product space:

$$A \times B = R \subset X \times Y, \tag{2.28}$$

with R having a membership function:

$$\mu_R(x,y) = \mu_{A \times B}(x,y) = \min\left(\mu_A(x), \mu_B(y)\right). \tag{2.29}$$

From Equation (2.29) we note that fuzzy relations are sets themselves. These relations can be expressed as matrices.

Example 2.9

As an example, consider the color of a fruit and its relation to indicate how ripe the fruit is. Let's choose a banana, and relate green, yellow, and brown to unripe, ripe, and spoiled. The first fuzzy set is the color of the fruit: A = {*green, yellow, brown*}, and B is accordingly B = {*unripe, ripe, spoiled*}. A matrix with the corresponding relationships can be composed as shown in Table 2.4.

Table 2.4 Fuzzy relation example for bananas and ripeness

R (*color, ripeness*)	*Unripe*	*Ripe*	*Spoiled*
Green	1	0.5	0
Yellow	0.5	1	0.25
Brown	0	0.5	1

Equation (2.29) details the computation of a Cartesian product.

Example 2.10

To illustrate this, consider two discrete sets A and B, where

$$A = \frac{0.25}{x_1} + \frac{0.55}{x_2} + \frac{0.95}{x_3} \text{ and } B = \frac{0.35}{y_1} + \frac{0.85}{y_2}.$$

The fuzzy Cartesian product R of sets A and B is hence

$$R = A \times B = \begin{matrix} & y_1 & y_2 \\ x_1 & \\ x_2 & \\ x_3 & \end{matrix} \begin{bmatrix} 0.25 & 0.25 \\ 0.35 & 0.55 \\ 0.35 & 0.85 \end{bmatrix}.$$

We can expand this operation to fuzzy compositions. We will look at two different compositions: the fuzzy MAX-MIN composition and the fuzzy MAX-PRODUCT composition. These two types can be shown by the following example. Suppose R is a fuzzy relation on the Cartesian space $X \times Y$, S is a fuzzy relation on $Y \times Z$, and T is a fuzzy relation on $X \times Z$. Then the fuzzy MAX-MIN composition is defined by

$$T = R \circ S, \qquad (2.30)$$

where

$$\mu_T(x,z) = \bigvee_{y \subset Y} \{\mu_R(x,y) \wedge \mu_S(y,z)\}. \qquad (2.31)$$

In Equation (2.31) the $\vee$ operator indicates "maximum" or "max," while $\wedge$ means "minimum" or "min." To compute the MAX-PRODUCT the membership function for T is calculated as

$$\mu_T(x,z) = \bigvee_{y \in Y} \{\mu_R(x,y) \bullet \mu_S(y,z)\}. \qquad (2.32)$$

Example 2.11

Let's illustrate these two computations with an example. Suppose we have three fuzzy sets: $X = \{x_1, x_2\}$, $Y = \{y_1, y_2\}$, and $Z = \{z_1, z_2, z_3\}$. Let's also define the following fuzzy relations:

$$R = \begin{array}{c} \\ x_1 \\ x_2 \end{array} \begin{array}{cc} y_1 & y_2 \\ \begin{bmatrix} 0.7 & 0.6 \\ 0.8 & 0.3 \end{bmatrix} \end{array} \text{ and } S = \begin{array}{c} \\ y_1 \\ y_2 \end{array} \begin{array}{ccc} z_1 & z_2 & z_3 \\ \begin{bmatrix} 0.8 & 0.5 & 0.4 \\ 0.1 & 0.6 & 0.6 \end{bmatrix} \end{array}.$$

Our objective is to find the relation between X and Z – that is, the fuzzy set T which relates the elements in set X with the elements in set Z. In order to find this relationship, we will employ both fuzzy composition methods:

Using MAX-MIN:

$T = R \circ S$: using x_1 and z_1 of R and S, we find the following degree of membership:

$$\mu_T(x_1, z_1) = \max\{\min(0.7, \ 0.8), \min(0.6, \ 0.1)\} = \max\{0.7, \ 0.1\} = 0.7.$$

Repeating this for every x and z combination will yield T:

$$T = \begin{array}{c} \\ x_1 \\ x_2 \end{array} \begin{array}{ccc} z_1 & z_2 & z_3 \\ \begin{bmatrix} 0.7 & 0.6 & 0.6 \\ 0.8 & 0.5 & 0.4 \end{bmatrix} \end{array}.$$

Using MAX-PRODUCT:

$T = R \circ S$: using again x_1 and z_1 of R and S, we find the following degree of membership:

$$\mu_T(x_1, z_1) = \max\{\min(0.7 \times 0.8), \min(0.6 \times 0.1)\} = \max\{0.56 \times 0.06\} = 0.56.$$

For x_2 and z_2 of R and S we find

$$\mu_T(x_2, z_2) = \max\{\min(0.8 \times 0.5), \min(0.3 \times 0.6)\} = \max\{0.4 \times 0.18\} = 0.4.$$

Continuing with all x and z combinations, we will arrive at

$$T = \begin{matrix} x_1 \\ x_2 \end{matrix} \begin{bmatrix} \overset{z_1}{0.56} & \overset{z_2}{0.36} & \overset{z_3}{0.35} \\ 0.64 & 0.4 & 0.32 \end{bmatrix}.$$

There exists a number of other compositions, such as MAX-MAX, MIN-MIN, MAX-AVERAGE, etc.; however, the primary idea is the same: We want to relate the elements in set X with the element in set Z using a fuzzy relation R that relates set X with set Y and a fuzzy relation S that relates set Y with set Z. The computation is simplified by finding the fuzzy relation T using the composition of relation R and S (i.e., $T = R \circ S$). The resulting relationship can be given in graphical form, as shown in Figure 2.17 for the MAX-MIN composition.

The MATLAB code given in Appendix D.1 is a simple implementation of the MAX-MIN composition computations for the above example.

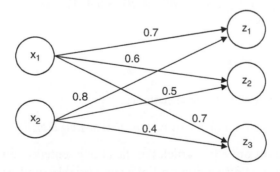

Figure 2.17 Fuzzy set X in relation to fuzzy set Z using T, which is computed using the MAX-MIN composition.

2.7   Linguistic Variables

Suppose we have a fuzzy set defined by its membership function $\mu_A(x)$ as shown in Figure 2.6 (i.e., a triangular membership function). We would like to alter the characteristics of this membership function to include attributes easily found in the English language: for example, *very*, *somewhat*, *a little*, etc. The triangular membership function may represent the degree of membership to the set of "Old" people, as introduced in Section 2.4. Then, including attributes to this membership function, one can express *very Old*, *somewhat Old*, etc. These so-called hedges are expressed mathematically and alter the shape of the membership function. Table 2.5 lists a set of hedges, their corresponding mathematical operators, and the effect on the shape of the membership function.

There are many more operators one could define and use, with specific outcomes for the shape of the membership function and its linguistic implication. The triangular membership function in the above discussion is only a placeholder for any other membership function that can be operated on using hedges.

Having defined hedges, we can introduce another concept that will be of use when we discuss fuzzy reasoning in the next section: fuzzy conjunctions. Consider two fuzzy sets expressed with two membership functions $\mu_A(x)$ and $\mu_B(x)$, and recalling the AND operator, which is the same as the min operator, we can state the following:

$$A \wedge B \triangleq \min(A, B). \tag{2.33}$$

Suppose $A \wedge B = C$, we can call C the conjunction of the quantity A and B. Let's illustrate this graphically. Figure 2.18 depicts two membership functions $\mu_A(x)$ and $\mu_B(x)$ as triangular membership functions. Suppose A is 1.5 and B is 2.25; we want to find $A \wedge B = C$. To find C, we first determine the degree of membership for each set – that is, for A we will have $\mu_A(x) = 0.4$ and for $\mu_B(x) = 0.6$. Applying the fuzzy AND operator: $A \wedge B = \min(A, B) = 0.4$. Hence, the conjunction C of set A and B is equals to 0.4.

2.8   Fuzzy Reasoning

Consider the following English sentence:

If temperature is cold and electricity is cheap then heating is high.

We may have left out a few articles in the above sentence, but it is still readable. We can revisit our discussion on linguistic variables and expand on it by also including linguistic values. Let's comment the above sentence with these identifiers as follows:

Table 2.5 Hedges and their operators

Hedge	Operator	Effect
A little	$\mu_A(x)^{1.4}$	
Slightly	$\mu_A(x)^{1.8}$	
Very	$\mu_A(x)^2$	
Extremely	$\mu_A(x)^3$	
Very very	$\mu_A(x)^4$	
Somewhat	$\mu_A(x)^{0.5}$	

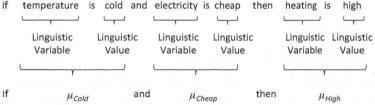

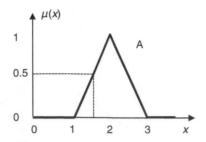 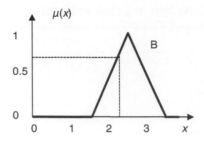

Figure 2.18 Example of conjunction $A \wedge B = C$ for $A = 1.5$ and $B = 2.25$.

Before discussing the sentence structure and the above indicated membership functions, we can define in more detail the linguistic variables used in our sentence. A linguistic variable is given by five elements:

1. Name of the variable: for example, "*Temperature.*"
2. Term set of its linguistic values or linguistic terms: for example, {*very cold, cold, more or less cold, hot, very hot,* ...}.
3. Universe of discourse: for example, we can define the range of the linguistic variable as between 32 and 100.
4. Syntactic rule: this generates the terms in the term set.
5. Semantic rule: this associates each linguistic value A with its meaning $M(A)$, where $M(A)$ is the fuzzy set in the universe of discourse.

Example 2.12

To illustrate the linguistic variable and value definition, consider the following MATLAB code defining a linguistic variable called "cost" with the term set of T (*cost*) = {*cheap, affordable, expensive*}. We will use a universe of discourse of [0, 10].

```
% MATLAB example for linguistic variables
cost = (0:0.1:10)';
cheap = sigmf(cost, [-2 2]);
affordable = gaussmf(cost, [3 5]);
expensive = sigmf(cost, [2 8]);
plot(cost, [cheap affordable expensive]);
text(1.25, 0.89, 'Cheap')
text(4.4, 0.89, 'Affordable')
text(8.2, 0.89, 'Expensive')
ylabel('Membership Grade')
xlabel('Cost')
title('MATLAB Linguistic Variables Example')
```

The graphical depiction of the linguistic variable and its values is given in Figure 2.19.

Cheap, affordable, and *expensive* are the primary terms of the linguistic variable *cost*. We can now alter these by fuzzy operations such as NOT, which is a negation. We also can employ the hedges introduced in the previous section and create the following values: *very cheap, more or less cheap, extremely cheap,* among others.

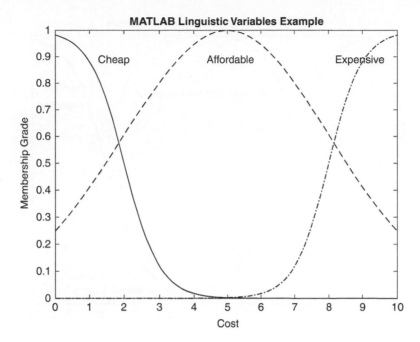

Figure 2.19 Example of linguistic variable *cost* with linguistic values of *cheap*, *afford-able*, and *expensive*.

To construct these composite linguistic values, we alter our MATLAB code as follows:

```
% MATLAB example for composite linguistic values
cost = (0:0.1:10)';
cheap = sigmf(cost, [-2 2]);
affordable = gaussmf(cost, [3 5]);
expensive = sigmf(cost, [2 8]);
more_or_less_cheap = cheap.^0.5;
not_cheap_and_not_expensive = min(1-cheap, 1-expensive);
expensive_but_not_too_expensive = min(expensive, 1-expensive.^2);
extremely_expensive = expensive.^0.75;
together = [more_or_less_cheap not_cheap_and_not_expensive
  expensive_but_not_too_expensive extremely_expensive];
plot(cost, together);
text(0.55, 0.89, 'More or Less Cheap')
text(3, 0.7, 'Not Cheap and Not Expensive')
text(8.2, 0.89, 'Extremely')
text(8.2, 0.84, 'Expensive')
text(7.1, 0.2, 'Expensive but')
text(7.2, 0.15, 'not too Expensive')
ylabel('Membership Grade')
xlabel('Cost')
title('MATLAB Linguistic Variables Example')
```

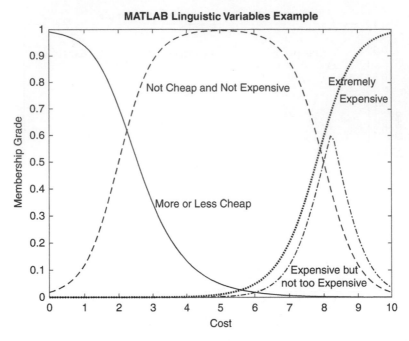

Figure 2.20 Composite linguistic variables created using simple fuzzy operations and hedges.

The corresponding membership functions are shown in Figure 2.20. Note that using hedges and simple negation we are able to create new values and membership functions.

Going back to our sentence from earlier, we notice that this is conditional logic using a set of statements with propositions that could be written in the form

IF X AND Y THEN Z,

where X and Y are variables called *antecedents* and Z is called the *consequence*. The sentence uses a simple "IF–THEN" rule (i.e., if X is true and Y is true, then Z will be true). We can use a standard notation to express this in short form:

$$A \rightarrow B \triangleq \text{If } x \text{ is } A \text{ then } y \text{ is } B. \tag{2.34}$$

In Equation (2.34), A and B are linguistic values defined on X and Y and A is an antecedent (or premise) and B is a consequence (or conclusion). To express the resulting set, we define the following:

$$\mu_{A \rightarrow B}(x,y) = \begin{cases} 1 & \mu_A(x) \leq \mu_B(y) \\ \mu_B(y) & \text{otherwise} \end{cases}. \tag{2.35}$$

We can also introduce truth tables to describe the IF–THEN statement (Table 2.6).

Table 2.6 IF–THEN rules truth table

A	B	$A \rightarrow B$
1	1	1
1	0	0
0	1	1
0	0	1

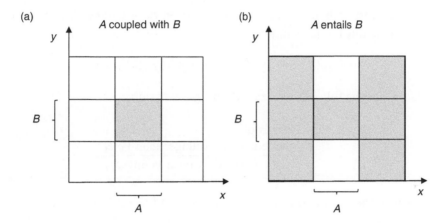

Figure 2.21 Fuzzy reasoning: IF–THEN rules with two possible interpretations: (a) A is coupled with B; (b) A entails B.

Now we have a choice to make: How do we interpret the resulting membership function? Suppose $\mu_R(x, y) = \mu_{A \rightarrow B}(x, y)$, we could imply that R is the outcome of A being coupled with B, or we could state that R is the result of A entailing B. To illustrate these two possible interpretations, consider Figure 2.21.

We can express each interpretation mathematically as well. Consider A is coupled with B: From Figure 2.21 and recalling Equation (2.15), we see that we could utilize the min operator to accomplish the rule IF A THEN B:

$$R = A \rightarrow B = \mu_R(x, y) = \min\{\mu_A(x), \mu_B(y)\}. \tag{2.36}$$

For the second option, Figure 2.21(b) allows us to construct a number of different equations to describe the resulting membership function. For example, we can state

$$R = A \rightarrow B \equiv \neg A \cup B. \tag{2.37}$$

Equation (2.37) is called the *material implication* of A entails B. Another way to construct the set given in Figure 2.21(b) is to use

$$R = A \rightarrow B \equiv \neg A \cup (A \cap B). \tag{2.38}$$

In the literature, Equation (2.38) has been labeled the *propositional calculus* implementation of A entails B. There are a number of other implementations that describe the second interpretation of the IF–THEN rule, which we will not review here. However, we can note the corresponding membership function for the given interpretations above as

$$\mu_R(x,y) = \max\{1 - \mu_A(x), \mu_B(y)\}, \tag{2.39}$$

which corresponds to the material implication of A entails B. For the propositional calculus implementation of A entails B, we write

$$\mu_R(x,y) = \max\{1 - \mu_A(x), \min\{\mu_A(x), \mu_B(y)\}\}. \tag{2.40}$$

Consider the following fuzzy reasoning statement:

$$\frac{\text{Peter is old}}{\text{Natalie is much younger than Peter}}.$$
$$\overline{\text{Natalie is (\{old\} \{much younger\})}}$$

Let's read these statements line by line. The first line could be given as a fuzzy set for "old":

$$\text{Age (Peter)} = \text{old}.$$

The second line can be given as a fuzzy relation:

$$\text{Age (Natalie), Age (Peter)} = \text{much younger}.$$

The last line is a fuzzy composite relation that can be read as

$$\text{Age (Natalie)} = \text{old} \circ \text{much younger}.$$

In a generalized form, we state this as

$$\frac{\begin{array}{l}\text{Rule: if } x \text{ is } A \text{ then } y \text{ is } B \\ \text{Fact: } x \text{ is } A'\end{array}}{\text{Conclusion: } y \text{ is } B'}.$$

This is a single rule with a single antecedent. Here, A' and B' are fuzzy sets related to A and B. We can also accommodate multiple antecedents:

$$\frac{\begin{array}{l}\text{Rule: if } x \text{ is } A \text{ and } y \text{ is } B \text{ then } z \text{ is } C \\ \text{Fact: } x \text{ is } A' \text{ and } y \text{ is } B'\end{array}}{\text{Conclusion: } z \text{ is } C'}.$$

This is depicted in Figure 2.22.

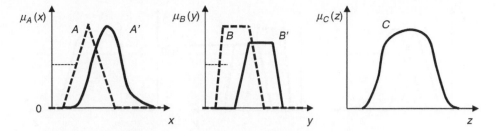

Figure 2.22 Fuzzy reasoning: a single rule with multiple antecedents.

Figure 2.22 is not quite complete; we have not included what C' is – that is, what is our conclusion? From our discussion above, the conclusion is that z is C'. How do we compute C'? Let's use the introduced mathematical tools to define C'.

We have $R = A \times B \to C$, which is written as

$$\mu_R(x,y,z) = \mu_{(A \times B) \times C}(x,y,z) = \mu_A(x) \wedge \mu_B(y) \wedge \mu_C(z)$$

and

$$
\begin{aligned}
\mu_{C'}(z) &= \max_{x,y}\left\{ \min\left\{ \mu_{A',B'}(x,y), \mu_R(x,y,z) \right\} \right\} \\
&= \vee_{x,y}\left\{ \mu_{A',B'}(x,y) \wedge \mu_R(x,y,z) \right\} \\
&= \vee_{x,y}\left\{ \mu_{A'}(x) \wedge \mu_{B'}(y) \wedge \mu_A(x) \wedge \mu_B(y) \wedge \mu_C(z) \right\} \\
&= \left[\vee_x \left(\mu_A(x) \wedge \mu_{A'}(x) \right) \right] \wedge \left[\vee_y \left(\mu_B(y) \wedge \mu_{B'}(y) \right) \right] \wedge \mu_C(z)
\end{aligned}
$$

The last equation means that we will use the AND operator on A and A' and take the maximum of the resulting membership function. We do the same for B and B' and take the maximum of the resulting membership function. Both maxima are compared using the MIN operator and the resulting value is used to perform an α-cut on C to obtain C'. The α value is called the *firing strength*:

$$\mu_{C'}(z) = \underbrace{\left[\vee_x \left(\mu_A(x) \wedge \mu_{A'}(x) \right) \right] \wedge \left[\vee_y \left(\mu_B(y) \wedge \mu_{B'}(y) \right) \right] \wedge \mu_C(z)}_{\text{firing strength}}. \tag{2.41}$$

Figure 2.22 can now be updated to indicate the conclusion C', as shown in Figure 2.23.

Looking at Equation (2.41) more carefully, we recognize that the consequence is computed as

$$C' = (A' \times B') \circ (A \times B \to C). \tag{2.42}$$

What if we have multiple rules and multiple antecedents? To entertain multiple rules, we can extend Figure 2.22 by adding for each rule a new row representing

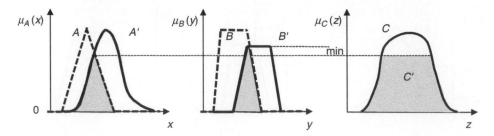

Figure 2.23 Fuzzy reasoning: computation of firing strength and consequence.

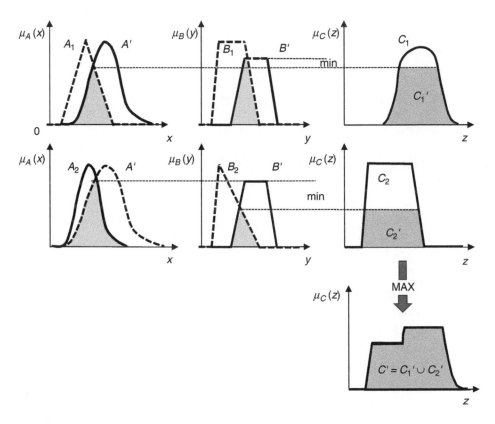

Figure 2.24 Fuzzy reasoning: multiple rules with multiple antecedents.

the corresponding fuzzy reasoning. The question then becomes how do we compute the conclusion. Let's look at the example given in Figure 2.24. Here we have two rules:

Rule 1: if x is A_1 and y is B_1 then z is C_1
Rule 2: if x is A_2 and y is B_2 then z is C_2
Fact: x is A' and y is B'

Conclusion: z is C'

The conclusion now can be computed in different ways. For example, we could compute:

$$C' = \left(A' \times B' \right) \circ (R_1 \cup R_2)$$
$$= [\left(A' \times B' \right) \circ R_1] \cup [\left(A' \times B' \right) \circ R_2] \cdot$$
$$= C_1' \cup C_2'$$

Hence, the conclusion is computed by using the maximum of each consequence.

2.9   Conclusions

In this chapter we made the transition from crisp sets to fuzzy sets. For a crisp set, an element is either a member or not a member of a particular set. For a fuzzy set, an element can have varying degrees of membership to one or more sets, but also not be a member of a particular set. Such degree of membership accommodates the notion of uncertainty and ambiguity, and allows us to bring such description into computations. The degree of membership an element takes on to a fuzzy set is described by its membership function. A membership function can take on many different forms, such as bell-shaped, triangular shaped, trapezoidal shape, etc. To perform fuzzy logic we work with these membership functions to express implications of belonging or exclusion using logical operations such as AND, OR, and NOT.

A fuzzy set has properties such as being normal if at least one element reaches the degree of membership of 1.0 and defines what kind of support it provides by the number or range of elements that have a nonzero membership degree. We also found that computations with fuzzy sets lead to new sets, which may retain the property of fuzziness. Some of the computations are simple cuts that remove the elements above a certain membership grade, called α-cut; others use the traditional UNION and INTERSECTION operations we find in crisp set theory. For fuzzy set operations that are labeled with the UNION operator, we use the *max* operator, identifying the maximum degree of membership, while the INTERSECTION operator is equivalent to the MIN operator, where we assemble the result by retaining the elements with the minimum degree of membership along the variable.

We also looked extensively at fuzzy relations, which is a mapping from some Cartesian space to the unit interval [0 1]. To quantify the strength of mapping we use the resulting degree of membership. The computation of finding a fuzzy relation is done by employing the Cartesian product of the fuzzy sets we are relating.

Using linguistic variables we can express simple verbal descriptions mathematically, and when used with fuzzy logic we gain the ability to compute inferences, construct rules and instructions. Additionally, hedges along with linguistic variables allow us to emphasize or de-emphasize descriptions mathematically in the framework of fuzzy logic. Our discussion on linguistic variables allowed us to introduce fuzzy reasoning. Here we utilize simple descriptions which we map mathematically using membership functions and linguistic variables to express things such as: IF temperature is cold AND electricity is cheap THEN heating is high. The words *temperature*, *electricity*, and *heating* in this sentence are fuzzy linguistic variables, while *cold*, *cheap*, and *high* are linguistic variables. A linguistic variable may have a number of different values, each being characterized by a membership function. For example, temperature can be hot, cold, or pleasant.

In fuzzy logic, conditional statements are captured with IF–THEN rules, which relate antecedents with consequences. As an example, IF A THEN B implies that A and B are fuzzy sets defined by their respective membership functions and A is the antecedent while B is the consequence. As such, IF–THEN rules may involve only partial truth – that is, the antecedent has a degree of membership, and hence the consequence(s) are true to some degree as well.

We will use this material to construct fuzzy inference systems and fuzzy logic-based control systems in the next chapter. Later chapters will make use of this fundamental treatise on fuzzy logic when we utilize some of the strengths of fuzzy inference to build hybrid systems that learn and adapt to different environments.

SUMMARY

In the following, a number of key concepts of fuzzy logic covered in this chapter are summarized.

Membership Function

A membership function assesses the degree of belonging of a variable to a set. For fuzzy sets a membership function may take on any value between 0 and 1. For a crisp set, a membership function will yield only a binary result, either 0 for not belonging or 1 for belonging. For fuzzy logic systems the membership function establishes a degree of truth.

UNION and INTERSECTION Operators

The UNION operator combines two sets by collecting all elements in both sets, excluding common elements from being counted more than once. For two sets A and B the UNION is defined as:

For a crisp set: $A \cup B = \{x | x \in A \text{ or } x \in B\}$

For a fuzzy set: $A(x) \cup B(x) = \max[A(x), B(x)]$.

The intersection of two or more sets is computed by collecting all the common elements of both sets. Algebraically, we can express the intersection of two sets A and B by:

For a crisp set: $A \cap B = \{x | x \in A \text{ and } x \in B\}$

For a fuzzy set: $A(x) \cap B(x) = \min[A(x), B(x)]$.

Fuzzy Relation

A fuzzy relation is a mapping from the Cartesian space $A \times B$ to the interval [0 1]. The mapping strength is given by the membership function of the relation for the ordered pairs:

$$R = \left\{ \left((a, b), \mu_R(a, b) \right) \middle| (a, b) \in A \times B \right\}.$$

Cartesian Product

The Cartesian product between two fuzzy sets A and B is the fuzzy relation R computed by

$$A \times B = R \subset X \times Y.$$

R represents a membership function defined by

$$\mu_R(x, y) = \mu_{A \times B}(x, y) = \min\left(\mu_A(x), \mu_B(y) \right).$$

MAX-MIN Composition

The MAX-MIN composition of two relations is defined by

$$T = R \circ S \text{ with } \mu_T(x, z) = \bigvee_{y \in Y} \{\mu_R(x, y) \wedge \mu_S(y, z)\}$$

where the $\vee$ operator indicates "maximum" or "max" and $\wedge$ means "minimum" or "min."

MAX-PRODUCT

The MAX-PRODUCT can be computed using two relations R and S as

$$T = R \circ S \text{ where } \mu_T(x, z) = \bigvee_{y \in Y} \{\mu_R(x, y) \bullet \mu_S(y, z)\}$$

Linguistic Variable and Values

Linguistic variables are given by their values, being words or sentences. Examples of a linguistic variable and its value is the variable *Age* with values of {*young, middle-aged, old*}.

Hedges

A hedge is a mathematical operation on a fuzzy set that modifies the severity of its shape using adverbs. Examples of a hedge are *very*, *more*, *less*, *almost*.

Antecedent and Consequence

In fuzzy logic IF–THEN statements, the *if* part represents the antecedent, while the *then* part corresponds to the consequence.

Fuzzy Reasoning

Fuzzy reasoning, also referred to as approximate reasoning, is a process to infer conclusions from fuzzy sets and fuzzy IF–THEN rules.

REVIEW QUESTIONS

1. What is the difference between a crisp set and a fuzzy set?
2. What does a degree of membership of 1.0 indicate if the membership function represents the linguistic variable *hot* compared to a degree of membership of 0.2?
3. How do you draw the membership function for a crisp set?
4. What is an interval-valued fuzzy set?
5. Why would one use a bell membership function over a triangular membership function?
6. What is an α-cut?
7. What is the difference between an α-cut and a strong α-cut?
8. When is a fuzzy set called normal?
9. What is the support of a fuzzy set?
10. What is the difference between the support and the core of a fuzzy set?
11. What is the Cartesian product of the two crisp sets $A = \{a, b\}$ and $B = \{9, 10\}$?
12. How do you find the fuzzy relation of two fuzzy sets?
13. Given two fuzzy sets A and B, how do you compute A AND B?
14. Given two fuzzy sets A and B, how do you compute A OR B?
15. How do you call R when R is computed as $R = A \times B$?
16. What do you compute when using the fuzzy MAX-PRODUCT?
17. Suppose you have a membership function $\mu(x)$ representing the linguistic value *deep*; how do you form the linguistic value *very deep* using $\mu(x)$?
18. Given two fuzzy sets A and B, what is the conjunction C of A and B?
19. Identify in the following sentence which word is the linguistic variable and which one is the linguistic value: IF exam is hard THEN grade is low.
20. What are the five elements of a linguistic variable?

21. Suppose A and B are linguistic values defined on X and Y, respectively. How do you read in plain English the statement $A \rightarrow B$?
22. Suppose A is true and B is false; what is $A \rightarrow B$?
23. What is the material implication, given two sets where A entails B?
24. Suppose sets A, B, and C have membership functions $\mu_A(x)$, $\mu_B(y)$, and $\mu_C(z)$; how do you express $R = A \times B \rightarrow C$?
25. How is the firing strength related to the α-cut?

PROBLEMS

2.1 Consider a universe X representing a collection of elements s; i.e., $\mu_A : A \rightarrow [0,1]$ where μ_A is the membership function of the fuzzy set A. Suppose $X = \{1,2,3,4,5,6,7,8,9,10\}$ and $A =$ "The element close to 4." Develop a discrete membership function (integer values for x) to represent A. Use a triangular membership function.

2.2 Suppose you want to model the grade in your intelligent controls course using a fuzzy set. Hence, $X = [0,4]$, which is your grade point average for the course. Define a continuous membership function to express a "good" grade for the course.

2.3 Consider the following fuzzy set B:

$$B = 1/1 + 0.9/2 + 0.8/3 + 0.7/4 + 0.6/5 + 0.5/6 + 0.25/7 + 0/8 + 0/9 + 0/10.$$

Find:

a. The support of B, supp(B).
b. For $\alpha = 0.1$, what is B_α?
c. For $\alpha = 0.4$, what is B_α?
d. For $\alpha = 0.75$, what is B_α?

2.4 Write a MATLAB function that generates a terminated ramp function with a support supp(A) = [5,10] and a core of core(A) = [6.5, 10]. Plot the function and verify the support and core.

2.5 Suppose you are given two fuzzy sets A and B:

$$A = 0.25/a_1 + 0.89/a_2 \text{ and } B = 0.29/b_1 + 0.48/b_2 + 1.0/b_3.$$

Find the fuzzy relation between these two sets.

2.6 Suppose $A = B = R^+$ and $C =$ "a is greater than b." The fuzzy relation C can be represented by the membership function

$$\mu_C(a,b) = \begin{cases} \dfrac{b-a}{a+b+2} & \text{for } b > a \\ 0 & \text{for } b \le a. \end{cases}$$

Suppose $A = \{3,4,5\}$ and $B = \{3,4,5,6,7\}$, find the fuzzy relation C.

2.7 Consider two fuzzy relations, $A \times B$ and $B \times C$, where $A = \{1,2,3\}$, $B = \{q,r,s,t\}$, and $C = \{m,n\}$. Suppose

$$R_1 = \begin{bmatrix} 0.1 & 0.3 & 0.5 & 0.7 \\ 0.4 & 0.2 & 0.8 & 0.9 \\ 0.6 & 0.8 & 0.3 & 0.2 \end{bmatrix}$$

represents that "*a* is relevant to *b*," and

$$R_2 = \begin{bmatrix} 0.9 & 0.1 \\ 0.2 & 0.3 \\ 0.5 & 0.6 \\ 0.7 & 0.2 \end{bmatrix}$$

represents that "*b* is relevant to *c*."

a. Find the relationship of "a is relevant to c" using the MAX-MIN composition.
b. Find the relationship of "a is relevant to c" using the MAX-PRODUCT composition.

2.8 Consider the fuzzy set A,

$$A = 0.2/a_1 + 0.9/a_2$$

and the fuzzy set B,

$$B = 0.3/b_1 + 0.5/b_2 + 1.0/b_3.$$

Find the relation $R = A \times B$.

2.9 Given two fuzzy relations R_1 and R_2,

$$R_1 = \begin{bmatrix} 0.6 & 0.8 \\ 0.7 & 0.9 \end{bmatrix} \text{ and } R_2 = \begin{bmatrix} 0.3 & 0.1 \\ 0.2 & 0.8 \end{bmatrix},$$

show that $R_1 \circ R_2 \neq R_2 \circ R_1$ using both the MAX-MIN as well as the MAX-PRODUCT rule.

2.10 Write a MATLAB routine that produces a graph of the Gaussian membership function for a variable *approximately 6*, where $\mu_A : X \to [0, 10]$.

2.11 Suppose

$$\mu_A(x) = \begin{cases} \dfrac{x-3}{3}, & 3 \leq x \leq 6 \\ \dfrac{9-x}{3}, & 6 \leq x \leq 9 \end{cases}, \quad \mu_B(x) = \begin{cases} \dfrac{x-2}{3}, & 2 \leq x \leq 5 \\ \dfrac{8-x}{3}, & 5 \leq x \leq 8 \end{cases}.$$

Find

$$\max[\min\{\mu_A(x), \mu_B(x)\}].$$

2.12 Suppose two discrete fuzzy sets A and B are defined as

$$A = 0.75/2 + 0.98/3 + 0.29/4 \text{ and } B = 0.35/2 + 1.0/3 + 0.65/4 + 0.18/5.$$

a. Apply the hedges *A little*, *Very*, and *Very very* to set A and plot these in MATLAB.
b. Apply the hedges *Slightly*, *Extremely*, and *Somewhat* to set B and plot these in MATLAB.
c. Suppose A defines the linguistic term *Small* and B defines the linguistic term *Large*. Compute the phrase *not very small* and *not extremely large*.

2.13 For the two sets given in Problem 2.12, find the fuzzy relation R that expresses: IF A' THEN B'.

2.14 Consider the fuzzy relation R given by the Cartesian product $X \times Y$, which represents the rule: IF x is A THEN y is B. What is the output B if

$$A = 0.3/x_1 + 1.0/x_2 + 0.6/x_3$$

and

$$R = \begin{bmatrix} 0.19 & 0.82 & 0.29 & 0.89 \\ 0.39 & 0.91 & 0.11 & 0.39 \\ 0.21 & 0.51 & 0.31 & 0.49 \end{bmatrix} ?$$

2.15 Consider a control system that acts on discrete sensor data – that is, the sensor reads the physical phenomenon and categorizes it to be one of six distinct values $\{1.0, 1.5, 2.0, 2.5, 3.0, 3.5\}$. Suppose you are in the process of designing a fuzzy logic controller and you defined two fuzzy membership functions A and B labeled as *less* and *more*, respectively. A graphical representation of A and B is shown in Figure 2.18.

a. Define the discrete fuzzy set A_d as *less* and the discrete fuzzy set B_d as *more*.
b. Find $A_d \cup B_d$, $A_d \cap B_d$, and $\overline{A}_d$.
c. Use MATLAB to solve part (b) of this problem.

2.16 In this problem we explore some of the tasks required in designing a fuzzy logic controller. Two sensors are utilized to compute the percentage of actuation signal needed to make corrections on an item of plant. Sensor A measures discrete values of temperature $A = \{18, 20, 22, 24\}$ and sensor B measures discrete values of deflection $B = \{-0.5, -0.25, 0.0, 0.25, 0.5, 0.75\}$. Let's define a fuzzy set *Medium Temperature* and a fuzzy set *Medium Deflection*, using discrete triangular membership functions:

Medium Temperature: $T_M = 0.4/20 + 1.0/22 + 0.6/24$
Medium Deflection: $D_M = 0.2/(-0.25) + 1.0/0.0 + 0.65/0.25 + 0.3/0.5$.

Hence, we can make the following statement: IF temperature is Medium THEN deflection is Medium (i.e., if T_M then D_M). This can be evaluated by computing $R = T_M \times D_M$. If one expands on this reasoning by including an implication fuzzy set,

Medium Actuation: $A_M = 0.2/(15) + 0.8/35 + 1.0/60 + 0.25/80$,
where the set determines the percentage of actuation (i.e., 0–100%), we can evaluate the following statement:

IF temperature is T_M, THEN deflection is D_M, else actuation is A_M

by computing

$$\mu_R(t = temperature, d = deflection) = \max\left[\{\mu_T(t) \wedge \mu_D(d)\}, \{(1 - \mu_T(t)) \wedge \mu_A(d)\}\right].$$

Determine μ_R and comment on the implication.

2.17 In Problem 2.16, suppose a new antecedent T_M^* is introduced as *Almost Medium Temperature*:

Almost Medium Temperature: $T_M^* = 0.1/18 + 0.8/20 + 1.0/22 + 0.2/24$.

What will be the consequence D_M^*?

2.18 Suppose a fuzzy system involves two rules:
Rule 1: IF x is A_1 AND y is B_1 THEN z is C_1
Rule 2: IF x is A_2 AND y is B_2 THEN z is C_2
Fact: x is A' AND y is B'

Conclusion: z is C'
Compute the conclusion according to

$$C' = (A' \times B') \circ (R_1 \cup R_2) = \left[(A' \times B') \circ R_1\right] \cup \left[(A' \times B') \circ R_2\right].$$

Use the following membership functions:

$$\mu_{A_1}(x) = \begin{cases} \dfrac{x - 1}{3}, & 1 \leq x \leq 4 \\ \dfrac{7 - x}{3}, & 4 \leq x \leq 7 \end{cases}$$

$$\mu_{B_1}(y) = \begin{cases} \dfrac{y - 3}{3}, & 3 \leq y \leq 6 \\ \dfrac{9 - y}{3}, & 6 \leq y \leq 9 \end{cases}$$

$$\mu_{C_1}(z) = \begin{cases} \dfrac{z - 0}{3}, & 0 \leq z \leq 3 \\ \dfrac{6 - z}{3}, & 3 \leq z \leq 6 \end{cases}$$

$$\mu_{A_2}(x) = \begin{cases} \dfrac{x - 4}{3}, & 4 \leq x \leq 7 \\ \dfrac{10 - x}{3}, & 7 \leq x \leq 10 \end{cases}$$

$$\mu_{B_2}(y) = \begin{cases} \dfrac{y - 2}{3}, & 2 \leq y \leq 5 \\ \dfrac{8 - y}{3}, & 5 \leq y \leq 8 \end{cases}$$

$$\mu_{C_2}(z) = \begin{cases} \dfrac{z - 2}{3}, & 2 \leq z \leq 5 \\ \dfrac{8 - z}{3}, & 5 \leq z \leq 8 \end{cases}$$

2.19 Using the membership functions and rules defined in Problem P2.18, evaluate graphically what C will be for an input of $x = 3.4$ and $y = 7.1$.

2.20 Consider the room temperature control problem described in Section 2.1.

Develop three rules for how the heating and cooling system should behave for each influence parameter – that is, outside temperature, number of people in the room, doors, and windows. Include the description of the actuation, the membership functions for each influence parameter, and the actuation. Utilizing the rules developed, draw out graphically the fuzzy inference system similar to Figure 2.24 and comment on your choice of membership functions.

REFERENCES

[1] Novák V, Perfilieva I, Močkoř J (1999). *Mathematical Principles of Fuzzy Logic*. Dordrecht: Kluwer Academic.

[2] Pelletier FJ (2000). Review of metamathematics of fuzzy logics. *Bulletin of Symbolic Logic*, 6(3): 342–346.

[3] Zadeh LA (1965). Fuzzy sets. *Information and Control*, 8(3): 338–353.

3 Fuzzy Inference and Fuzzy Logic Control

3.1 Fuzzy Inference: Mamdani Fuzzy Inference

So far we have been able to construct a fuzzy reasoning system using membership functions which allow us to define rules and their consequences, and aggregate them to find an overall conclusion. However, we need to be able to move back from the "fuzzy" world into the real world – that is, to have a crisp outcome to implement such systems in control applications. Consider the conclusion of the fuzzy reasoning system shown in Figure 2.23. The conclusion is given by a single membership function. If this conclusion represents the amount of heating, we would need to apply a crisp value to the actuator. In our temperature controller system example the membership function C' is useless as it does not represent a number the actuator understands. Hence, we need to defuzzify the conclusion of our fuzzy reasoning system. There are a number of different implementations of such a fuzzy system, as well as a number of different aggregation and defuzzification methods. One of the more popular fuzzy inference systems (FIS) implementations is the Mamdani Fuzzy Inference System (MFIS). The MFIS was first developed in 1975 by Professor Ebrahim Mamdani, who devised a fuzzy system controlling a steam engine and a boiler. The MFIS can be constructed in five steps: (1) fuzzification of the input variables; (2) applying the fuzzy operation; (3) applying implication; (4) applying the aggregation method; and (5) defuzzification.

Consider Figure 3.1, which represents all five steps of an MFIS for a temperature- and pressure-based control system.

The first step of fuzzification is done by applying the crisp input values for the pressure and the temperature and determining the degree of membership of each appropriate fuzzy set. In the example shown in Figure 3.1, the input pressure is 34 N/m^2 and the input temperature is 58°C. There are three rules:

R1: IF *pressure* is low OR *temperature* is low THEN *activation* is high.
R2: IF *pressure* is medium OR *temperature* is medium THEN *activation* is average.
R3: IF *pressure* is high OR *temperature* is high THEN *activation* is low.

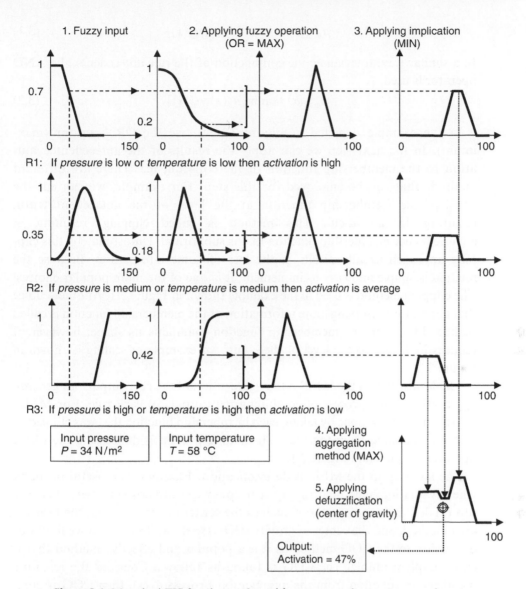

Figure 3.1 Mamdani FIS for three rules, with *pressure* and *temperature* as input.

Accordingly, the input pressure of 34 N/m² yields a degree of membership for the fuzzy *pressure* set of R1 a value of 0.7, for R2 $\mu_P(P = 34) = 0.35$, and for R3 $\mu_P(P = 34) = 0$. The same process of fuzzification is done for the input variable *temperature*, resulting in $\mu_T(P = 58) = 0.2$ for R1, $\mu_T(P = 58) = 0.18$ for R2, and $\mu_T(P = 58) = 0.42$ for R3.

The second step is to take the fuzzified inputs and apply them to the antecedents of the fuzzy rules. If a fuzzy rule has several antecedents, as shown in Figure 3.1, the fuzzy operator AND or OR is utilized to obtain a single number representing the result of the antecedent evaluation. The OR operator allows us to evaluate the disjunction of the rule antecedents:

$$\mu_{A \cup B}(x) = \max\{\mu_A(x), \mu_B(x)\}. \tag{3.1}$$

In a similar way, to evaluate the conjunction of the rule antecedents, the AND operator is used:

$$\mu_{A \cap B}(x) = \min\{\mu_A(x), \mu_B(x)\}. \tag{3.2}$$

In the example depicted in Figure 3.1 we use the OR operator (maximum). In the next step we can apply the results of the antecedent evaluation to the membership function of the consequence. There are different methods that can be employed for this step. For example, we can cut the consequence membership function at the level of the antecedent truth, resulting in an α-cut. This method is called clipping, resulting in a consequence membership function that is subnormal. Unfortunately, this clipping approach usually results in the loss of some information. Because the method is simple and easy to implement, it is one of the more popular choices. The clipping method is used in the example shown in Figure 3.1. Another choice that allows for preserving more information of the membership function is called scaling. In scaling the membership function maintains its shape; however, it reduces its height to the level of the α-cut. An example of scaling is shown in Figure 3.2.

The fourth step in the MFIS is the aggregation of the rule outputs. In general, aggregation is the process of combining the outputs of each rule. For MFIS, aggregation can be done by taking each membership function that was clipped or scaled and adding them together. The adding process is done using the MAX operator, as shown in Figure 3.1.

The last step in the MFIS is defuzzification. Fuzziness is a useful property when evaluating rules, but to apply an output we need a crisp number. There are several defuzzification methods such as the center of area method, the bisector area method, and the center of gravity (COG) method. In this text we limit our discussion to the COG method as it is a popular and effective method that is easily implementable. The COG is found as follows: Consider the resulting membership function from the aggregation process $\mu_R(x)$, then COG is computed as

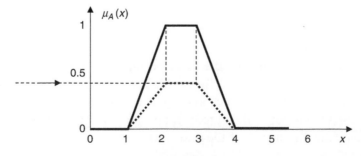

Figure 3.2 Mamdani fuzzy inference: scaling of the consequence membership function.

$$COG = \frac{\int x\mu_R(x)dx}{\int \mu_R(x)dx}. \tag{3.3}$$

In the example shown in Figure 3.1 the resulting crisp value using the COG method is 47.

Designing a MFIS in MATLAB can be done in a number of different ways. In this text, we will utilize MATLAB's Fuzzy Logic Toolbox GUI as it is very efficient and allows for exporting the resulting fuzzy logic design into Simulink.

Example 3.1

To start the process, type in the command window of MATLAB `fuzzyLogic Designer`. If you are using an older version of MATLAB the command is simply `fuzzy`. This will open the FIS Editor window, shown in Figure 3.3.

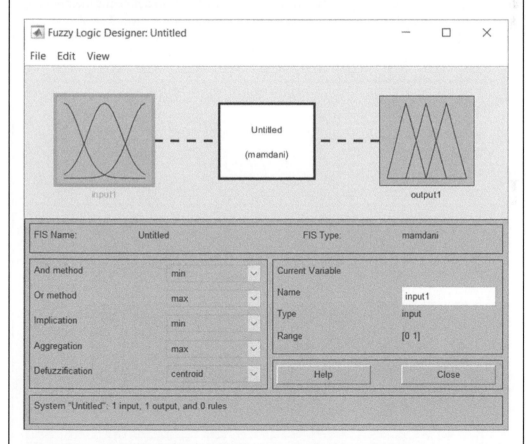

Figure 3.3 Mamdani fuzzy inference: MATLAB implementation – FIS Editor.

Suppose we want to create an MFIS for the following two-input, single-output model using four rules:

R1: IF X is small AND Y is small THEN Z is negative large.
R2: IF X is small AND Y is large THEN Z is negative small.
R3: IF X is large AND Y is small THEN Z is positive small.
R4: IF X is large AND Y is large THEN Z is positive large.

Let's also assume the universe of discourse for the three variables is

$$X = [-10, 10], \ Y = [-10, 10], \text{ and } Z = [-10, 10].$$

We will start creating the MFIS by first defining the inputs. As the FIS system we are building has two inputs, we will add one input to the default FIS system. Adding and removing variables is done by selecting **Edit > Add Variable > Input**. This will create a second input box labeled **input2**. Before proceeding with defining each input's set of membership functions, we first change the names of the input to the ones given by the set of rules. Click on the **input1** box and edit the name field **input1** to X. Repeat the same for the **input2** box to change the name to Y. Now we are ready to define the membership functions for X and Y. To do so, double-click the box for X, which will open the Membership Function Editor, as shown in Figure 3.4.

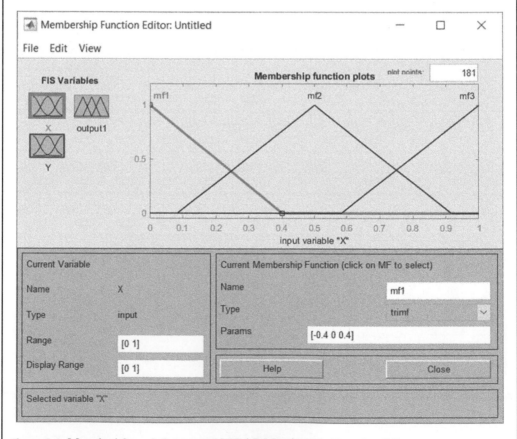

Figure 3.4 Mamdani fuzzy inference: MATLAB Membership Function Editor.

First, we will define the universe of discourse for X. Under the **Current Variable** panel in the Membership Function Editor, change the default range [0 1] to [−10 10]. The **Display Range** is automatically altered after you hit enter. From the rule set given above, we notice that X has two membership functions: *small* and *large*. Currently, the default set of membership functions has three components, as shown in Figure 3.4, labeled as *mf1*, *mf2*, and *mf3*. To change this we first delete the current set of membership functions. This is done by selecting one at a time from the **Membership Function Plots** panel and using the delete button on the keyboard. Next, we add two membership functions by selecting under **Edit > Add MFs** ... A new window will appear that allows us to select how many and what type of membership function we would like to create. For this example, we choose one `zmf` and one `smf` membership function. Each membership function has its set of parameters (see Table 2.2) and can be edited in the **Current Membership Function** panel. In addition, the names of the membership functions can also be edited. We will use *small* and *large* and change the parameters to [−3 9] and [−9 3], respectively. The membership functions for the variable X should look like the one shown in Figure 3.5.

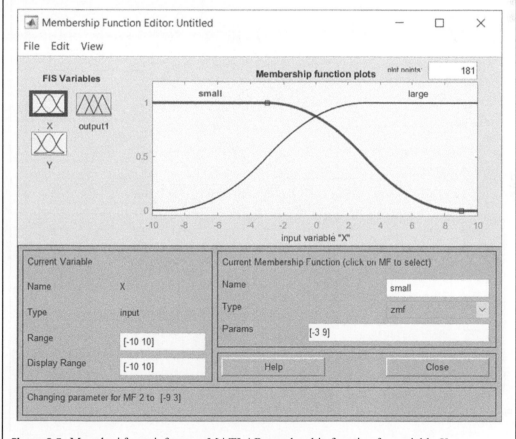

Figure 3.5 Mamdani fuzzy inference: MATLAB membership function for variable X.

We repeat this process for the variable Y, which also has two membership functions called *small* and *large*. We utilize the same type of membership function zmf for small with parameters [−5 5] and smf for large with parameters [−5 5]. The resulting graph of the two membership functions for the variable Y is shown in Figure 3.6.

For the output variable Z we construct four membership functions: *negative large*, *negative small*, *positive small*, and *positive large*. We use gbellmf for all three membership functions and the following parameters, respectively: [1.667 2.5 0], [1.667 2.5 3.33], [1.667 2.5 6.667], and [1.667 2.5 10].

To construct the MFIS that connects the inputs and output we use the Rule Editor, which can be found from the main FIS Editor using **Edit > Rules** ... and is shown in Figure 3.7.

To construct the set of rules – in this example we have four rules – we select the linguistic value of each input variable and the connection (AND or OR), and

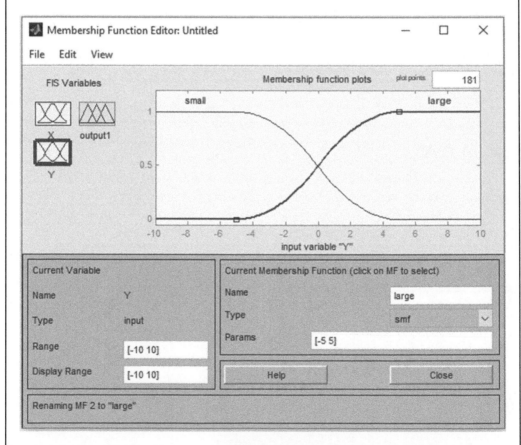

Figure 3.6 Mamdani fuzzy inference: MATLAB membership function for variable Y.

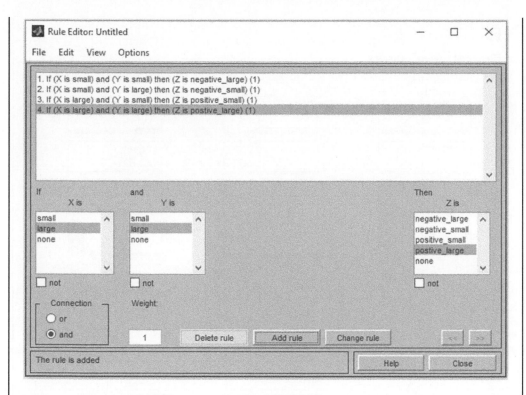

Figure 3.7   Mamdani fuzzy inference: MATLAB Rule Editor.

associate them with the output. There are some other choices that can be included in the rule definitions, such as negating (NOT) and adding weights to a rule. After adding each of the four rules, the Rule Editor should look as shown in Figure 3.7. With the addition of the rules we have constructed an MFIS. To see how it performs, we can open **View > Rules** and find the Rule Viewer. The Rule Viewer is MATLAB's version of Figure 3.1. However, it is interactive. For example, in the **Input** panel one can provide the input quantities (crisp values) for X and Y, and immediately obtain the resulting crisp output. The default aggregation method is MAX, and the defuzzification method corresponds to the CENTROID method. These settings can be changed in the main FIS Editor. Figure 3.8 shows the final MFIS system using the Rule Viewer and an input of $X = 0.5$ and $Y = 2.5$.

To use the created FIS in MATLAB or Simulink you will need to save the given file with a *.fis extension. To save an FIS file, use **File > Export > To File**. When in a MATLAB environment (command window or *.m file application) the FIS file is loaded using the following command:

```
fis = readfis('filename');
```

To generate an output with a given input (i.e., $X = -2$ and $Y = 3$) use the following command in MATLAB:

```
out = evalfis(fis,[-2 3])
```

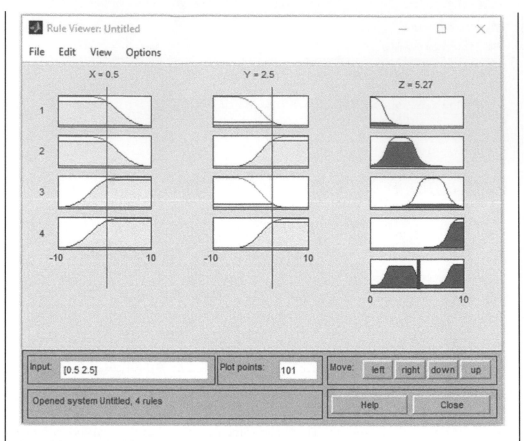

Figure 3.8   Mamdani fuzzy inference: MATLAB Rule Viewer window.

The created FIS can also be imported into Simulink. To do so use the Fuzzy Logic Controller block from the Fuzzy Logic Toolbox in Simulink and enter the *.fis file created previously.

3.2   Fuzzy Inference: Sugeno Fuzzy Inference

Aggregation and defuzzification for Mamdani-style inference systems are generally not very computationally efficient. To bypass this shortcoming, Michio Sugeno proposed a singleton as a membership function for the rule consequent. A fuzzy singleton is a fuzzy set with a membership function that is unity at a single point on the universe of discourse and zero everywhere else. The Sugeno-style fuzzy system is based on the following format:

Rule 1: IF x is A_1 and y is B_1 THEN z is C_1

    IF x is A
    AND y is B
    THEN z is $f(x, y)$.

Here, x, y, and z are the linguistic variables, A and B are the fuzzy sets with the universe of discourses X and Y, respectively, and $f(x, y)$ is a function of x and y. Figure 3.9 shows the implementation of the Sugeno FIS using the temperature–pressure example from the MFIS discussion.

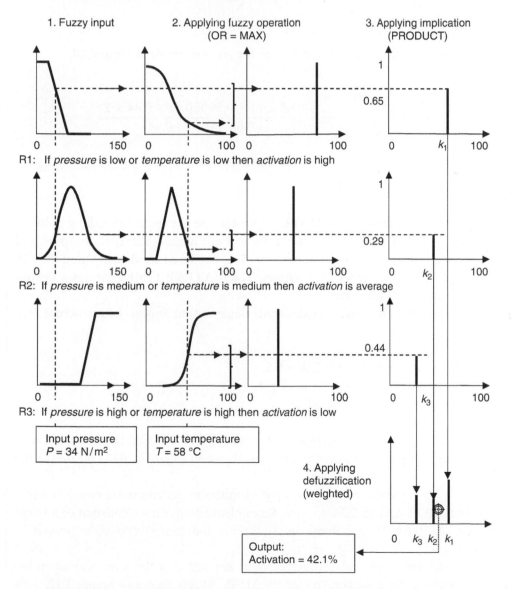

Figure 3.9 Sugeno FIS for three rules with pressure and temperature as input.

Four steps are involved in creating a Sugeno FIS output. The first two steps are identical to the MFIS. However, we note that the aggregation and defuzzification methods differ from the MFIS approach. The singleton for the activation membership function is reduced in height based on the firing strength of the rule and the defuzzification is performed using a weighted average. This weighted average can be computed by

$$\text{WA} = \frac{\sum_{i=1}^{n} \mu(k_i) \times k_i}{\sum_{i=1}^{n} \mu(k_i)}. \tag{3.4}$$

Consider the weighted average for the given example in Figure 3.9:

$$\text{WA} = \frac{\sum_{i=1}^{n} \mu(k_i) \times k_i}{\sum_{i=1}^{n} \mu(k_i)} = \frac{0.65 \times 71 + 0.29 \times 0.42 + 0.44 \times 27}{0.65 + 0.29 + 0.44} = 42.1\%.$$

Example 3.3

To implement a Sugeno FIS in MATLAB we can use the same FIS Editor as introduced in the previous section. However, prior to implementing the inputs, outputs, and rules, we need to set the FIS to Sugeno type. This can be done when creating the FIS structure: go to **File > New FIS > Sugeno**. The MATLAB FIS Editor has a default setting for the MFIS structure.

Suppose we want to create a single-input, single-output Sugeno FIS structure with three rules in MATLAB:

R1: IF X is small THEN $Y = 0.25X + 6.5$.
R2: IF X is medium THEN $Y = -0.4X + 5$.
R3: IF X is large THEN $Y = X - 1.25$.

Using `gbellmf()` membership functions for small, medium, and large on a universe of discourse [0 10], the membership functions shown in Figure 3.10 are created in the FIS Editor.

For the rule implementation, the output membership functions are now just equations. The MATLAB FIS Editor allows for implementing either a constant or a linear equation. As the rules are linear in X, we use the second choice, as shown in Figure 3.11.

Having defined input and output, the rules are added in the same way as in the previous section where we constructed the MFIS. This is shown in Figure 3.12, with equations substituting for the output membership functions. When using the Rule Viewer, as depicted in Figure 3.13, we can see that the output for each rule is a singleton.

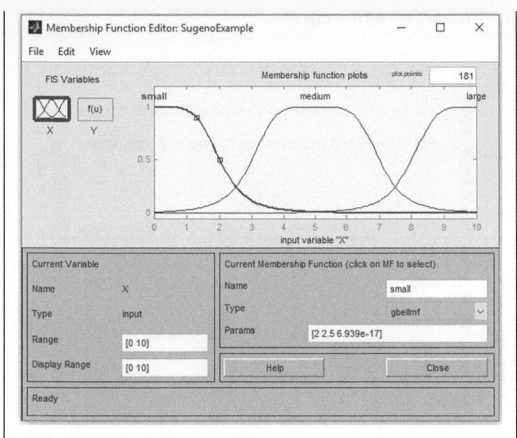

Figure 3.10 Sugeno FIS: input membership function using `gbellmf()`.

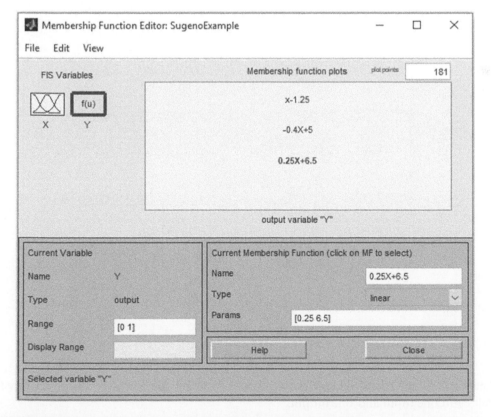

Figure 3.11 Sugeno FIS: output membership function as linear equations.

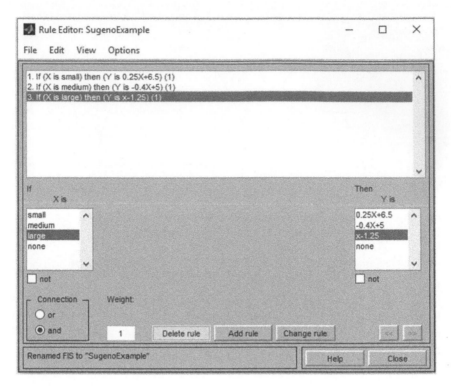

Figure 3.12 Sugeno FIS: rule implementation.

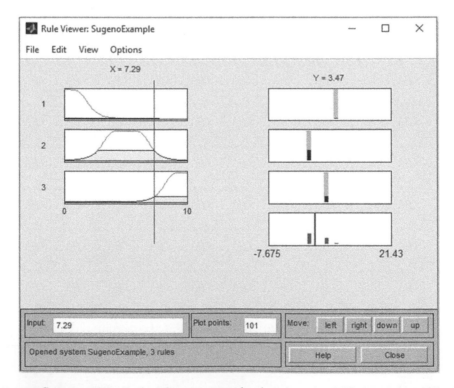

Figure 3.13 Sugeno FIS: Rule Viewer example showing the FIS structure with three rules and one input and one output.

Having introduced two different FIS structures, we have a choice to make on which structure to use for which application. So far, we know Sugeno is simpler to implement as it requires fewer computational resources. However, are there other criteria to be considered? From the literature, MFIS structures are considered to be able to capture expert knowledge. The MFIS seems to allow a more intuitive and human-like description of the process to be modeled. For Sugeno structures, the efficient implementation is utilized often for problems entailing optimization and adaptive techniques. Such techniques are found in control systems, especially complex control systems such as dynamic nonlinear system control and adaptive control.

How should we tune FIS? There are a few simple steps one can take to improve the performance of an FIS:

1. Determine if the universe of discourse – or range – for the input and the output is sufficient. For example, if the simulations result in saturation of either variable it may be beneficial to extend the range.
2. Starting with just a few fuzzy sets is usually advantageous; however, the performance is somewhat rough, and a review of the fuzzy sets with additional sets over the universe of discourse may yield smoother performance.
3. When a number of fuzzy sets covers the universe of discourse, make sure to provide for sufficient overlap between adjacent fuzzy sets. As a rough guide, have an overlap of their bases of 25–50% between triangle to triangle or trapezoidal to triangle fuzzy sets.
4. Increasing the number of rules may also help with the performance. Sometimes this requires creating more linguistic values and membership functions using hedges, for example.
5. Incorporating weights for each rule allows for emphasizing the importance of some rules over others.

You may find other FIS structures in the literature, such as the Larsen, Takagi, or Kang FIS; however, for most applications Mamdani and Sugeno are quite suitable. In the next section we present fuzzy logic-based controllers, and with that we will have a chance to introduce the Takagi–Sugeno FIS controller.

3.3 Fuzzy Logic Control

Let's revisit our room temperature controller, as depicted in Figure 2.1. The controller block will now house our fuzzy logic system, which is directing the heating and cooling system. The decision-making process is done through a set of conditional IF–THEN statements derived from the knowledge base. The input to the controller block is fuzzified and the output is defuzzified. From the outside, nothing distinguishes the fuzzy logic-based controller block from a regular hard

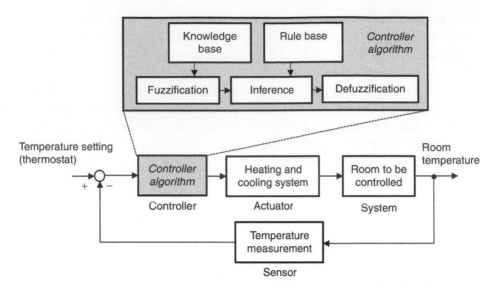

Figure 3.14 Block diagram of temperature control of a room.

computing controller: it is a block in our block diagram with peripheral input and output data streams. The inner workings, however, are not based on the dynamics of the system described in some mathematical form, but in terms of the rules we impose and the way we apply an implication and a defuzzification. Some conditions, however, are shared with regular controllers. For example, the system needs to be observable and controllable and a solution to the control problem must exist. Additionally, stability and optimality are not addressed explicitly in fuzzy logic-based controllers, as these types of topics are still open issues to be addressed in research on fuzzy logic controllers.

Figure 3.14 incorporates the details of the fuzzy logic portion of the fuzzy logic controller into our original control diagram for the temperature control problem. The question now becomes how we design the fuzzy logic system so that it correctly interacts with the parameters and inner workings of a feedback control system. We can summarize the process in 10 steps:

Step 1: For a given plant, identify the input, output, and state variables.

Step 2: For each variable, determine the range and partition the universe of discourse into a number of fuzzy subsets, each having a linguistic name.

Step 3: Choose a membership function for each fuzzy subset.

Step 4: Define the rule base by using each input, state, and output fuzzy subset.

Step 5: Use scaling factors for the input and output variables to normalize the variables to be in some defined interval.

Step 6: Fuzzify the input variables to the controller.

Step 7: Apply fuzzy approximate reasoning to determine the output of each rule.

Step 8: Use aggregation of the fuzzy outputs.

Step 9: Use defuzzification to compute a crisp output.

Step 10: Apply crisp output to the actuator and sense the system output to evaluate error.

To illustrate these steps the temperature control of a room is implemented using a step-by-step approach.

Example 3.4

Consider a classroom at a school in which there is a heating unit, an air-conditioning unit, several windows, and two doors. We want to design a fuzzy logic controller to maintain the air temperature at some comfortable level, regardless of any influences such as doors or windows being open or closed. The schematic of the system and controller interaction is shown in Figure 3.14.

Using fuzzy logic controllers we can craft two simple rules to achieve the above stated control goal:

- If the temperature is higher than 69°F, then start cooling.
- If the temperature is lower than 69°F, then start heating.

To incorporate these two simple rules into a fuzzy logic controller, we follow the steps listed above:

Step 1: For a given plant, identify the input, output, and state variables.

The plant is the room, the input is the air flow either from the heater or the air-conditioner, the state is the temperature in the room, and the output is the state (i.e., the temperature).

Step 2: For each variable, determine the range and partition the universe of discourse into a number of fuzzy subsets, each having a linguistic name.

Let's define the error in temperature as

$$e = T_o - T_m,$$

where T_o is the desired temperature and T_m is the measured temperature. In order to fuzzify the error, let's define five fuzzy sets:

- Negative Big (NB)
- Negative Small (NS)
- Zero (O)
- Positive Small (PS)
- Positive Big (PB).

According to Step 2, we also need to identify the universe of discourse. We can do this by associating different levels of error. For example, $-3°F$ is associated with NB by a degree of membership of 1.0 and 0 for PB. Doing this for each of the fuzzy sets, we can establish Table 3.1.

The same approach can be utilized to fuzzify the input. Let's define five fuzzy sets for the input r: NB, NS, O, PS, and PB. For the control input range we will use the ranges as specified in Table 3.2.

Table 3.1 Error level and fuzzy sets

MF degree		Range						
		−3	−2	−1	0	+1	+2	+3
Fuzzy	PB	0	0	0	0	0	0.5	1.0
Sets	PS	0	0	0	0	1.0	0.5	0
	O	0	0	0.5	1.0	0.5	0	0
	NS	0	0.5	1.0	0	0	0	0
	NB	1.0	0.5	0	0	0	0	0

Table 3.2 Input level and fuzzy sets

MF degree		Range								
		−4	−3	−2	−1	0	+1	+2	+3	+4
Fuzzy	PB	0	0	0	0	0	0	0	0.5	1.0
Sets	PS	0	0	0	0	0	0.5	1.0	0.5	0
	O	0	0	0	0.5	1.0	0.5	0	0	0
	NS	0	0.5	1.0	0.5	0	0	0	0	0
	NB	1.0	0.5	0	0	0	0	0	0	0

Step 3: Choose a membership function for each fuzzy subset.

Inspecting Table 3.1 and 3.2, we notice that Step 3 has been accomplished by the discrete values chosen for the degree of membership to each fuzzy set. The membership functions for both the error and the input are all simple triangular membership functions – although discrete.

Step 4: Define the rule base by using each input, state, and output fuzzy subset.

As stated at the beginning of the example, we heat when the temperature is below our target temperature T_o and we cool when the room temperature is above T_o. Using the grading given by the discrete membership functions and the corresponding fuzzy sets, we can construct the following simple fuzzy logic rules:

Rule 1: IF e = NB THEN r = NB
Rule 2: IF e = NS THEN r = NS
Rule 3: IF e = 0 THEN r =0
Rule 4: IF e = PS THEN r = PS
Rule 5: IF e = PB THEN r = PB.

Step 5: Use scaling factors for the input and output variables to normalize the variables to be in some defined interval.

In this example we do not need to scale the input and output as we have not defined the implementation of the given control system. If the implementation utilizes a specific voltage level for a certain measurement and a certain voltage level for a particular control action, these ranges would be used to normalize the variables.

Before moving on to Step 6, let's express the FIS we have constructed so far mathematically. By utilizing the membership functions and fuzzy sets, we can establish the relationship between the error e and the input r using the Cartesian product between the two:

$$R = \{NB(e) \times NB(r)\} \cup \{NS(e) \times NS(r)\} \cup$$

$$\{O(e) \times O(r)\} \cup \{PS(e) \times PS(r)\} \cup \{PB(e) \times PB(r)\}.$$

To compute each term, consider the first Cartesian product between $NB(e)$ and $NB(r)$. Using each corresponding row of NB as shown in Figure 3.15, we can compute the product:

MF degree		Range						
		−3	−2	−1	0	+1	+2	+3
Fuzzy Sets	PB	0	0	0	0	0	0.5	1.0
	PS	0	0	0	0	1.0	0.5	0
	O	0	0	0.5	1.0	0.5	0	0
	NS	0	0.5	1.0	0	0	0	0
	NB	1.0	0.5	0	0	0	0	0

$NB(e)$ ⟹

MF degree		Range							
		−4	−3	−2	−1	0	+1	+2	+3	+4
Fuzzy Sets	PB	0	0	0	0	0	0	0	0.5	1.0
	PS	0	0	0	0	0	0.5	1.0	0.5	0
	O	0	0	0	0.5	1.0	0.5	0	0	0
	NS	0	0.5	1.0	0.5	0	0	0	0	0
	NB	1.0	0.5	0	0	0	0	0	0	0

$NB(r)$ ⟹

Figure 3.15 Computation of the relationship between e and r.

$$NB(e) \times NB(r) = \begin{bmatrix} 1.0 \\ 0.5 \\ 0 \\ 0 \\ 0 \\ 0 \\ 0 \end{bmatrix} \times \begin{bmatrix} 1.0 & 0.5 & 0 & 0 & 0 & 0 & 0 & 0 & 0 \end{bmatrix}$$

$$= \begin{bmatrix} 1.0 & 0.5 & 0 & 0 & 0 & 0 & 0 & 0 & 0 \\ 0.5 & 0.50 & 0 & 0 & 0 & 0 & 0 & 0 & 0 \\ 0 & 0 & 0 & 0 & 0 & 0 & 0 & 0 & 0 \\ 0 & 0 & 0 & 0 & 0 & 0 & 0 & 0 & 0 \\ 0 & 0 & 0 & 0 & 0 & 0 & 0 & 0 & 0 \\ 0 & 0 & 0 & 0 & 0 & 0 & 0 & 0 & 0 \\ 0 & 0 & 0 & 0 & 0 & 0 & 0 & 0 & 0 \end{bmatrix}.$$

Moving one row up in each table in order to compute the Cartesian product between $NS(e)$ and $NS(r)$, we find:

$$NS(e) \times NS(r) = \begin{bmatrix} 0 \\ 0.5 \\ 1.0 \\ 0 \\ 0 \\ 0 \\ 0 \end{bmatrix} \times \begin{bmatrix} 0 & 0.5 & 1.0 & 0.5 & 0 & 0 & 0 & 0 & 0 \end{bmatrix}$$

$$= \begin{bmatrix} 0 & 0 & 0 & 0 & 0 & 0 & 0 & 0 & 0 \\ 0 & 0.5 & 0.5 & 0.5 & 0 & 0 & 0 & 0 & 0 \\ 0 & 0.5 & 1.0 & 0.5 & 0 & 0 & 0 & 0 & 0 \\ 0 & 0 & 0 & 0 & 0 & 0 & 0 & 0 & 0 \\ 0 & 0 & 0 & 0 & 0 & 0 & 0 & 0 & 0 \\ 0 & 0 & 0 & 0 & 0 & 0 & 0 & 0 & 0 \\ 0 & 0 & 0 & 0 & 0 & 0 & 0 & 0 & 0 \end{bmatrix}.$$

Repeating this for the other terms:

$$O(e) \times O(r) = \begin{bmatrix} 0 & 0 & 0 & 0 & 0 & 0 & 0 & 0 & 0 \\ 0 & 0 & 0 & 0.5 & 0.5 & 0.5 & 0 & 0 & 0 \\ 0 & 0 & 0 & 0.5 & 1.0 & 0.5 & 0 & 0 & 0 \\ 0 & 0 & 0 & 0.5 & 0.5 & 0.5 & 0 & 0 & 0 \\ 0 & 0 & 0 & 0 & 0 & 0 & 0 & 0 & 0 \\ 0 & 0 & 0 & 0 & 0 & 0 & 0 & 0 & 0 \\ 0 & 0 & 0 & 0 & 0 & 0 & 0 & 0 & 0 \end{bmatrix}$$

$$PS(e) \times PS(r) = \begin{bmatrix} 0 & 0 & 0 & 0 & 0 & 0 & 0 & 0 & 0 \\ 0 & 0 & 0 & 0 & 0 & 0 & 0 & 0 & 0 \\ 0 & 0 & 0 & 0 & 0 & 0 & 0 & 0 & 0 \\ 0 & 0 & 0 & 0 & 0 & 0 & 0 & 0 & 0 \\ 0 & 0 & 0 & 0 & 0 & 0.5 & 1.0 & 0.5 & 0 \\ 0 & 0 & 0 & 0 & 0 & 0.5 & 0.5 & 0.5 & 0 \\ 0 & 0 & 0 & 0 & 0 & 0 & 0 & 0 & 0 \end{bmatrix},$$

and

$$PB(e) \times PB(r) = \begin{bmatrix} 0 & 0 & 0 & 0 & 0 & 0 & 0 & 0 & 0 \\ 0 & 0 & 0 & 0 & 0 & 0 & 0 & 0 & 0 \\ 0 & 0 & 0 & 0 & 0 & 0 & 0 & 0 & 0 \\ 0 & 0 & 0 & 0 & 0 & 0 & 0 & 0 & 0 \\ 0 & 0 & 0 & 0 & 0 & 0 & 0 & 0 & 0 \\ 0 & 0 & 0 & 0 & 0 & 0 & 0 & 0.5 & 0.5 \\ 0 & 0 & 0 & 0 & 0 & 0 & 0 & 0.5 & 1.0 \end{bmatrix},$$

which yields

$$R = \begin{bmatrix} 1.0 & 0.5 & 0 & 0 & 0 & 0 & 0 & 0 & 0 \\ 0.5 & 0.5 & 0.5 & 0.5 & 0 & 0 & 0 & 0 & 0 \\ 0 & 0.5 & 1.0 & 0.5 & 1.0 & 0.5 & 0 & 0 & 0 \\ 0 & 0 & 0 & 0.5 & 0.5 & 0.5 & 0 & 0 & 0 \\ 0 & 0 & 0 & 0.5 & 0.5 & 0.5 & 1.0 & 0.5 & 0 \\ 0 & 0 & 0 & 0 & 0 & 0.5 & 0.5 & 0.5 & 0.5 \\ 0 & 0 & 0 & 0 & 0 & 0 & 0 & 0.5 & 1.0 \end{bmatrix},$$

which is the relationship between the error and the control input.

Step 6: Fuzzify the input variables to the controller.

Use the given fuzzy sets to obtain the fuzzified input.

Step 7: Apply fuzzy approximate reasoning to determine the output of each rule.
Step 8: Use aggregation of the fuzzy outputs.

To make a fuzzy decision, we can use the fuzzy composition operator:

$$r = e \circ R.$$

Suppose the error e is NB:

$$e = [1.0 \quad 0.5 \quad 0 \quad 0 \quad 0 \quad 0 \quad 0].$$

Then, we can compute the fuzzy controller value as

$$r = e \circ R = [1.0 \quad 0.5 \quad 0 \quad 0 \quad 0 \quad 0 \quad 0] \circ \begin{bmatrix} 1.0 & 0.5 & 0 & 0 & 0 & 0 & 0 & 0 & 0 \\ 0.5 & 0.5 & 0.5 & 0.5 & 0 & 0 & 0 & 0 & 0 \\ 0 & 0.5 & 1.0 & 0.5 & 1.0 & 0.5 & 0 & 0 & 0 \\ 0 & 0 & 0 & 0.5 & 0.5 & 0.5 & 0 & 0 & 0 \\ 0 & 0 & 0 & 0.5 & 0.5 & 0.5 & 1.0 & 0.5 & 0 \\ 0 & 0 & 0 & 0 & 0 & 0.5 & 0.5 & 0.5 & 0.5 \\ 0 & 0 & 0 & 0 & 0 & 0 & 0 & 0.5 & 1.0 \end{bmatrix}$$

$$= [1.0 \quad 0.5 \quad 0.5 \quad 0.5 \quad 0 \quad 0 \quad 0]$$

Step 9: Use defuzzification to compute a crisp output.

Our computation included the fuzzy rules as expressed in R. For the case of $e = \begin{bmatrix} 1.0 & 0.5 & 0 & 0 & 0 & 0 & 0 \end{bmatrix}$, we computed r to be $r = \begin{bmatrix} 1.0 & 0.5 & 0.5 & 0.5 & 0 & 0 & 0 \end{bmatrix}$, which needs to be defuzzified. We can choose from a number of different methods to accomplish the defuzzification. For example, using the maximum degree of membership means we use the highest value in r and find the corresponding value in Table 3.2. In this example the maximum value of r is 1.0 at the -4 range level, hence $r = -4$.

Step 10: Apply crisp output to the actuator and sense system output to evaluate the error.

The steps for the temperature controller can also be implemented in MATLAB or Simulink. The MATLAB implementation is given in Appendix D.2. In the given program, we made use of MATLAB's `addrule()` function. This function allows for adding a rule list, which is an indexed form of the given set of rules. In the above example the rule list is a matrix with the following entries:

$$\text{rule list} = \begin{bmatrix} 1 & 1 & 1 & 1 \\ 2 & 2 & 1 & 1 \\ 3 & 3 & 1 & 1 \\ 4 & 4 & 1 & 1 \\ 5 & 5 & 1 & 1 \end{bmatrix}.$$

Each line in the rule list corresponds to and expresses one rule. The first entry in the row corresponds to the input variable-associated membership function, while the second entry corresponds to the membership function associated with the output variable. The third entry indicates the weighting for this rule, and the last entry specifies if we use an AND (=1) or an OR (=2) for the rule. Hence, the first row in the above rule list is read as follows:

IF *error* is Negative Big THEN r is Negative Big, with a weighting of 1.0 – that is, IF $e =$ NB THEN $r =$ NB.

Rule 1: IF $e =$ NB THEN $r =$ NB.
Rule 2: IF $e =$ NS THEN $r =$ NS.
Rule 3: IF $e =$ 0 THEN $r =$ 0.
Rule 4: IF $e =$ PS THEN $r =$ PS.
Rule 5: IF $e =$ PB THEN $r =$ PB.

Figure 3.16 depicts the implemented membership function for the input and output.

The MATLAB program shown above also generates the Rule Viewer, which is depicted in Figure 3.17. The Rule Viewer allows for changing the input and seeing the corresponding output. You can verify the computation given above by sliding the ruler to $e = -3$ and finding the r will be -4.

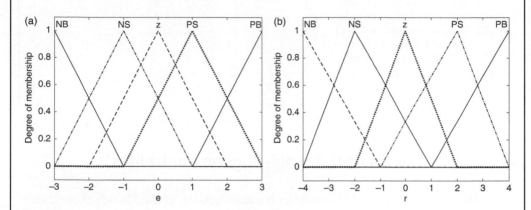

Figure 3.16 Membership functions for the input (a) and output (b).

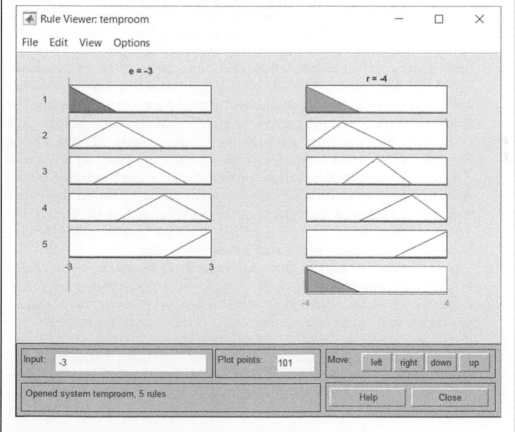

Figure 3.17 MATLAB's Rule Viewer for the given example.

In MATLAB's command window the implemented rules and the control action r can be called up and verified, as shown below:

```
FuzzyRoomTemperature
Fuzzy Rules Implemented
   1. If (e is NB) then (r is NB) (1)
   2. If (e is NS) then (r is NS) (1)
   3. If (e is Z) then (r is Z) (1)
   4. If (e is PS) then (r is PS) (1)
   5. If (e is PB) then (r is PB) (1)
control_r= -4 -2 -2 0 2 2 4
r = -4
```

3.4   Fuzzy Logic Control, Takagi–Sugeno Controllers

The Takagi–Sugeno (TS) controller is a modification of the Mamdani controller. The alteration is found at the output of the controller, which is now defined as a function of the inputs instead of fuzzy sets. Hence, TS controllers have the rule base consisting of antecedents of the linguistic type and implications are piecewise linear crisp outputs. By changing the output to crisp functions, defuzzification is no longer needed and each fuzzy subspace becomes a linear input–output relation. For Mamdani-based controllers a known disadvantage of the linguistic-based inference is that they do not contain – in an explicit form – the objective knowledge about the system [1]. Tomohiro Takagi and Michio Sugeno made two key observations [2]: (1) Complex technological processes may be described in terms of interacting yet simpler subprocesses. This is the mathematical equivalent of fitting a piecewise linear equation to a complex curve. (2) The output variable(s) of a complex physical system (complex in the sense that it can take a number of input variables to produce one or more output variables) can be related to the system's input variable in a linear manner provided the output space can be subdivided into a number of distinct regions. Not surprisingly, the TS fuzzy model structure has been extensively employed to identify the structures and parameters of plants and to control nonlinear systems.

If we compare the Mamdani inference system to the TS system, we notice that the Mamdani inference system is good at capturing expert knowledge and at facilitating an intuitive description of the knowledge. However, the Mamdani inference system employs the computation of a two-dimensional shape by integrating across a continuously varying function, which may represent a high cost in terms of computational load, especially if one implements a Mamdani inference system on a microcontroller. Consider that, for every rule, one has to find the membership function for the corresponding linguistic variables for both the antecedents and the consequents. Then, during the composition and defuzzification process, for every rule in the inference system, one has to compute the membership functions for the consequents. Additionally, if the relationship between the input and output is inherently nonlinear, it becomes more difficult to identify the membership functions for the linguistic variables in the consequent.

In comparison to the Mamdani inference system, the TS system allows for a reduced computational load due to having crisp output functions. Considering the nonlinear relationship between an input and an output (for a nonlinear system), the TS system makes use of piecewise linear approximations. For conventional control systems, complex nonlinear systems have been described in the control literature by a set of subsystems, each subsystem being linear. To operate the controller, a switching function is utilized that selects the appropriate linear subsystem in the vicinity of the operating point. Unfortunately, the switching may cause problems, as it includes artificial frequencies in the response that are not part of the real system.

In 1985 Takagi and Sugeno proposed that, in order to develop a simple mathematical instrument for calculating fuzzy implications, one needs to formulate a fuzzy partition of the fuzzy input space. In each of the defined fuzzy subspaces, a linear input–output relation is formed. The output of the fuzzy reasoning is then given by the values computed by some implications that were applied to the input.

A symbolic representation of the utilization of fuzzy subspaces for dealing with nonlinear relationships is given in Figure 3.18. In this method, the linear equations and their corresponding coefficients are trained by a set of measured data points, analogous to the learning and training phase of a neural network. The linear equation outputs are then defuzzified by evaluating the weighted average of the consequents using the degree of membership.

TS fuzzy implications are noted in the following form:

$$\text{R: IF } x_1 \text{ is } \mu_{A_1}(x_1), \ x_2 \text{ is } \mu_{A_2}(x_2), \ldots, x_n \text{ is } \mu_{A_n}(x_n) \ \text{ THEN } y = g(x_1, x_2, \ldots, x_n).$$

As a summary, the overall TS fuzzy system is shown in Figure 3.19, excluding the linear approximation scheme as detailed above.

3.4.1 Design of TS Controllers

Having the basic concept for a TS system-based controller, the question becomes how to design such controllers systematically and, more importantly, how to assess stability concerns. In any control task, stability analysis is of paramount importance. Fuzzy control systems are generally nonlinear systems, and hence one cannot draw from the abundant set of stability analysis tools developed for linear systems. However, the TS fuzzy model, as discussed above, represents a system of linear local dynamics for an overall nonlinear system. The TS model allows for obtaining an overall model by fuzzy blending of these regional linear models. Sufficient

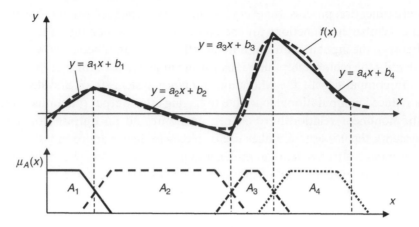

Figure 3.18 Nonlinear function approximation using subspaces with linear functions and corresponding fuzzy sets.

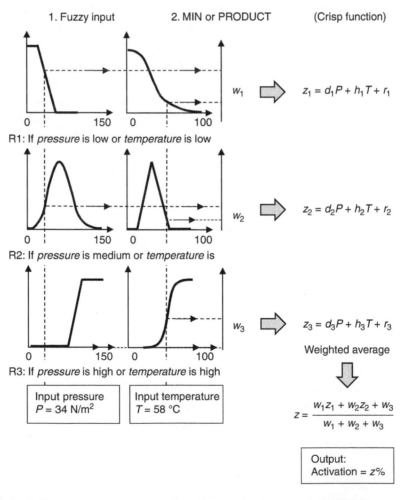

Figure 3.19 Takagi–Sugeno controller using a crisp function as implication.

conditions for the stability of a TS fuzzy control system were developed by Tanaka and Sugeno in 1992 [3]. The sufficient conditions involve a suitable Lyapunov function that satisfies the stability of all fuzzy subsystems. There are numerous works in the literature where extensions and refinements have been presented that build on the proposed sufficient conditions stated by Tanaka and Sugeno. In this section we introduce one particular approach for the systematic controller design, the modeling of a dynamical system using the TS fuzzy structure, and stability analysis based on a TS fuzzy controller along with the sufficient conditions proposed by Tanaka and Sugeno.

We express each subsection of the overall dynamics by a discrete-time linear local dynamic model in state-space form as

$$x(k+1) = \mathbf{A}_i x(k) + \mathbf{B}_i u(k),$$
$$y(k) = \mathbf{C}_i x(k), \tag{3.5}$$

where $x \in \mathbb{R}^{n \times 1}$ is the state vector, $u \in \mathbb{R}^{n_i \times 1}$ is the input vector, $y \in \mathbb{R}^{n_o \times 1}$ is the output vector, $\mathbf{A}_i \in \mathbb{R}^{n \times n}$ is the state matrix, $\mathbf{B}_i \in \mathbb{R}^{n \times n_i}$ is the input matrix, and $\mathbf{C}_i \in \mathbb{R}^{n_o \times n}$ is the output matrix.

Note that continuous-time models have the same form and are introduced in the examples later in this section. We can form the ith model rule of a TS fuzzy system as:

IF $z_1(k)$ is M_{i1} and $z_2(k)$ is M_{i2} and ... and $z_p(k)$ is M_{ip}

THEN $\begin{array}{l} x(k+1) = \mathbf{A}_i x(k) + \mathbf{B}_i u(k), \\ y(k) = \mathbf{C}_i x(k) \end{array}$ for $i = 1, 2, \ldots, r$ $\tag{3.6}$

In Equation (3.6) the variables $z_i(k)$ are premise variables and M_{ij} are the fuzzy sets for each of the r subsystems. The premise variables may represent the states of the local dynamic systems. Collecting all p premise variables, we can form the vector $z(k)$ that contains $z_i(k)$ for $i = 1, 2, \ldots, r$. The output of the fuzzy system for a pair of state and input data at discrete time k is found as

$$x(k+1) = \frac{\displaystyle\sum_{i=1}^{r} w(z(k))\{\mathbf{A}_i x(k) + \mathbf{B}_i u(k)\}}{\displaystyle\sum_{i=1}^{r} w(z(k))}, \tag{3.7}$$

$$y(k) = \frac{\sum_{i=1}^{r} w(z(k))\mathbf{C}_i x(k)}{\sum_{i=1}^{r} w(z(k))}.$$ (3.8)

Note how the output is computed by the weighted average of the consequents. In particular, the weights are computed as

$$w_i(z(k)) = \prod_{j=1}^{p} M_{ij}(z_j(k)),$$ (3.9)

where $M_{ij}(z_j(k))$ is the grade of membership of $z_j(k)$ for M_{ij}.

Our first objective is to design a controller using the TS framework. Consider having a full state feedback system (i.e., no observer is necessary as all states are measured by the system to be controlled). A simple state feedback control scheme is given by computing the input $u(k)$ as a function of the state variable $x(k)$ and some constant feedback gain F:

$$u(k) = -\mathbf{F}x(k).$$ (3.10)

A depiction of the control system interacting with the full state-space system is shown in Figure 3.20.

The block with q^{-1} in Figure 3.20 represents the backshift operator, sometimes referred to as the lag operator. In case not all states are measured, an observer can be utilized to estimate the missing state values. Figure 3.21 shows a possible implementation.

Any common state observer can be used. Observers and state-space control are not reviewed in this book, but a great number of standard controls textbooks cover these topics in detail [4].

To incorporate the TS structure to define a TS state-space controller, we can combine Equations (3.6) with Equations (3.7–3.9), which results in the following ith control rule:

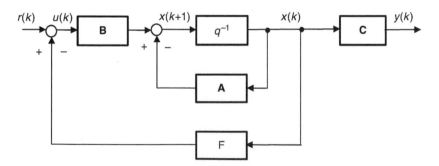

Figure 3.20 Full state feedback control system.

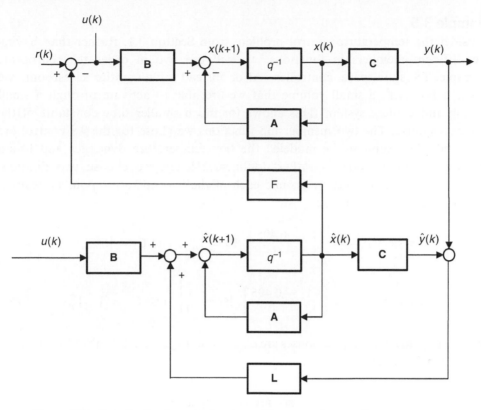

Figure 3.21 State feedback control system with a state observer **L**.

IF $z_1(t) = M_{i1}$ and $z_{i2}(t) = M_{i2}$ and $\ldots z_{ip}(t) = M_{ip}$
THEN $u(k) = -\mathbf{F}_i x(k)$.

Since we have r rules, we use the weighted average computation as shown above to find the overall fuzzy controller:

$$u(k) = -\frac{\sum\limits_{i=1}^{r} w_i(z(k))\mathbf{F}_i x(k)}{\sum\limits_{i=1}^{r} w_i(z(k))}. \tag{3.11}$$

Equation (3.11) is equivalent to the parallel distributed compensation (PDC) method. In order to realize a PDC, a nonlinear system is modeled using a TS fuzzy model. Then, each control rule can be drawn from the corresponding rule of the TS fuzzy model. To illustrate the PDC we will revisit our temperature control problem with some minor modifications below.

Example 3.5

Consider the temperature control problem from Section 3.1. Rather than having five different membership functions, we will use two in order to incorporate a simple TS state-space controller. Also, instead of an entire classroom, we model a box with a small volume that we are able to activate through a small heating and cooling system. This allows for much smaller time constants of the system response. The two membership functions we chose for the TS control are *Cold* and *Hot*. Suppose we modeled the box temperature dynamics and found that the resulting model is rather nonlinear. Hence, we choose two different operating points and linearize around each of the two operating points, resulting in two state-space equations:

$$\text{System 1: } \mathbf{A}_1 = \begin{bmatrix} 1 & -0.495 \\ 1 & 0 \end{bmatrix}, \mathbf{B}_1 = \begin{bmatrix} 1 \\ 0 \end{bmatrix}, \text{ and } \mathbf{C}_1 = \begin{bmatrix} 1 & 1 \end{bmatrix}.$$

$$\text{System 2: } \mathbf{A}_2 = \begin{bmatrix} -1 & -0.495 \\ 1 & 0 \end{bmatrix}, \mathbf{B}_2 = \begin{bmatrix} 1 \\ 0 \end{bmatrix}, \text{ and } \mathbf{C}_2 = \begin{bmatrix} 1 & 1 \end{bmatrix}.$$

The two systems' unit step responses are depicted in Figure 3.22. Both subsystems are stable and slightly underdamped.

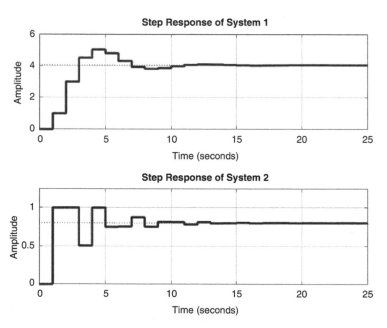

Figure 3.22 Unit step response of each subsystem, exhibiting small overshoot and sufficient damping.

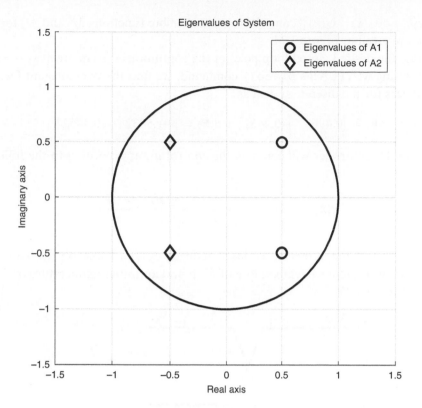

Figure 3.23 Pole location of both discrete subsystems.

The damping ratio for system 1 is approximately 0.912 and for system 2 it is 0.989. The natural frequency of the first system is 0.86 rad/s and for the second system 2.39 rad/s. The stability is also indicated by the pole location of the discrete-time model in Figure 3.23. Each of the system's poles are well within the unit circle.

Even though the response is close to being critically damped, a reasonable control objective would be to have the temperature not oscillate if a change in the reference temperature is made. This requires that the damping ratio is unity or larger. A damping ratio of unity implies critically damped and is desirable as it is the fastest response without oscillation. However, the modeling of the subsystems may not be completely accurate, and the inaccuracies combined with a controller that targets a damping ratio of exactly 1.0 may result in an underdamped system. Hence, we design a controller which drives the overall system just past critically damped into the overdamped region of the complex plane.

The TS controller will have two rules:

Rule 1: IF $x_2(k)$ is M_1, THEN $x(k + 1) = \mathbf{A}_1 x(k) + \mathbf{B}_1 u(k)$,
Rule 2: IF $x_2(k)$ is M_2, THEN $x(k + 1) = \mathbf{A}_2 x(k) + \mathbf{B}_2 u(k)$,

where $x(k) = [x_1(k) \quad x_2(k)]^T$, and the two membership functions M_1 and M_2 for *Cold* and *Hot* are given in Figure 3.24.

Suppose the desired closed-loop poles of the continuous system are at $s_1 = -1$ and $s_2 = -2$. Using MATLAB's `place()` command, we find the two constant feedback gain matrices for the discrete system as

$$F_1 = [0.4968 \quad -0.4452] \text{ and } F_2 = [-1.5032 \quad -0.4452].$$

Hence, our TS controller will compute the control input according to the following formula:

$$u(k) = -\frac{w_1(z(k))F_1x(k) + w_2(z(k))F_2x(k)}{w_1(z(k)) + w_2(z(k))}.$$

The simulation of the above closed-loop feedback system based on the TS control framework for a reference temperature of 67°F and an initial temperature of 72°F is shown in Figure 3.25.

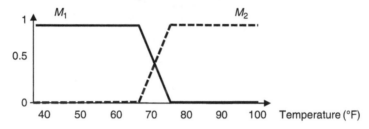

Figure 3.24 Membership function for the example problem.

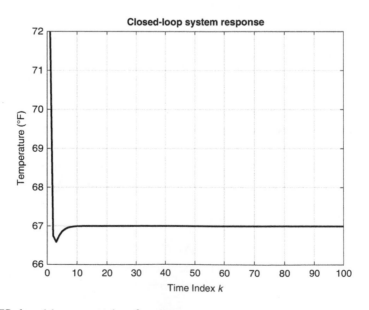

Figure 3.25 TS closed-loop control performance.

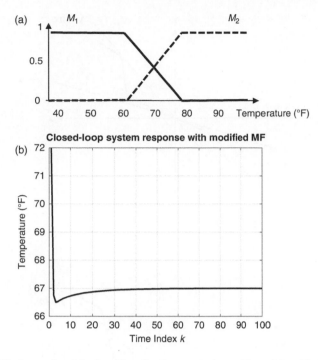

Figure 3.26 Modified membership function for the example problem (a), with the corresponding closed-loop response (b).

In Figure 3.25 the response corresponds to the membership function as shown in Figure 3.24. Note that the response will exhibit a different form with different membership functions and membership function parameters. For example, if we change the membership functions such that the corner points of the Z-type (M_1) and S-type (M_2) membership functions were moved out further by five units, as shown in Figure 3.26(a), the response will change slightly, as shown in Figure 3.26(b).

It is easy to imagine that the form and type of membership function has quite some influence on the performance of the system. In addition, the blending of the two subsystems using the TS framework along with different initial conditions may cause instability, even though the controller designed using the linear system-based approach makes both subsystems stable [5]. Hence, it is usually advisable to compute the feedback gain matrix F for the global system and its performance. This requires some understanding of stability analysis of TS control systems, which will be presented later in this section.

The above example used two linear subsystems. In this section we briefly outline how to deal with nonlinear systems. There is an extensive literature on this topic in regard to TS modeling and controller design and the

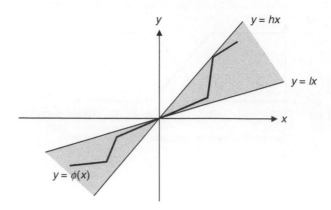

Figure 3.27 Global sector nonlinearity.

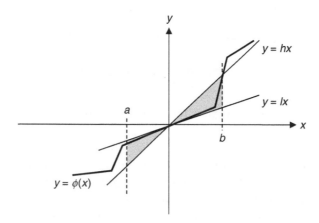

Figure 3.28 Local sector nonlinearity.

implications for stability. We do not aim to provide a comprehensive overview and detailed design approach for this type of topic. However, the following section does include some important concepts for systems entailing nonlinear elements.

The first concept is the *sector nonlinearity* for model constructions [6]. Consider a function $y = \phi(x)$. The function $\phi : \mathbb{R} \rightarrow \mathbb{R}$ is considered to be in sector $[l, h]$ if for all $x \in \mathbb{R}$, $y = \phi(x)$ lies between lx and hx. Graphically, we can represent this as shown in Figure 3.27.

Equivalently, this can be stated as a quadratic inequality as

$$(y - hx)(y - lx) \leq 0 \,\forall\, x, \; y = \phi(x). \tag{3.12}$$

We can expand on this concept and introduce local sector nonlinearity, which is shown in Figure 3.28.

Example 3.6

A traditional controls example for studying marginally stable systems is the ball-on-a-beam problem. Details on the mathematical description of the linear approximate system using servos and a cam to change the beam angle to prevent the ball falling off the beam are given on MATLAB's Control Tutorials webpage [7]. This system is also considered by Wang and Mendel [8], where the servos are replaced by a control input that directly affects the acceleration of the angle of the beam. In this example, we are following the approach as detailed in *Fuzzy Control Systems: LMI-Based Design* [9], with the objective to develop a TS model representing the dynamics of the ball-on-a-beam system. The system configuration is shown schematically in Figure 3.29.

The system in state-space form [10] is given as

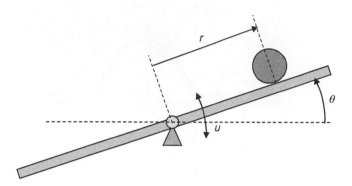

Figure 3.29 Ball-on-a-beam problem with direct actuation of the acceleration of the angle of the beam.

$$
\begin{Bmatrix} \dot{x}_1 \\ \dot{x}_2 \\ \dot{x}_3 \\ \dot{x}_4 \end{Bmatrix} = \begin{bmatrix} x_2(t) \\ B\left(x_1(t)x_4^2(t) - g\sin(x_3(t))\right) \\ x_4(t) \\ 0 \end{bmatrix} + \begin{Bmatrix} 0 \\ 0 \\ 0 \\ 1 \end{Bmatrix} u(t).
$$

The state variables $\vec{x}(t) = [x_1(t) \quad x_2(t) \quad x_3(t) \quad x_4(t)]^T = [r(t) \quad \dot{r}(t) \quad \theta(t) \quad \dot{\theta}(t)]^T$ are the position r, the velocity of the ball, the angle of the beam, and the velocity of the angle of the beam, respectively. The constant B is given by

$$
B = \frac{M}{\left(\dfrac{J_b}{R^2 + M}\right)},
$$

where M is the mass of the ball, J_b is the moment of inertia of the beam, R is the radius of the ball, and g is the gravitational constant. From the above state-space description of the dynamic system we recognize that we have a nonlinear system. To proceed with the section nonlinear approach we define the nonlinearities with new variables: $z_1 = \sin(x_3(t))$ and $z_2 = x_1 x_4^2$.

The change in notation allows us to rewrite the state-space equations as

$$\begin{Bmatrix} \dot{x}_1 \\ \dot{x}_2 \\ \dot{x}_3 \\ \dot{x}_4 \end{Bmatrix} = \begin{bmatrix} x_2(t) \\ B(z_2 - gz_1) \\ x_4(t) \\ 0 \end{bmatrix} + \begin{Bmatrix} 0 \\ 0 \\ 0 \\ 1 \end{Bmatrix} u(t).$$

A common linearization approach for such a system is to assume small angles; hence, $\sin(\theta) \simeq \theta$. However, we will not restrict the angle to some small range in order to have linearized this term. Hence, let's assume the beam can rotate from $-90°$ to $+90$: $x_3 \in [-\pi/2, \pi/2]$. For this range, the global section nonlinearity can be depicted as in Figure 3.30.

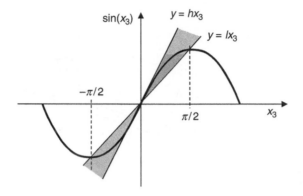

Figure 3.30 Global section nonlinearity for the ball-on-a-beam problem for the state variable $\theta(t)$.

From the figure we can determine that z_1 is bounded by the coefficients l and h:

$$\left| \frac{2}{\pi} x_3(t) \right| \leq |\sin(x_3(t))| \leq |x_3(t)|.$$

Similarly, we can define a range for z_2:

$$-dx_4 \leq x_1 x_4^2 \leq dx_4,$$

where d is to be selected based on the physical parameters of the system. Recalling Equation (3.9) and noting that the maximum of z_1 is 1.0 and the minimum of z_1 is -1, we have

$$\max(z_1(t)) = 1 \text{ and } \min(z_1(t)) = -1.$$

Analogously,

$$\max(z_2(t)) = -d \text{ and } \min(z_2(t)) = d.$$

From the maximum and minimum values, z_1 and z_2 can be represented as

$$z_1 = \sin(x_3(t)) = M_{11}(z_1(t)) \cdot 1 + M_{12}(z_1(t)) \cdot (-1),$$

$$z_2 = x_1(t)x_4^2(t) = M_{21}(z_2(t)) \cdot d + M_{22}(z_2(t)) \cdot (-d),$$

where

$$M_{11}(z_1(t)) + M_{12}(z_1(t)) = 1,$$

$$M_{21}(z_2(t)) + M_{22}(z_2(t)) = 1.$$

This allows us to compute the respective membership functions M_{ij}:

$$M_{11}(z_1) = \frac{z_1(t) + 1}{2}, \ M_{12}(z_1) = \frac{1 - z_1(t)}{2} \ \text{ and}$$

$$M_{21}(z_2) = \frac{z_2(t) + d}{d}, \ M_{22}(z_2) = \frac{d - z_2(t)}{d}.$$

Knowing that M_{1j} is a function of z_1, which varies between -1 and $+1$, we name the two membership functions *Positive* and *Negative*. For z_2, let's use *Large* and *Small* as the respective membership functions.

Hence, our TS model is given as follows:

Model Rule No. 1
IF $z_1(t)$ is Positive and $z_2(t)$ is Large THEN

$$\dot{\vec{x}}(t) = \mathbf{A}_1 + \begin{Bmatrix} 0 \\ 0 \\ 0 \\ 1 \end{Bmatrix} u(t).$$

Model Rule No. 2
IF $z_1(t)$ is Positive and $z_2(t)$ is Small THEN

$$\dot{\vec{x}}(t) = \mathbf{A}_2 + \begin{Bmatrix} 0 \\ 0 \\ 0 \\ 1 \end{Bmatrix} u(t).$$

Model Rule No. 3
IF $z_1(t)$ is Negative and $z_2(t)$ is Large THEN

$$\dot{\vec{x}}(t) = \mathbf{A}_3 + \begin{Bmatrix} 0 \\ 0 \\ 0 \\ 1 \end{Bmatrix} u(t).$$

Model Rule No. 4

IF $z_1(t)$ is Negative and $z_2(t)$ is Small THEN

$$\dot{\vec{x}}(t) = \mathbf{A}_4 + \begin{Bmatrix} 0 \\ 0 \\ 0 \\ 1 \end{Bmatrix} u(t).$$

We composed the state matrices $\mathbf{A}_i$ using the maximum and minimum values for z_1 and z_2, resulting in

$$\mathbf{A}_1 = \begin{bmatrix} x_2(t) \\ B(d-g) \\ x_4(t) \\ 0 \end{bmatrix}, \mathbf{A}_2 = \begin{bmatrix} x_2(t) \\ B(-d-g) \\ x_4(t) \\ 0 \end{bmatrix}, \mathbf{A}_3 = \begin{bmatrix} x_2(t) \\ B(d+g) \\ x_4(t) \\ 0 \end{bmatrix}, \text{ and } \mathbf{A}_4 = \begin{bmatrix} x_2(t) \\ B(g-d) \\ x_4(t) \\ 0 \end{bmatrix}.$$

The membership functions can be drawn as shown in Figure 3.31.

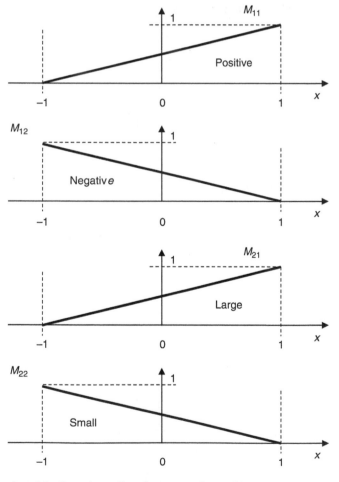

Figure 3.31 Membership functions for the example problem.

Accordingly, the defuzzification is computed first by finding the weights for each implication:

$$w_1 = M_{11}(z_1(t)) \times M_{21}(z_2(t))$$
$$w_2 = M_{11}(z_1(t)) \times M_{22}(z_2(t))$$
$$w_3 = M_{21}(z_1(t)) \times M_{21}(z_2(t))$$
$$w_4 = M_{21}(z_1(t)) \times M_{22}(z_2(t))$$

and

$$\dot{x}(t) = \sum_{i=1}^{4} w_i(z(t))\mathbf{A}_i + \begin{Bmatrix} 0 \\ 0 \\ 0 \\ 1 \end{Bmatrix} u(t).$$

Clearly, the above example problem is dynamically unstable. For such a system a suitable controller may be fashioned to accomplish some control objective while keeping the system stable. This brings us back to our stability problem of TS systems, regardless of whether it is a model or a control system. The following is a brief introduction to the stability analysis of such systems using Lyapunov stability methods. There is much recent literature addressing the stability problem of TS system. Our treatment here focuses only on one such method – the Lyapunov-based stability method.

Consider a simple open-loop system of the form

$$x(k+1) = \sum_{i=1}^{r} h_i(z(k))\mathbf{A}_i(k), \tag{3.13}$$

where

$$h_i(z(k)) = \frac{w_i(z(k))}{\sum_{i=1}^{r} w_i(z(k))}. \tag{3.14}$$

According to Tanaka and Sugeno, a sufficient stability condition for a TS system can be stated as follows.

Theorem 3.1 [3,10]

The equilibrium of a fuzzy system given in Equation (3.7) is globally asymptotically stable if there exists a common positive definite matrix $\mathbf{P}$ such that

$$\mathbf{A}_i^T \mathbf{P} \mathbf{A}_i - \mathbf{P} < \mathbf{0}, \text{ for } i = 1, 2, \ldots, r. \tag{3.15}$$

This means a common matrix $\mathbf{P}$ has to exist for all subsystems. Theorem 3.1 can be applied to any nonlinear system that can be approximated by a piecewise linear function if the stability condition of Equation (3.15) is met. One can extend this criterion such that if there exists a common positive definite matrix $\mathbf{P}$, then all $\mathbf{A}_i$ matrices are stable. Note, Theorem 3.1 is referring to the sufficient condition, which means that there is a possibility that one may not find a $\mathbf{P} > 0$ even though all $\mathbf{A}_i$ matrices are stable matrices. The opposite is true too – that is, a fuzzy system may be globally asymptotically stable even if no $\mathbf{P} > 0$ is found.

Theorem 3.1 is based on Lyapunov stability theory for nonlinear systems. Lyapunov provided two methods for stability analysis of nonlinear systems. The first method is in essence described by the movement of its solution over time. In particular, if we have a solution to a nonlinear system that starts out near an equilibrium point of that system and stays close to that equilibrium point – say, point x_e – for all times, then this point x_e is considered Lyapunov stable. Defining asymptotically stable points using Lyapunov will require that if x_e is Lyapunov stable and all solutions in the vicinity around x_e converge over time to x_e, then x_e is considered an asymptotically stable point. The second stability method proposed by Lyapunov utilizes the energy balance of a dynamic nonlinear system. Specifically, if the system attains more energy over time than it loses, the system is considered unstable. If the system loses more energy over time than enters the system, then the system is stable. To quantify the energy of the system a Lyapunov function V is required to be constructed. This function is similar to the potential function in classical dynamics. The difficulty of this method is finding a suitable Lyapunov function V for the system to be analyzed.

Theorem 3.1 is based on this Lyapunov function V in the form $V(x(k)) = x^T(k)\mathbf{P}x(k)$, which means that rather than finding the Lyapunov function V, we need to find the positive definite matrix $\mathbf{P}$ for the system described in Equation (3.13). This problem of finding $\mathbf{P}$ is an ongoing topic in the research community and various approaches and methods have been proposed recently. For now, we want to expand Theorem 3.1 to a more conventional control system, such as the one defined by Equation (3.11).

Theorem 3.2 [3]

The equilibrium of a fuzzy control system as given by Equation (3.11) is considered globally asymptotically stable if there exists a common positive definite matrix $\mathbf{P}$ such that

$$\{\mathbf{A}_i - \mathbf{B}_i\mathbf{F}_j\}^T\mathbf{P}\{\mathbf{A}_i - \mathbf{B}_i\mathbf{F}_j\} - \mathbf{P} < 0 \qquad (3.16)$$

for $h_i(z(k)) \cdot h_j(z(k)) \neq 0 \ \forall k$ and $i, j = 1, 2, 3, \ldots, r$.

Using Equations (3.14) and (3.15) in Equation (3.11), the state vector of the system can be expressed as

$$x(k+1) = \sum_{i=1}^{r} \sum_{j=1}^{r} h_i(z(k)) h_j(z(k)) \{\mathbf{A}_i - \mathbf{B}_i \mathbf{F}_j\} x(k). \tag{3.17}$$

Defining

$$G_{ij} = \frac{\{\mathbf{A}_i - \mathbf{B}_i \mathbf{F}_j\} + \{\mathbf{A}_j - \mathbf{B}_j \mathbf{F}_i\}}{2} \text{ for } h_i(z(k)) \times h_j(z(k)) = 0 \ \forall z(k). \tag{3.18}$$

Equation (3.17) can be restated using Equation (3.18) as follows [5]:

$$x(k+1) = \left\{ \sum_{i=1}^{r} h_i(z(k)) h_i(z(k)) \{\mathbf{A}_i - \mathbf{B}_i \mathbf{F}_i\} x(k) \right.$$

$$\left. +2 \sum_{i=1}^{r} \sum_{i<j}^{r} h_i(z(k)) h_j(z(k)) \mathbf{G}_{ij} x(k) \right\}. \tag{3.19}$$

This leads to the following stability theorem.

Theorem 3.3

Global asymptotic stability exists for a fuzzy control system as given by Equation (3.17) if there exists a common positive definite matrix $\mathbf{P}$ satisfying

$$\{\mathbf{A}_i - \mathbf{B}_i \mathbf{F}_i\}^T P\{\mathbf{A}_i - \mathbf{B}_i \mathbf{F}_i\} - \mathbf{P} < \mathbf{0} \text{ for } i = 1, 2, 3, \ldots, r \tag{3.20}$$

and

$$\mathbf{G}_{ij}^T \mathbf{P} \mathbf{G}_{ij} - \mathbf{P} < \mathbf{0} \text{ for } i < j \leq r \text{ such that } h_i(z(k)) \times h_j(z(k)) = 0 \ \forall z(k). \tag{3.21}$$

Example 3.7

Consider the following two state-space systems:

$$\mathbf{A}_1 = \begin{bmatrix} 0.9 & -0.3 \\ 1.0 & 0 \end{bmatrix}, \ \mathbf{A}_2 = \begin{bmatrix} 0.6 & 0.4 \\ 1.0 & 0 \end{bmatrix}, \ \mathbf{B}_i = \begin{bmatrix} 1 \\ 0 \end{bmatrix} \text{ for } i = 1, 2.$$

Suppose we design a fuzzy controller of the following form:

Rule 1: IF $x_2(k)$ is M_1 THEN $x(k+1) = \mathbf{A}_1 x(k) + \mathbf{B}_1 u(k)$,
Rule 2: IF $x_2(k)$ is M_2, THEN $x(k+1) = \mathbf{A}_2 x(k) + \mathbf{B}_2 u(k)$,

where $x(k) = [x_1(k) \quad x_2(k)]^T$, and the two membership functions M_1 and M_2 are identical to the one depicted in Figure 3.24. Consider the following positive definite matrix $\mathbf{P}$:

$$\mathbf{P} = \begin{bmatrix} 3 & -1 \\ -1 & 1 \end{bmatrix}.$$

Checking the condition of Theorem 3.1 (i.e., Equation (3.15)), we have

$$\mathbf{A}_1^T\mathbf{P}\mathbf{A}_1 - \mathbf{P} < 0 : \begin{Bmatrix} -1.6352 \\ -0.4648 \end{Bmatrix} \text{ and } \mathbf{A}_2^T\mathbf{P}\mathbf{A}_2 - \mathbf{P} < 0 : \begin{Bmatrix} -2.3700 \\ -0.2700 \end{Bmatrix},$$

implying that the equilibrium is globally asymptotic.

For Theorem 3.2 we need to design a controller. Suppose we follow the previous example and design a state feedback controller using pole placement (with the desired closed-loop poles at -1 and -2 in the s-plane), resulting in the following two constant discrete-time feedback matrices:

$$\mathbf{F}_1 = [0.3968 \quad -0.2502] \text{ and } \mathbf{F}_2 = [0.0968 \quad -0.3502].$$

We do satisfy Equation (3.16):

$$\{\mathbf{A}_i - \mathbf{B}_i\mathbf{F}_j\}^T\mathbf{P}\{\mathbf{A}_i - \mathbf{B}_i\mathbf{F}_j\} - \mathbf{P} < 0.$$

Hence, the equilibrium of the closed-loop state feedback fuzzy control system is considered globally asymptotically stable.

For Theorem 3.3 we compute $\mathbf{G}_{1,2}$:

$$\mathbf{G}_{1,2} = \begin{bmatrix} 0.5032 & -0.0498 \\ 1.0000 & 0 \end{bmatrix}.$$

With this we can test the conditions given in Equations (3.20):

$$\{\mathbf{A}_1 - \mathbf{B}_1\mathbf{F}_1\}^T\mathbf{P}\{\mathbf{A}_1 - \mathbf{B}_1\mathbf{F}_1\} - \mathbf{P} = \begin{Bmatrix} -0.7541 \\ -0.0066 \end{Bmatrix} < 0,$$

$$\{\mathbf{A}_2 - \mathbf{B}_2\mathbf{F}_2\}^T\mathbf{P}\{\mathbf{A}_2 - \mathbf{B}_2\mathbf{F}_2\} - \mathbf{P} = \begin{Bmatrix} -0.7541 \\ -0.0066 \end{Bmatrix} < 0,$$

and (3.21):

$$\mathbf{G}_{ij}^T\mathbf{P}\mathbf{G}_{ij} - \mathbf{P} = \begin{Bmatrix} -2.7786 \\ -0.4607 \end{Bmatrix} < 0.$$

Hence, the system is globally asymptotically stable.

3.5   Conclusions

In this chapter we built FIS using the concepts of fuzzy logic. An FIS utilizes fuzzy logic to map an input to its output. To accomplish this mapping the FIS engages four distinct components: a rule base composed of a set of distinct fuzzy rules, a fuzzification stage, an inference engine, and the defuzzification process. The rule base is composed of a set of IF–THEN rules that may be connected with OR and AND logic operators. The fuzzification step is responsible for converting crisp inputs into fuzzy sets. The inference engine utilizes the fuzzified input to determine a fuzzy output based on the rule base. To be able to use the fuzzy output as a signal, we employ a defuzzification process that generates a crisp output.

However, the defuzzification process can be implemented in different ways. For example, the MFIS utilizes the COG or similar computations to find the crisp output. In contrast, the Sugeno FIS uses a weighted average computation to find the crisp output. For the TS FIS, linear functions are used instead of membership functions when computing the implications. In our discussion of the Mamdani and Sugeno FIS we discovered that MATLAB provides easy implementation for both systems by the use of the FIS App.

Fuzzy logic controllers make use of FIS to compose the controller in a closed-loop feedback system. A particular implementation of fuzzy logic inference to controls is the TS controller. Here, we can accommodate nonlinear relations by associating membership functions to linear subsections. By having linear subsections we can utilize the rich library of linear control and linear control theory. However, since the overall structure remains nonlinear, the question of the resulting stability needs to be addressed in more detail. For this, we utilized Lyapunov theory to find ways to characterize stability or the lack of it in fuzzy controller systems.

SUMMARY

In the following, a number of key concepts of fuzzy inference and fuzzy logic control systems covered in this chapter are summarized.

Mamdani Fuzzy Inference

The MFIS maps inputs to outputs by incorporating linguistic control rules, which are generated using simple logic and experience gained through observations by the fuzzy inference designer. The MFIS in particular shapes the output locally by evaluating each individual rule without necessarily affecting the other rules, and aggregates the output by retaining much of the nature of each rule composition.

Defuzzification

Since the MFIS output is given by a membership function, a defuzzification process is required. The process takes on a fuzzy input and converts this into a crisp output that is scalable for actuator activation. Some of the common defuzzification methods are the COG method, the centroid method, the weighted average method, the mean-max membership method, and the center of sums method, to mention a few.

Sugeno Fuzzy Inference

The Sugeno FIS is composed similarly to the Mamdani inference system, with the exception of the implication and defuzzification process. The implication is computed using simple linear equations, while the defuzzification can be accomplished using a simple weighted average of each implication of each rule. The Sugeno FIS is less computational expensive than the MFIS because of the missing aggregation computation and the use of simple linear equations.

Fuzzy Logic Control

Fuzzy logic controllers use an FIS to integrate human reasoning and experimental observations in a heuristic fashion to determine and shape the output of the dynamic systems. Fuzzy logic controllers can be applied without detailed knowledge of the process dynamics. They are also suitable for controlling systems that have unknown process dynamics, have considerable nonlinear elements, and that change over time.

Takagi–Sugeno Fuzzy Logic Control

The TS controller is a controller that utilizes the Sugeno fuzzy inference model. Therefore, the rule implications are evaluated using functions based on the inputs to the controller. The function evaluations are weighted by the rule implication level. Since no defuzzification is needed, the control action is computed directly by the weighted average of the implication product.

Sector Nonlinearity

The sector nonlinearity bounds the nonlinear system using inequalities. These bounds can be expressed locally, resulting in local sector nonlinearities, or globally, which gives rise to the global sector nonlinearity. Mathematically, we can express this using a function $y = \phi(x)$. The function $\phi : \mathbb{R} \to \mathbb{R}$ is considered to be in sector $[l, h]$ if for all $x \in \mathbb{R}$, $y = \phi(x)$ lies between lx and hx.

Lyapunov Function

Lyapunov functions are useful in determining the stability and the equilibrium points of dynamic systems, including nonlinear systems.

Asymptotic Stability

Given a system has an equilibrium point, asymptotic stability implies that any initial condition applied to this system and located in a close neighborhood of the equilibrium point yields a solution that converges asymptotically to that equilibrium point. Asymptotic stability implies that the corresponding Lyapunov function has a negative definite derivative.

REVIEW QUESTIONS

1. Why do we need to apply defuzzification at the end of an inference system that is used for controlling a dynamic system?
2. What are the five steps in constructing a Mamdani fuzzy logic controller?
3. How do we evaluate the disjunction of a fuzzy rule?
4. How do we evaluate the conjunction of a fuzzy rule?
5. What is clipping?
6. Why do we sometimes use scaling rather than clipping?
7. What is computed with the center of gravity method?
8. What is the difference between a Mamdani and a Sugeno FIS?
9. What is a singleton when used in an FIS?
10. How do we perform defuzzification for a Sugeno FIS?
11. What information about the plant is necessary to design a fuzzy logic controller?
12. Why do we scale and normalize input and output quantities in a fuzzy logic controller? Give an example.
13. In MATLAB, can an entire rule base be represented by a numeric matrix?
14. For a TS controller, are the subsystems considered linear?
15. Why are observer matrices used in state-space models?
16. What is the local sector nonlinearity?
17. How is stability defined for a discrete-time linear system using the system's poles?
18. Define what asymptotically stable means for a linear system and a specific set of initial conditions.
19. Considering Lyapunov stability theory, is a system that gains more energy than it loses over time deemed to be stable or unstable?
20. How do you determine if a system is globally asymptotically stable?
21. Partitioning a nonlinear system into a set of linear systems, and each individual linear system is found to be stable, is it a valid statement that, since all subsystems are stable, the overall nonlinear system is stable too? If not, why?
22. For a TS state-space controller, does the shape of the selected membership functions have any influence on the performance of the closed-loop system?
23. What kind of aggregation method is used for the TS controller?

PROBLEMS

3.1 Suppose $A = B = R^+$ and $C =$ "a is greater than b." The fuzzy relation C can be represented by the membership function

$$\mu_C(a, b) = \begin{cases} \dfrac{b - a}{a + b + 2} & \text{for } b > a \\ 0 & \text{for } b \leq a \end{cases}.$$

Suppose $A = \{3, 4, 5\}$ and $B = \{3, 4, 5, 6, 7\}$. Find the fuzzy relation C.

3.2 Consider two fuzzy relations, $A \times B$ and $B \times C$, where $A = \{1, 2, 3\}$, $B = \{q, r, s, t\}$, and $C = \{m, n\}$. Suppose

$$R_1 = \begin{bmatrix} 0.1 & 0.3 & 0.5 & 0.7 \\ 0.4 & 0.2 & 0.8 & 0.9 \\ 0.6 & 0.8 & 0.3 & 0.2 \end{bmatrix} \text{ represents that "a is relevant to b," and}$$

$$R_2 = \begin{bmatrix} 0.9 & 0.1 \\ 0.2 & 0.3 \\ 0.5 & 0.6 \\ 0.7 & 0.2 \end{bmatrix} \text{ represents that "b is relevant to c."}$$

a. Find the relationship of "a is relevant to c" using the MAX-MIN composition.
b. Find the relationship of "a is relevant to c" using the MAX-PRODUCT composition.

3.3 Consider the fuzzy set A

$$A = 0.2/a_1 + 0.9/a_2$$

and the fuzzy set B

$$B = 0.3/b_1 + 0.5/b_2 + 1.0/b_3.$$

Find the relation $R = A \times B$.

3.4 Suppose you constructed a single-input Sugeno fuzzy model, represented by

$$\begin{cases} \text{If } A \text{ is dark then } B = 0.2A + 6.5 \\ \text{If } A \text{ is medium then } B = 0.75A + 5 \\ \text{If } A \text{ is light then } B = 2A - 3 \end{cases}.$$

Construct membership functions for dark, medium, and light as follows:
a. The membership functions are such that the antecedents result in crisp rules.
b. The membership functions are such that the antecedents result in fuzzy rules.

3.5 Consider designing a fuzzy logic-based autopilot for an airplane. The autopilot's objective is to maintain the speed, altitude, and direction of the plane.
The control for the altitude is accomplished by activating the elevator. The
control for the constant direction is given by activating the rudder. The
control of the speed is given by activating the propeller speed. The schematic
control diagram for the different objectives is found in Figure P3.1. Each
system is a one-input, one-output system.

   a. In your control design, incorporate the rate of change information of
      the controlled quantity. Define the linguistic values for each variable,
      define a set number of rules, and implement this fuzzy logic controller
      in Simulink (use Mamdani fuzzy logic). Once implemented, tune your
      system so that it regulates the controlled output accurately in the
      presence of disturbances of 10% of each input (use a sinusoidal dis-
      turbance input with magnitude of 10% of the nominal set-point and
      a frequency close (within 10%) to the first fundamental frequency of the
      system).

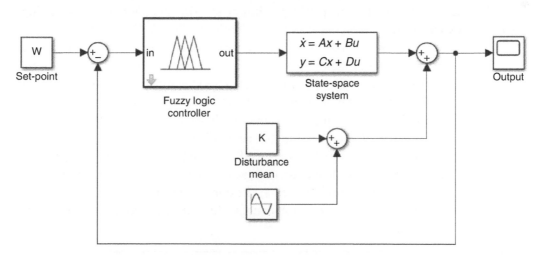

Figure P3.1 Control schematics for autopilot implementation.

   b. Using the Control Systems Toolbox, design a **PID** controller and simu-
      late the performance of the PID controller. Compare the simulation
      results between the PID controller and the fuzzy logic controller using
      standard control performance criteria.
   c. Write a short (very concise) report, include all your assumptions, compu-
      tations, design rules, and results. Note, the open-loop system dynamics is
      given by the following transfer function:

$$G(s) = \frac{3.125}{(s^2 + 0.625s + 1.5625)}.$$

3.6 Consider the TS system given by the following three implications:

Rule 1: IF x_1 is small and x_2 is small, THEN $y = x_1 + x_2$.

Rule 2: IF x_1 is medium and x_2 is small, THEN $y = 2x_1$.

Rule 3: IF x_2 is high, THEN $y = 3x_2$.

a. Develop membership functions for *small*, *medium*, and *high*. Use triangular-shaped membership functions and plot them for a universe of discourse of $0 \leq x_1 \leq 5$ and $0 \leq x_2 \leq 10$.

b. What is the output of the system if $x_1 = 3$ and $x_2 = 4$?

3.7 Consider Problem 3.6.

a. Design a TS controller and compare it to the two controllers developed in Problem 3.5.

b. Analyze your designed controller and determine if the TS controller is globally asymptotically stable.

3.8 Following direction from Appendix G and information from Appendices E and F, use the propellor–pendulum system to conduct the following experiments.

a. Investigate how the change in the angle of the pendulum as measured by the potentiometer is processed using five membership functions: negative big (NB), negative small (NS), neutral (N), positive small (PS), and positive big (PB). For this, create an FIS system that entails these five membership functions for the input and provides as output the degree of membership to each of these membership functions. Connect the pendulum system as described in Appendix G and provide an initial angular deflection to have the pendulum swing freely. Record the angle information as well as the corresponding voltage information in a table. For example, you can use the following table to record your experiment.

Pot. meter voltage (V)	Angle (deg)	Degree of membership NB	Degree of membership NS	Degree of membership N	Degree of membership PS	Degree of membership PB

b. Extend your fuzzy system by including a defuzzification method. This method should compute one crisp value as output which is based on the degree of membership the measured angle of the pendulum provides. Use the COG method and vary the angle between the maximum and minimum while recording the voltage at equal distance (angle).

c. Design a Mamdani fuzzy logic controller to regulate the pendulum at a set angle. Choose an angle that the actuator (propellor and motor) can easily reach.

d. Design a Sugeno fuzzy logic controller with the same objectives as given in (c).

3.9 Consider the propellor–pendulum system given in Appendix G. Design a TS controller using the following steps.

a. Model the system using your dynamics background. Make sure you linearize the model around an operating point of interest. Hint: Use the logarithmic decrement method for estimating the damping coefficient.

b. Verify your model by simulating it in MATLAB or Simulink and comparing the results with the physical systems response.

c. Use the local sector method and define a set of state-space models for each sector. Use pole placement to compute a constant feedback controller gain and simulate the closed-loop response for each sector.

d. Use the results from (c) and design a TS controller. Compare the results of your controller by using a PID block from Simulink and repeat the same experiment as with the TS controller.

REFERENCES

[1] Yager RR, Filev DP (1994). *Essentials of Fuzzy Modeling and Control*. Wiley, New York.

[2] Takagi T, Sugeno M (1985). Fuzzy identification of systems and its applications to modelling and control. *IEEE Transactions on Systems, Man and Cybernetics*, 15(1): 116–132.

[3] Tanaka K, Sugeno M (1992). Stability analysis and design of fuzzy control systems. *Fuzzy Sets and Systems*, 45: 135–156.

[4] Friedland B (2001). *Control System Design, an Introduction to State-Space Methods*. McGraw-Hill, New York.

[5] Kazuo Tanaka K, Wang HO (2001). *Fuzzy Control Systems Design and Analysis*, Wiley, New York.

[6] Kawamoto S, Tada K, Ishigame A, Taniguchi T (1992). An approach to stability analysis of second order fuzzy systems. *IEEE International Conference on Fuzzy Systems*. 1427–1434.

[7] MATLAB (n.d.). Ball & beam: system modeling. http://ctms.engin.umich.edu/CTMS/index.php?example=BallBeam & section=SystemModeling.

[8] Wang LX, Mendel JM (1992). Fuzzy basis functions, universal approximation, and orthogonal least-squares learning, *IEEE Transaction on Neural Networks*, 3(5): 807–814.

[9] Seidi M, Hajiaghamemar H, Segee B (2012). *Fuzzy Control Systems: LMI-Based Design*. Intech, London.

[10] Tanaka K, Sugeno M (1990). Stability analysis of fuzzy systems using Lyapunov's direct method. Proc*eedings of* NAFIPS, 133–136.

4 Optimization: Hard Computing

4.1 Introduction

Optimization plays a large role in controls and system operations, so much so that there exists an entire field termed *optimal control*. In addition, optimization can be found in estimation theory, commonly used in controls and system identification. However, this chapter is a preparation for using optimization in intelligent controls. One of the many applications of optimization algorithms within intelligent control is found in artificial neural networks and deep learning. We treat optimization with those application in mind and cover two types of optimizations: hard computing, the topic of this chapter; and soft computing, which is treated in the next chapter. We will limit ourselves to applicable optimization methods commonly found in intelligent controls without spending much time on treating optimization as its own field.

Mathematically speaking, the optimization problem in a general form can be stated as

$$\text{minimize } f_o(x), \tag{4.1}$$

$$\text{subject to } g_i(x) \le b_i, \text{ where } i = 1, \ldots, m, \tag{4.2}$$

where $f_o : \mathbb{R}^n \to \mathbb{R}$ is the objective function, $x = [x_1 \quad x_2 \quad \ldots \quad x_n]$ are the design variables, and $g_i : \mathbb{R}^n \to \mathbb{R}$ are the inequality constraints. Note that the design variables must be linearly independent. A so-called feasible solution refers to a solution that satisfies all the constraints. An optimal solution is one that has met the objective function and results in its minimum or maximum while remaining feasible. The search space is enclosed by the constraints and is called the feasible domain. Notice that we do not distinguish between the maximum or the minimum in this treatment. The reason for this indifference is that if a point x^* in the search space corresponds to the minimum value of the objective function $f_o(x)$, then the same point also is the corresponding maximum value of the negative of the objective function – that is, $-f_o(x)$. This is shown in Figure 4.1. An optimization problem involving multiple objective functions is commonly referred to as a multi-objective optimization problem. However, the present treatment of optimization methods does not cover multi-objective optimization algorithms.

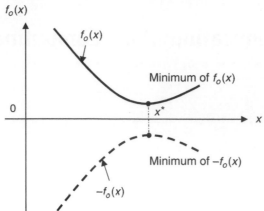

Figure 4.1 Minimum and maximum value of a function.

We can also distinguish between different categories of constraints. One category captures the behavioral constraints, which represent limitations on the behavior or performance of the system, and the side constraints, which represent physical limitations on the variables. The latter category of constraints can be further categorized into hard and soft constraints – that is, constraints that strictly limit the search field and constraints that are preferred not to be crossed. Hence, a candidate optimum point can be in the feasible region, can be an acceptable solution on the boundary defined by a hard constraint, or lie beyond a boundary defined by a soft constraint.

To find the optimum, an objective function is utilized. Suppose the objective function $f_o(x) = c$, where c is a constant, then the locus of all points satisfying this equation forms a hypersurface in the search space. For each value of the constant c, there are different families of surfaces. These surfaces are referred to as objective function surfaces. Graphically they correspond to isoclines which illustrate the search space with information that is similar to that seen on geographical maps (i.e., indication of mountains and valleys).

4.2   Review of Relevant Mathematics

The objective of this section is to review some basic mathematics that is necessary for hard computing optimization algorithms. A more in-depth treatment of these topics is found in [1].

Let's start with some basic concepts from linear algebra.

Definition 4.1
A matrix $\mathbf{A}$ is positive definite if all its eigenvalues are positive. More precisely, all values of λ that satisfy the equation

$$|\mathbf{A} - \vec{\lambda}\mathbf{I}| = 0 \qquad (4.3)$$

need to be positive.

Here, $\mathbf{I}$ is the identity matrix and $\vec{\lambda}$ are the eigenvalues. Extending Definition 4.1, we can also define *negative definite* if the eigenvalues of $\mathbf{A}$ are strictly negative.

In MATLAB one has at least two choices for testing a matrix $\mathbf{A}$ for being positive definite or not: using the `eig(A)` function or the `chol(A)` function. The `eig(A)` function simply computes the eigenvalues of the matrix $\mathbf{A}$, which then can be inspected for their magnitudes. However, there exists a more computationally efficient method by using `chol(A)`, which is the Cholesky decomposition. The `chol()` function allows for an optional second output argument which is zero if $\mathbf{A}$ is positive definite. If $\mathbf{A}$ is not positive definite, then this output will be a positive integer:

```
[~,p]=chol(A)
```

In this MATLAB script, p is the second argument indicating if the matrix $\mathbf{A}$ is positive definite or not. Note, if some of the eigenvalues of $\mathbf{A}$ are positive and some are zero, the matrix is referred to as *positive semidefinite*.

Definition 4.2

A matrix $\mathbf{A} \in \mathbb{R}^{n \times n}$ is called *nonsingular* or invertible if there exists a matrix $\mathbf{B} \in \mathbb{R}^{n \times n}$ such that

$$\mathbf{AB} = \mathbf{I}_n = \mathbf{BA}. \qquad (4.4)$$

Matrix $\mathbf{B}$ is the inverse of matrix $\mathbf{A}$. If $\mathbf{A}$ does not have an inverse, $\mathbf{A}$ is called *singular*.

An easy test of Definition 4.2 is to find the determinant of the matrix $\mathbf{A}$. If the determinant is not equal to zero, the matrix is nonsingular. However, this is not recommended when using MATLAB, as MATLAB employs a numerical approach that results in an ambiguous result (e.g., if the determinant of matrix $\mathbf{A}$ is 2.5489e-12, would this indicate a numerical error or a singularity of the matrix $\mathbf{A}$?). This leads us to Definition 4.3.

Definition 4.3

A matrix $\mathbf{A} \in \mathbb{R}^{n \times n}$ has rank $r \geq 1$ if and only if $\mathbf{A}$ has an $r \times r$ submatrix with nonzero determinant, while any matrix that has $r + 1$ or more rows and contains $\mathbf{A}$ has a determinant of zero.

In MATLAB the function `rank(A)` is used to find the rank of a matrix. If the matrix is full rank (i.e., $r = n$) we have a nonsingular matrix. An additional way to check for singularity in MATLAB is the `cond(A)` function, which finds the condition number of the matrix $\mathbf{A}$. As a rule of thumb, any condition number above 1e15 indicates that $\mathbf{A}$ is numerically singular. In addition to

Example 4.1

Examples of matrix surface plots that correspond to their properties of definiteness are given in Figure 4.2.

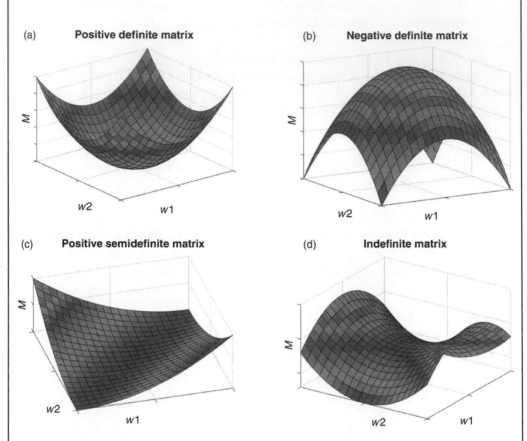

Figure 4.2 Examples of (a) positive definite matrix, (b) negative definite matrix, (c) positive semidefinite matrix, and (d) indefinite matrix.

rank() and det(), a rather useful MATLAB function is svd(), which is in effect used for computing the condition number of a matrix or the rank of a matrix. svd(A) will compute the singular values of **A** and we can examine if there is a singular value that is relatively small compared to the largest singular value of the matrix to indicate singularity or the degree of this matrix being ill-conditioned.

Applying the above-listed definitions to an optimization problem, we can state the following:

$$\text{Minimize } f(\vec{x}) = c^T \vec{x}, \tag{4.5}$$

$$\text{subject to } g(\vec{x}): \mathbf{A} \vec{x} = \vec{b} \tag{4.6}$$

$$\text{and } x_i \geq 0. \tag{4.7}$$

Note that the existence of a solution to the above problem is dependent on **A**: If the rows of **A** are linearly independent, then there exists a unique solution to the system of equations. However, if the determinant of **A** is zero (i.e., **A** is singular), then we have either no solution or an infinite number of solutions.

When discussing multivariable optimization problems we will come across some specialty matrices, which we review here briefly.

Consider $f : \mathbb{R}^n \to \mathbb{R}$ represents a function with an input vector $\vec{x} \in \mathbb{R}^n$ that produces an output function $f(\vec{x}) \in \mathbb{R}$. Suppose all second partial derivatives of $f(\vec{x})$ exist and are continuous over the domain of the function, then we can define the Hessian matrix of $f(\vec{x})$ as

$$
\mathbf{H}_f = \begin{bmatrix}
\dfrac{\partial^2 f}{\partial x_1^2} & \dfrac{\partial^2 f}{\partial x_1 \partial x_2} & \cdots & \dfrac{\partial^2 f}{\partial x_1 \partial x_2} \\[2ex]
\dfrac{\partial^2 f}{\partial x_1 \partial x_2} & \dfrac{\partial^2 f}{\partial x_2^2} & \cdots & \dfrac{\partial^2 f}{\partial x_1 \partial x_2} \\[2ex]
\vdots & \vdots & \ddots & \vdots \\[2ex]
\dfrac{\partial^2 f}{\partial x_1 \partial x_2} & \dfrac{\partial^2 f}{\partial x_1 \partial x_2} & \cdots & \dfrac{\partial^2 f}{\partial x_2^2}
\end{bmatrix}.
\tag{4.8}
$$

Note, the Hessian matrix $\mathbf{H}_f$ is a symmetric matrix. We can extend this definition to the Jacobian matrix by noting that the Hessian matrix of a function $f(\vec{x})$ is the Jacobian matrix of the gradient of the function $f(\vec{x})$ – that is, $\mathbf{H}_f(f(\vec{x})) = \mathbf{J}(\nabla f(\vec{x}))$.

4.3   Classical Optimization Techniques

Consider Figure 4.3 and note that the objective function $f_o(x)$ with one variable has a relative or so-called local minimum at $x = x^L$ if $f_o(x^L) \leq f_o(x^L + h)$ for some sufficiently small value of h. Also, the function $f_o(x)$ is said to have a global or absolute minimum at $x = x^*$ if $f_o(x^*) \leq f_o(x)$ for all x in the domain over which $f_o(x)$ is defined.

The necessary condition for a minimum point is as follows: If a function $f_o(x)$ is defined in an interval given by $\alpha \leq x^* \leq \beta$, with $x = x^*$ being a relative minimum, and $\dfrac{df_o(x)}{dx}$ exists as a finite number at $x = x^*$, then $\dfrac{df_o(x^*)}{dx} = 0$. In essence, we state that the necessary condition for a relative minimum is given by the function having an inflection point (also called a stationary point) at $x = x^*$. The sufficient condition for a minimum point is that at $x = x^*$ we have

Figure 4.3 Local and global minimum of a function.

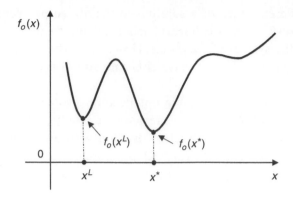

$$\frac{df_o(x^*)}{dx} = \frac{d^2f_o(x^*)}{dx^2} = \cdots = \frac{d^{(n-1)}f_o(x^*)}{dx^{(n-1)}} = 0 \qquad (4.9)$$

$$\text{and } \frac{d^{(n)}f_o(x^*)}{dx^{(n)}} \neq 0. \qquad (4.10)$$

Then,

$$f_o(x^*) \text{ is a minimum value of } f_o(x) \text{ if } \frac{d^{(n)}f_o(x^*)}{dx^{(n)}} > 0 \text{ and } n \text{ is even,} \qquad (4.11)$$

$$f_o(x^*) \text{ is a maximum value of } f_o(x) \text{ if } \frac{d^{(n)}f_o(x^*)}{dx^{(n)}} < 0 \text{ and } n \text{ is even, and} \qquad (4.12)$$

$$f_o(x^*) \text{ is neither a minimum nor maximum if } n \text{ is odd.} \qquad (4.13)$$

Example 4.2

Consider the single variable function $f(x)$ given by

$$f(x) = 12x^5 - 45x^4 + 40x^3 + 5.$$

Suppose we want to find the maximum and the minimum values of $f(x)$. Hence, we find the following derivatives: $\dfrac{df(x)}{dx} = 60x^4 - 180x^3 + 120x^2$ with $f(x) = 0$ at $x = 0, 1, 2$. Taking the second derivative yields:

$$\frac{d^2f(x)}{dx^2} = 240x^3 - 540x^2 + 240x.$$

Note that at $x = 1$ the second derivative of x is –60. Therefore, $x = 1$ is a relative maximum and $f(x^* = 1) = f_{max} = 12$. At $x = 2$ the second derivative of x is 240. Hence, $x = 2$ is a relative minimum and $f(x^* = 2) = f_{min} = -11$.

At $x = 0$ we have the second derivative of x equals to 0, which implies we need to test the next derivative:

$$\frac{d^3f(x)}{dx^3} = 720x^2 - 1{,}080x + 240.$$

Hence, $\left.\dfrac{d^3f(x)}{dx^3}\right|_{x=0} = 240$, which implies that $x = 0$ is neither a maximum nor a minimum; however, $x = 0$ is an inflection point.

This example can also be solved using MATLAB. As the problem is a single equation with a single variable x, we can utilize MATLAB's 1D solver `fzero()`. A more robust solution may be obtained using MATLAB's Optimization Toolbox's function `fsolve()`. Using `fzero()` for the above problem will result in the following MATLAB code:

```
[x fval exitflag]=fzero(@(x)(12*x^5-45*x^4+40*x^3+5),2)
```

In this code the "2" is our initial guess, preceded by the actual function we want to investigate. The @ handle is used to pass the objective function to the `fzero()` function, indicating that (x) is the variable to be adjusted. Note, the `fzero()` function can fail at a discontinuity, so we need to have some warning. Therefore, we include the return variable `exitflag` which indicates success if its value is 1.

We can extend the above discussion to multivariable optimization problems with no constraints. The corresponding necessary condition for $\vec{x} \in \mathbb{R}^n$ can be stated as follows.

Necessary condition: If $f_o(\vec{x})$ has an extreme point – either minimum or maximum – at $\vec{x} = \vec{x}^*$ and if the first partial derivative of $f_o(\vec{x})$ exists at $\vec{x}^*$, then

$$\frac{\partial f_o(\vec{x}^*)}{\partial x_1} = \frac{\partial f_o(\vec{x}^*)}{\partial x_2} = \cdots = \frac{\partial f_o(\vec{x}^*)}{\partial x_n} = 0. \tag{4.14}$$

Furthermore, we can update the sufficient condition for a multivariable optimization problem as follows:

Sufficient condition: A sufficient condition for a stationary point $\vec{x} = \vec{x}^*$ to be either a minimum or maximum is that the Hessian matrix of $f_o(\vec{x}^*)$ evaluated at $\vec{x}^*$ is

$$\text{positive definite when } \vec{x}^* \text{ is a relative minimum point,} \tag{4.15}$$

$$\text{negative definite when } \vec{x}^* \text{ is a relative maximum point.} \tag{4.16}$$

This problem can also be solved with MATLAB. If you have the Optimization Toolbox, a GUI will guide you through it. However, for this instance we will utilize the function optimset(). It is used to create an optimization option structure that takes on the parameters and options for the optimization problem.

Example 4.3

As an example, we want to find the maximum, minimum, and inflection points of the following function:

$$f(x_1, x_2) = x_1^3 + x_2^3 + 2x_1^2 + 4x_2^2 + 10.$$

To find the extreme points, we can apply the necessary conditions:

$$\frac{\partial f(\vec{x})}{\partial x_1} = 3x_1^2 + 4x_1 = 0 \text{ and } \frac{\partial f(\vec{x})}{\partial x_2} = 3x_2^2 + 8x_2 = 0.$$

From the two above equations, we find the following candidate solutions:

$$\vec{x} = (0,0), (0,-2.666), (-1.333, 0), (-1.333, -2.666).$$

We follow up by investigating each one of these candidate points by applying the sufficient condition, hence establishing the Hessian matrix of the problem:

$$\mathbf{H}_f(\vec{x}) = \begin{bmatrix} 6x_1 + 4 & 0 \\ 0 & 6x_2 + 8 \end{bmatrix}.$$

For this we have

$\mathbf{H}_f(\vec{x} = (0,0)) \ldots$ positive definite, hence relative minimum with

$\quad f(\vec{x}^* = (0,0)) = 10;$

$\mathbf{H}_f(\vec{x} = (0,-2.666)) \ldots$ indefinite, hence inflection (saddle) point with

$\quad f(\vec{x}^* = (0,-2.666)) = 15.481;$

$\mathbf{H}_f(\vec{x} = (-1.333, 0)) \ldots$ indefinite, hence inflection (saddle) point with

$\quad f(\vec{x}^* = (-1.333, 0)) = 7.185;$ and

$\mathbf{H}_f(\vec{x} = (-1.333, -2.666)) \ldots$ negative definite, relative maximum point with

$\quad f(\vec{x}^* = (-1.333, -2.666)) = 16.666.$

Example 4.4

The following code is given in two parts: one part is a function that describes the objective function, the second part is what one would type in the command window or as the main MATLAB file:

```
function f = fun(x)
f = x(1)^3+x(2)^3+2*x(1)^2+4*x(2)^2+10;

% Main function:
x0 = [0,0]; % also try other initial conditions
options = optimset('LargeScale','off');
[x,fval,exitflag,output] = fminunc('fun', x0, options);
x
fval
exitflag
```

Since constraint optimization for our future applications in this book is not employed, we leave out any discussion on imposing constraints on the optimization problem. For further treatment of these topics, see [2].

Instead, we will focus on five different types of optimization techniques commonly used in the learning process for neural networks. This learning process is generally formulated in terms of a minimization problem employing a loss function. The loss function acts as our objective function from the above treatment. The loss function is usually composed of an error term and an additional term, which we will later call the regularization term. The error term gauges how the neural network fits a given data set that we are trying to model, while the regularization term allows the neural network to remain general enough that new but similar data will fit the model. We will discuss this in much more detail in Chapter 6. For now, suppose we have a set of parameters that determine the neural network, and we want to find the optimum value based on a loss function (or objective function). Figure 4.4 depicts a two-dimensional problem with a quadratic loss function $f_o(w_1, w_2)$ representing the error, and two parameters labeled w_1 and w_2 representing the individual weights of the neural network.

Consider any point on the surface shown in Figure 4.4, containing the corresponding w_1 and w_2 values. At such a point we can compute the first and second derivatives of the objective function. For an n-dimensional problem, the first derivatives are assembled in the gradient vector, whose elements can be written as

$$\nabla_i f_o(\vec{w}) = \frac{\partial f_o(\vec{w})}{\partial w_i} \quad \text{for } i = 1, 2, \ldots, n. \tag{4.17}$$

We can compute the second derivatives of the objective function and assemble the corresponding Hessian matrix:

$$H_{i,j}(f_o) = \frac{\partial^2 f_o(\vec{w})}{\partial w_i \partial w_j} \quad \text{for } i, j = 1, 2, \ldots, n. \tag{4.18}$$

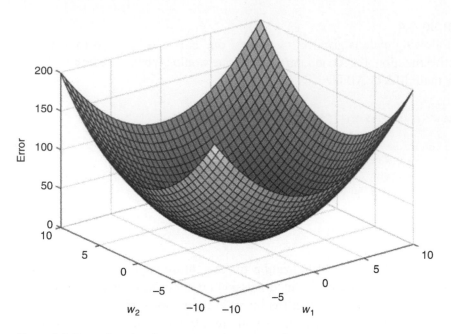

Figure 4.4 Error function for two parameters w_1 and w_2.

The Jacobian **J** is a matrix of all partial derivatives of that function:

$$\mathbf{J} = \frac{\partial(u_1, u_2, \ldots, u_m)}{\partial(x_1, x_2, \ldots, x_n)}. \tag{4.19}$$

Note that the Jacobian and the Hessian matrices are related by

$$\mathbf{H}(f_o) = \mathbf{J}(\nabla f_o). \tag{4.20}$$

MATLAB has some built-in functions to compute some of these terms. For example, the gradient can be computed using the function `gradient()`, the Jacobian is found by utilizing the MATLAB function `jacobian()`, and the Hessian matrix is found using the function `hessian()`.

Example 4.5

Consider the following function:

$$f(x, y) = 4x^2 - 4xy + 2y^2.$$

We can find the stated entities using the following MATLAB commands:

```
syms x y
f=4*(x.^2)-4*x*y+2*(y.^2)
```

```
gradient(f)
            ans =
            8*x - 4*y
            4*y - 4*x
jacobian(f)
            ans =
            [8*x - 4*y, 4*y - 4*x]
hessian(f,[x,y])
            ans =
            [8, -4]
            [-4, 4]
```

With these formulations we can now consider the five different optimization methods commonly used for neural network training having a parameter set $\vec{w}$, which are discussed in the following.

4.4   Gradient Descent Method

One of the simpler methods in neural network training is the gradient descent method, which is also often referred to as the steepest descent method. As the name indicates, it only requires the gradient information, hence it is a first-order method. The method begins at an initial point on the error function surface corresponding to an initial set of parameters $\vec{w}_o$. It moves stepwise until a stopping criterion is reached in the direction of the steepest descent, which can be expressed as:

$$\vec{w}_{i+1} = \vec{w}_i - \nabla f_o(\vec{w})\eta. \tag{4.21}$$

The parameter η is called the learning rate or sometimes referred to as the training rate in neural networks. The choice of η can have a large impact on the outcome of the optimization process. If η is too large, the step size becomes large and the optimum values may not be found – that is, the algorithm does not converge. A too small learning rate η may make the algorithm slow and computationally more expensive than necessary. Hence, there are methods where the learning rate is optimized based on the current location on the cost function surface. An indication of a good learning rate adaptation is given by plotting the iteration number against the objective function value. An exponential decline in the magnitude of the objective function indicates a proper learning rate selection. In simple algorithms this learning rate is just a constant. However, better implementations employ an optimization along the training direction at each step to find the optimum value of η. A stopping criterion can also be implemented. A sample algorithm for the implementation of the gradient descent method is given in Table 4.1.

Table 4.1 Gradient descent algorithm

Step	Description
Step 1	Choose a starting point $\vec{w}_o$.
	Choose a tolerance value for the objective function value to be used as a stopping criterion ε_f.
	Choose a tolerance value for the gradient value ε_g.
Step 2	Compute the objective function value $f_o(\vec{w})$ and its gradient vector $\nabla f_o(\vec{w})$.
	Set the search direction as $\vec{d}_i = -\nabla f_o(\vec{w}_i)$.
	Minimize $f_o(\vec{w}_{i+1})$ and determine η.
	Update the current search position $\vec{w}_{i+1} = \vec{w}_i - \nabla f_o(\vec{w}_i)\eta$.
	Test: If $|f_o(\vec{w}_{i+1}) - f_o(\vec{w}_i)| > \varepsilon_f$ or $\|\nabla f_o(\vec{w}_i)\| > \varepsilon_g$,
	then go to Step 2;
	else go to Step 3.
Step 3	$\vec{w}^* = \vec{w}_{i+1}, \quad f_o(\vec{w}^*) = f_o(\vec{w}_{i+1})$.

Example 4.6

An example of applying the gradient descent method for an optimization problem is given below.

Suppose you want to find the minimum value for x, y for a function $f(x, y) = 4x^2 - 4xy + 2y^2$. Let's randomly pick an initial starting point, $\vec{w}_o = (x = 2, y = 3)$. As a first step, we compute the gradient information:

$$\nabla f(x, y) = (8x - 4y, 4y - 4x) \text{ using } \vec{w}_o: \nabla f(2, 3) = (4, 4).$$

Then, our update is defined by Equation (4.21):

$$\vec{w}_1 = \vec{w}_o - \nabla f(\vec{w}_o) * \eta = (2, 3) - (4, 4)\eta.$$

Suppose we choose $\eta = 0.5$; we obtain $\vec{w}_1 = (0, 1)$. We repeat this process and compute the following values:

$$\vec{w}_2 = (0.4, 0.6), \vec{w}_3 = (0, 0.2) \dots \vec{w}^* = (0, 0).$$

We can achieve the same using MATLAB code to implement the gradient descent method. The MATLAB script is given in Appendix D.3.

The results are shown in Figure 4.5. Note that in the figure we plot the individual step results as the program cycles through its 30 iterations.

Using a different learning rate may result in a solution that does not converge. For example, using a learning rate of $\eta = 0.5$ will cause the MATLAB script not to converge. Another rather severe drawback of the

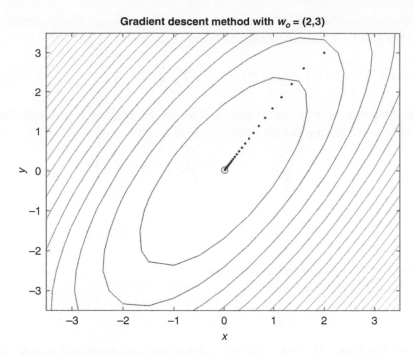

Figure 4.5 Graphical depiction of the gradient descent method for the example problem, including the tracing of intermediate steps.

algorithm is the requirement for many iterations. This is particularly the case when the objective function surface has a narrow valley structure. Here, even though the gradient is the direction in which the objective function decreases most rapidly, it does not yield the fastest convergence. However, the gradient descent method is particularly useful for very large neural networks, where the number of parameters can go into the thousands. For such applications the gradient descent method only needs to store the gradient vector, compared to other methods where the Hessian matrix is used (and therefore quadratically more parameters).

4.5 Newton's Method

Unlike the gradient descent method, Newton's method is a second-order algorithm as it utilizes the Hessian matrix. The purpose of the Hessian matrix is to find a more effective way to traverse the objective function surface with the goal of quickly finding the minimum point.

Suppose our objective function is a quadratic approximation at a point $\vec{w}_o$ and we use the Taylor series expansion to express it as

$$f_o(\vec{w}) \approx h(\vec{w}) = f_o\left(\widetilde{\vec{w}}\right) + \nabla f_o\left(\widetilde{\vec{w}}\right)^T \left(\vec{w} - \widetilde{\vec{w}}\right) + \frac{1}{2}\left(\vec{w} - \widetilde{\vec{w}}\right)^T \mathbf{H}_{f_o}\left(\widetilde{\vec{w}}\right)\left(\vec{w} - \widetilde{\vec{w}}\right). \quad (4.22)$$

Now we can attempt to minimize $h(\vec{w})$ by computing a point $\vec{w}$ for which $\nabla h(\vec{w}) = 0$. Because the gradient of $h(\vec{w})$ is

$$\nabla h(\vec{w}) = \nabla f_o\left(\widetilde{\vec{w}}\right)\mathbf{H}_{f_o}\left(\widetilde{\vec{w}}\right)\left(\vec{w} - \widetilde{\vec{w}}\right), \quad (4.23)$$

we can solve for

$$0 = \nabla h(\vec{w}) = \nabla f_o\left(\widetilde{\vec{w}}\right) + \mathbf{H}_{f_o}\left(\widetilde{\vec{w}}\right)\left(\vec{w} - \widetilde{\vec{w}}\right). \quad (4.24)$$

The corresponding solution for this is

$$\vec{w} = \widetilde{\vec{w}} - \mathbf{H}_{f_o}\left(\widetilde{\vec{w}}\right)^{-1}\nabla f_o\left(\widetilde{\vec{w}}\right). \quad (4.25)$$

Note, the term $-\mathbf{H}_{f_o}\left(\widetilde{\vec{w}}\right)^{-1}\nabla f_o\left(\widetilde{\vec{w}}\right)$ is a direction at step $\widetilde{\vec{w}}$ and is referred to as the Newton direction; in some literature it is called the Newton step. However, this step may very well lead to a maximum rather than a minimum. This can happen if the Hessian matrix is not positive definite. Hence, the function evaluation does not guarantee a reduction of the objective function at each iteration. To prevent such a move, Newton's method is modified by using a learning rate η, which results into the following update equation, constituting Newton's method:

$$\vec{w}_{i+1} = \vec{w}_i - \mathbf{H}_{f_o}\left(\vec{w}_i\right)^{-1}\nabla f_o\left(\vec{w}_i\right)\eta_i. \quad (4.26)$$

The learning rate in Equation (4.26) is indexed in case one chooses to utilize a line minimization to adapt to the current performance of the algorithm. However, it can also be set to a constant value. Also note that we still require that $\mathbf{H}_{f_o}\left(\widetilde{\vec{w}}\right)$ to be nonsingular at each iteration.

In general, Newton's method requires fewer steps to find the optimum in comparison to the gradient descent method. This comes at the cost of computing the Hessian and its inverse, which makes Newton's method more computationally expensive than the gradient descent method.

A sample implementation of Newton's method is given in Table 4.2.

Newton's method is implemented and given in Appendix D.4 using MATLAB.

Table 4.2 Newton's method

Step	Description
Step 1	Choose a starting point $\vec{w}_o$. Choose a tolerance value for the objective function value to be used as a stopping criterion ε_f. Choose a tolerance value for the gradient value ε_g.
Step 2	Compute the objective function value $f_o(\vec{w})$, its gradient vector $\nabla f_o(\vec{w})$, and the corresponding Hessian matrix $\mathbf{H}_{f_o}(\vec{w}_i)$. Set the search direction as $\vec{d}_i = -\mathbf{H}_{f_o}(\vec{w}_i)^{-1}\nabla f_o(\vec{w}_i)$. Determine η. Update the current search position $\vec{w}_{i+1} = \vec{w}_i - \nabla f_o(\vec{w}_i) * \eta$. Test: If $|f_o(\vec{w}_{i+1}) - f_o(\vec{w}_i)| > \varepsilon_f$ or $\|\nabla f_o(\vec{w}_i)\| > \varepsilon_g$, then go to Step 2; else go to Step 3.
Step 3	$\vec{w}^* = \vec{w}_{i+1},\ f_o(\vec{w}^*) = f_o(\vec{w}_{i+1})$.

Example 4.7

Solving the given problem from Section 4.4 using Equation (4.26) results in the following iteration – assuming the same learning rate of $\eta = 0.5$ and initial guess of $\vec{w}_o = (x = 2, y = 3)$:

$$\mathbf{H}_{f_o}\left(\tilde{\vec{w}}\right) = \begin{bmatrix} 8 & -4 \\ -4 & 4 \end{bmatrix} \text{ and } \nabla f_o\left(\tilde{\vec{w}}\right) = \begin{bmatrix} 8x - 4y \\ 4y - 4x \end{bmatrix}.$$

Iteration	x	y	$\nabla f_o(\vec{w})$
1	2	3	$[2\ 2]^{\mathrm{T}}$
2	1.0	1.5	$[1\ 1]^{\mathrm{T}}$
...	...	...	...
9	0.0039	0.0059	$[0.0076\ 0.0080]^{\mathrm{T}}$
10	0.0019	0.0029	

Example 4.8

To illustrate the performance difference between the gradient descent method and Newton's method, let's look at the following problem statement:

$$\text{Minimize } f_o(x, y) = 100\left(y - x^2\right)^2 + (1 - x)^2.$$

This function is called the Rosenbrock function. As many algorithms for optimization – hard computing and soft computing – are used and proposed, standard functions are utilized to compare the performance of these algorithms, such as the Rosenbrock function. This particular test function is a nonconvex unimodal function (i.e., it has one local minimum point which is also the global minimum point at coordinates $x = 1, y = 1$). The function's contour resembles a horseshoe, where the horseshoe is the valley. In this case, the valley is rather flat, with little gradient; however, away from this flat region are steep inclines. An algorithm that relies on the gradient for finding its updated solution and happens to choose an initial value that is in the steep part of the function may compute large step sizes that result in an overshoot of the minimum point. Because of these difficulties, the Rosenbrock function has been used extensively as a test function for optimization algorithms.

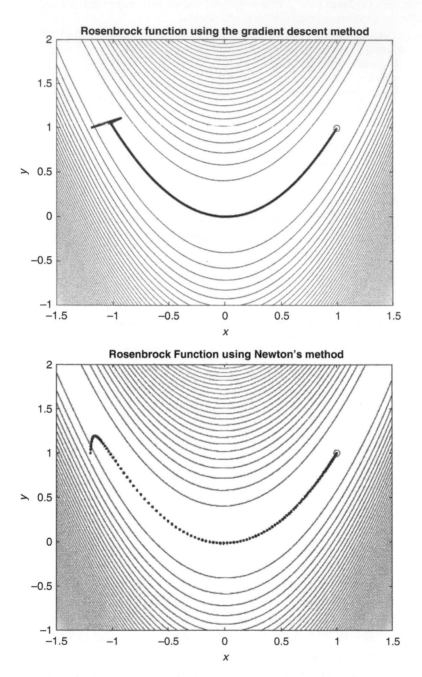

Figure 4.6 Comparison between the gradient descent method and Newton's method using the Rosenbrock test function.

To employ the gradient descent method and Newton's method, we compute the gradient and the Hessian matrix for the Rosenbrock function as

$$\nabla f_o(\vec{x}) = \begin{bmatrix} 100\left(-4xy + 4x^3\right) + (2x - 2) \\ 100\left(2y - 2x^2\right) + 2 \end{bmatrix},$$

$$\mathbf{H}_{f_o}(\vec{x}) = \begin{bmatrix} 100\left(12x^2 - 4y\right) + 2 & -400x \\ -400x & 200 \end{bmatrix}.$$

Utilizing the corresponding MATLAB code from this and the previous section, we can generate a contour plot and trace the incremental solutions approaching the minimum point, as shown in Figure 4.6.

Notice how each of the algorithms starts out their search: the gradient descent method utilizes a large step initially and "zig-zags" in the search space rather inefficiently, while Newton's method generates a smooth pathway to the optimum point from the start. The gradient descent method may result in a substantial decrease in the objective function initially, but generally is rather slow after that. This is shown in Figure 4.6 when comparing the steps taken to approach the minimum for both methods.

4.6   Conjugate Gradient Method

In the previous section we noted that Newton's method allows for a rather fast convergence to the optimum point compared to the gradient descent approach; however, this is at the cost of more computations as the Hessian matrix is involved. The conjugate gradient method is something in between Newton's method and the gradient descent approach. The motivation behind the conjugate gradient method comes from the desire to improve the convergence rate associated with the gradient descent method and to avoid the computation and inversion of the Hessian matrix as used by Newton's method. As the name indicates, the conjugate gradient method performs a search aligned with the conjugate directions, which are conjugated with respect to the Hessian matrix (two vectors are conjugate if they are orthogonal with respect to the inner product). This generally allows for a faster convergence of the algorithm than the gradient descent method. The recursive form of the conjugate gradient method is given by Equation (4.27):

$$\vec{w}_{i+1} = \vec{w}_i + \alpha_i \vec{d}_i. \tag{4.27}$$

Here, α_i is a parameter that controls the length of each step and $\vec{d}_i$ is the conjugate direction. The difference to the prior methods is given by the fact that $\vec{d}_i \neq \nabla f_o$ and that the step size governed by α_i is not constant. The question is, if $\vec{d}_i$ is not the gradient and the step size is not constant, how do we construct $\vec{d}_i$ while avoiding computing the Hessian matrix? It turns out that there are multiple propositions in use. For example, the implementation proposed by Polak and Ribière [3] is often employed for the training of neural networks and is listed below without the detailed derivation. We assume that the error surface is quadratic in nature. The procedure is as follows:

1. Compute the steepest direction $\Delta \vec{w}_i = -\nabla f_o(\vec{w}_i)$. $\hspace{2cm}$ (4.28)

2. Compute the gradient correction factor $\beta_i = \dfrac{\Delta \vec{w}_i^T (\Delta \vec{w}_i - \Delta \vec{w}_{i-1})}{\Delta \vec{w}_{i-1}^T \Delta \vec{w}_{i-1}}$. $\hspace{1cm}$ (4.29)

3. Update the conjugate direction $\vec{d}_i = \Delta \vec{w}_i + \beta_i \vec{d}_{i-1}$. $\hspace{2cm}$ (4.30)

4. Use line search to optimize $\alpha_i = \arg\min_{\alpha} f_o\left(\vec{w}_i + \alpha_i \vec{d}_i\right)$. $\hspace{1cm}$ (4.31)

5. Update the weight vector $\vec{w}_{i+1} = \vec{w}_i + \alpha_i \vec{d}_i$. $\hspace{2cm}$ (4.27)

Step 2 involves the computation of β_i, which is based on Polak and Ribière for Equation (4.29); however, there are a number of other implementations in use, such as

$$\text{Fletcher–Reeves: } \beta_i = \frac{\Delta \vec{w}_i^T \Delta \vec{w}_i}{\Delta \vec{w}_i^T \Delta \vec{w}_{i-1}}, \hspace{2cm} (4.32)$$

$$\text{Dai–Yuan: } \beta_i = \frac{\Delta \vec{w}_i^T \Delta \vec{w}_i}{-\vec{d}_{i-1}^T (\Delta \vec{w}_i - \Delta \vec{w}_{i-1})}. \hspace{1.5cm} (4.33)$$

Suppose we want to solve a set of equations for $\vec{w}$ that can be stated as

$$\mathbf{A}\vec{w} = \vec{b}.$$

The algorithm detailed by Equations (4.27)–(4.31) can be modified to accommodate this problem structure by first computing the residual vector

$$\Delta \vec{w}_o = \vec{b} - \mathbf{A}\vec{w}_o, \hspace{2cm} (4.34)$$

which we substitute for Equation (4.28) in this algorithm. Also, the initial search direction is found as $\vec{d}_o = \Delta \vec{w}_i$. With that we find the first α_o by computing $\alpha_1 = \frac{\Delta \vec{w}_o^T \Delta \vec{w}_o}{\vec{d}_o^T \mathbf{A} \vec{d}_o}$, which allows us to use Equation (4.27) to update the search position $\vec{w}$. We can repeat the computation by utilizing Equation (4.34), which

Example 4.9

Consider the MATLAB code in Appendix D.5 with the task to solve the following set of equations:

$$4x_1 - x_2 + x_3 = 12,$$

$$-x_1 + 4x_2 - 2x_3 = -1,$$

$$x_1 - 2x_2 + 4x_3 = 5.$$

Using an initial guess of $\vec{x}_o = [0 \ \ 0 \ \ 0]^T$, we obtain the correct solution of $\vec{x}^* = [3 \ \ 1 \ \ 1]^T$.

is used to find the updated gradient correction factor by employing either Equation (4.29), (4.32), or (4.33).

The corresponding MATLAB code given in Appendix D.5 using the conjugate gradient method will solve for the optimum values of $\vec{w}$ given an initial guess $\vec{w}_o$ and the set of equations contained in $\mathbf{A}$ and $\vec{b}$.

However, if the matrix $\mathbf{A}$ is ill-conditioned, the application of the conjugate gradient method may result in numerical errors.

4.7   Quasi-Newton Method

The development of the quasi-Newton method was motivated by the fact that Newton's method is computationally rather expensive as it requires the evaluation of the Hessian matrix and its inverse. However, the information provided by the Hessian matrix has shown to improve the convergence rate compared to other methods. Rather than computing the Hessian matrix, the quasi-Newton method attempts to approximate this matrix with a positive definite matrix, which is iteratively updated. As Newton's method in effect uses the inverse of the Hessian matrix, the quasi-Newton method may approximate the inverse of the Hessian matrix directly, hence reducing the computational costs. The approximation is accomplished by utilizing the gradient information from some or all of the previous iterations; for example:

$$\mathbf{H}_{i+1}(\vec{w}_{i+1} - \vec{w}_i) \approx \nabla f_o(\vec{w}_{i+1}) - \nabla f_o(\vec{w}_i). \tag{4.35}$$

Equation (4.35) is the result of applying the Taylor series expansion of $\nabla f_o(\vec{w}_{i+1})$ about $\nabla f_o(\vec{w}_i)$, which results in the given finite difference equation. It implies that two successive iterates $\vec{w}_i$ and $\vec{w}_{i+1}$ along with the gradients $\nabla f_o(\vec{w}_i)$ and $\nabla f_o(\vec{w}_{i+1})$ contain the curvature – Hessian – information. Hence, at each iteration one would prefer to choose the approximated Hessian matrix $\mathbf{B}_{i+1}$ to satisfy

$$\mathbf{B}_{i+1}\left(\nabla f_o(\vec{w}_{i+1}) - \nabla f_o(\vec{w}_i)\right) = \vec{w}_{i+1} - \vec{w}_i, \tag{4.36}$$

where $\mathbf{B}_i$ is a positive definite matrix approximating the Hessian information. Hence, the quasi-Newton method can be given by

$$\vec{w}_{i+1} = \vec{w}_i + \alpha_i \vec{d}_i, \tag{4.37}$$

where the direction $\vec{d}_i$ is given by

$$\vec{d}_i = -\mathbf{B}_i \nabla f_o(\vec{w}_i). \tag{4.38}$$

To iteratively converge to the Hessian information, we update at each iteration $\mathbf{B}_i$ by using a correction term $\mathbf{C}_i$ in Equation (4.36), which results into the following secant equation:

$$(\mathbf{B}_i + \mathbf{C}_i)\Big(\nabla f_o(\vec{w}_{i+1}) - \nabla f_o(\vec{w}_i)\Big) = \vec{w}_{i+1} - \vec{w}_i;$$

hence,

$$\mathbf{C}_k\Big(\nabla f_o(\vec{w}_{i+1}) - \nabla f_o(\vec{w}_i)\Big) = \vec{w}_{i+1} - \vec{w}_i - \mathbf{B}_i\Big(\nabla f_o(\vec{w}_{i+1}) - \nabla f_o(\vec{w}_i)\Big).$$

We did not specify the correction term $\mathbf{C}_i$ in the above equations. It turns out that there are many propositions for this term in the literature and in use. Many of these propositions are based on a basic form, which can be given as follows.

Using $q_i = \nabla f_o(\vec{w}_{i+1}) - \nabla f_o(\vec{w}_i)$ and $p_i = \vec{w}_{i+1} - \vec{w}_i$, $\mathbf{C}_i$ takes the form

$$\mathbf{C}(\gamma) = \frac{\vec{p}\vec{p}^T}{\vec{p}^T\vec{q}} - \frac{\mathbf{B}\vec{q}\vec{q}^T}{\vec{q}^T\mathbf{B}\vec{q}} + \gamma\kappa\vec{\mu}\vec{\mu}^T,$$

where $\kappa = \vec{q}^T\mathbf{B}\vec{q}$ and $\vec{\mu} = \frac{\vec{p}}{\vec{p}^T\vec{q}} - \frac{\mathbf{B}\vec{q}}{\kappa}$.

Hence, we can choose the value of $\gamma \in [0, 1]$, which gives some of the different choices for $\mathbf{C}_i$:

- Davidson–Fletcher–Powell (DFP) quasi-Newton method:

$$\mathbf{C}^{DFP} = \frac{\vec{p}\vec{p}^T}{\vec{p}^T\vec{q}} - \frac{\mathbf{B}\vec{q}\vec{q}^T\mathbf{B}}{\vec{q}^T\mathbf{B}\vec{q}}, \tag{4.39}$$

where $\gamma = 0$ is utilized.

- Broyden–Fletcher–Goldvarb–Shanno (BFGS) quasi-Newton method:

$$\mathbf{C}^{BFGS} = \frac{\vec{p}\vec{p}^T}{\vec{p}^T\vec{q}}\left\{1 + \frac{\vec{q}^T\mathbf{B}\vec{q}}{\vec{p}^T\vec{q}}\right\} - \frac{\mathbf{B}\vec{q}\vec{p}^T + \vec{p}\vec{q}^T\mathbf{B}}{\vec{p}^T\vec{q}}, \tag{4.40}$$

where $\gamma = 1$ is used.

A sample implementation of the DFP method is given in Table 4.3.
A sample implementation of the BFGS method is given in Table 4.4.

Table 4.3 The DFP method

Step	Description
Step 1	Choose a starting point $\vec{w}_o$. Choose a tolerance value for the objective function value to be used as a stopping criterion ε_f. Choose a tolerance value for the gradient value ε_g and an initial $\mathbf{B}_o$, which can be the identity matrix.
Step 2	Compute the objective function value $f_o(\vec{w})$, its gradient vector $\nabla f_o(\vec{w})$, and $\quad \vec{d}_i = -\mathbf{B}_i \nabla f_o(\vec{w}_i)$. Update $\vec{w}_{i+1} = \vec{w}_i + \alpha_i \vec{d}_i$. Minimize $f_o(\vec{w}_{i+1})$ and select α_i.
Step 3	Compute $\Delta \vec{w} = \vec{w}_i - \vec{w}_{i-1}$ and $\Delta g = \nabla f_o(\vec{w}_{i+1}) - \nabla f_o(\vec{w}_i)$. Update $\mathbf{B}_{i+1} = \mathbf{B}_i + \dfrac{\Delta \vec{w} \Delta \vec{w}^T}{\Delta \vec{w}^T \nabla g} - \dfrac{\mathbf{B}_i \nabla g \nabla g^T \mathbf{B}_i}{\nabla g^T \mathbf{B}_i \nabla g}$. Compute $\vec{d}_{i+1} = -\mathbf{B}_{i+1} \nabla f_o(\vec{w}_{i+1})$. Update $\vec{w}_{i+2} = \vec{w}_{i+1} - \alpha \vec{d}_{i+1}$. Determine α and minimize $f_o(\vec{w}_{i+2})$. Test: If $|f_o(\vec{w}_{i+2}) - f_o(\vec{w}_{i+1})| > \varepsilon_f$ or $\|\nabla f_o(\vec{w}_{i+1})\| > \varepsilon_g$, then go to Step 3; else go to Step 4.
Step 4	$\vec{w}^* = \vec{w}_{i+2}, \quad f_o(\vec{w}^*) = f_o(\vec{w}_{i+2})$.

Table 4.4 The BFGS method

Step	Description
Step 1	Choose a starting point $\vec{w}_o$. Choose a tolerance value for the objective function value to be used as a stopping criterion ε_f. Choose a tolerance value for the gradient value ε_g and an initial $\mathbf{B}_o$, which can be the identity matrix.
Step 2	Compute the objective function value $f_o(\vec{w})$, its gradient vector $\nabla f_o(\vec{w})$, and $\quad \vec{d}_i = -\mathbf{B}_i \nabla f_o(\vec{w}_i)$. Update $\vec{w}_{i+1} = \vec{w}_i + \alpha_i \vec{d}_i$. Minimize $f_o(\vec{w}_{i+1})$ and select α_i.
Step 3	Compute $\Delta \vec{w} = \vec{w}_i - \vec{w}_{i-1}$ and $\Delta g = \nabla f_o(\vec{w}_{i+1}) - \nabla f_o(\vec{w}_i)$. Update $\mathbf{B}_{i+1} = \mathbf{B}_i + \dfrac{\nabla g \nabla g^T}{\nabla g^T \Delta \vec{w}} + \dfrac{\nabla f_o(\vec{w}_i) \nabla f_o(\vec{w}_i)^T}{\nabla f_o(\vec{w}_i)^T \vec{d}_i}$. Compute $\vec{d}_{i+1} = -\mathbf{B}_{i+1}^T \nabla f_o(\vec{w}_{i+1})$. Update $\vec{w}_{i+2} = \vec{w}_{i+1} + \alpha \vec{d}_{i+1}$. Determine α and minimize $f_o(\vec{w}_{i+2})$. Test: If $|f_o(\vec{w}_{i+2}) - f_o(\vec{w}_{i+1})| > \varepsilon_f$ or $\|\nabla f_o(\vec{w}_{i+1})\| > \varepsilon_g$, then go to Step 3; else go to Step 4.
Step 4	$\vec{w}^* = \vec{w}_{i+2}, \quad f_o(\vec{w}^*) = f_o(\vec{w}_{i+2})$.

There are other methods that combine the DFP and BFGS methods by weighting each part to some degree; however, for training neural networks the BFGS method has shown to be superior to most other approaches.

Example 4.10

Using the Rosenbrock test function we can compare the performance of the two options detailed above for the quasi-Newton method using the given MATLAB function. The results are shown in Figure 4.7.

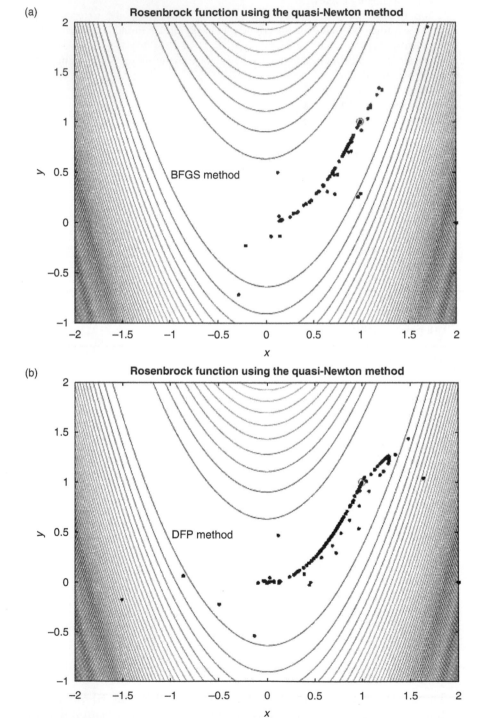

Figure 4.7 Comparison between quasi-Newton method and (a) BFGS and (b) DFP Hessian correction term for finding the optimum point for the Rosenbrock test function.

For this particular test function the quasi-Newton method employing the BFGS correction term resulted in having initially large step sizes, some of which are well outside the shown range of values in Figure 4.7. However, the BFGS-based simulation required 102 steps to find the optimum value of $x = 1, y = 1$, while the DFP-based simulation used 133 steps to achieve the same result with the same error tolerance of $e = 0.00005$.

The MATLAB function incorporating both the DFP and the BFGS methods is given in Appendix D.6.

4.8   Levenberg–Marquardt Method

The Levenberg–Marquardt method is also a rather popular learning algorithm used for neural network training. The literature sometimes refers to it as the damped least-squares method. It is a method that is particularly designed for an objective function that is composed of the sum of squared errors. Just like the quasi-Newton method, it does not employ the Hessian matrix; rather, it utilizes the Jacobian matrix and the gradient vector.

Suppose our objective function is expressed as a sum of squared errors:

$$f_o(\vec{e}) = \sum_{i=1}^{m} e_i^2 \text{ for } i = 0, 1, \ldots, m, \qquad (4.41)$$

where m represents the number of instances in the data set. The individual entries of the corresponding Jacobian matrix for the given objective function are determined by

$$J_{i,j}\left(f_o(\vec{w})\right) = \left[\frac{\partial e_i}{\partial w_j}\right] \text{ for } i = 1, \ldots, m \text{ and } j = 1, \ldots, n, \qquad (4.42)$$

where n represents the number of parameters. The associated gradient vector of the objective function can be evaluated as

$$\nabla f_o(\vec{e}) = 2\mathbf{J}^T e. \qquad (4.43)$$

With this, we can approximate the corresponding Hessian matrix as

$$\mathbf{H} \approx 2\mathbf{J}^T\mathbf{J} + \lambda\mathbf{I}. \qquad (4.44)$$

Here, λ is the damping factor that ensures the Hessian matrix is positive definite, and $\mathbf{I}$ is the identity matrix of appropriate dimensions. The parameter update process using the Levenberg–Marquardt algorithm is hence given by

Table 4.5 Levenberg–Marquardt method

Step	Description
Step 1	Choose a starting point $\vec{w}_o$. Choose a tolerance value for the objective function value to be used as a stopping criterion ε_f. Choose a tolerance value for the gradient value ε_g.
Step 2	Compute the objective function value $f_o(\vec{w})$ and its gradient vector $\nabla f_o(\vec{w})$. Set the search direction as $\vec{d}_i = -\left[\nabla f_o(\vec{w}_i)^T \nabla f_o(\vec{w}_i) + \lambda \mathbf{I}\right] \nabla f_o(\vec{e})$. Update the current search position $\vec{w}_{i+1} = \vec{w}_i - \vec{d}_i$. Test: If $f_o(\vec{w}_{i+1}) < f_o(\vec{w}_i)$, then change λ to $\lambda/2$; else change λ to 2λ. If $\lvert f_o(\vec{w}_{i+1}) - f_o(\vec{w}_i)\rvert > \varepsilon_f$ or $\lVert \nabla f_o(\vec{w}_{i+1})\rVert > \varepsilon_g$, then go to Step 2; else go to Step 3.
Step 3	$\vec{w}^* = \vec{w}_{i+1}$, $f_o(\vec{w}^*) = f_o(\vec{w}_{i+1})$.

$$\vec{w}_{i+1} = \vec{w}_i - \left(\mathbf{J}_i^T \mathbf{J}_i + \lambda_i \mathbf{I}\right)^{-1}\left(2\mathbf{J}_i^T e_i\right). \tag{4.45}$$

When we set the damping factor $\lambda = 0$, Equation (4.45) becomes Newton's method using the approximated Hessian matrix. However, if we select λ to be large, we in effect have the gradient descent method with a small learning rate. The Levenberg–Marquardt algorithm starts off using a large λ, ensuring that the first updates represent small steps in the gradient descent direction. However, if at any point during the iteration the algorithm results in a failure, then λ is increased by some magnitude. Otherwise, we reduce the magnitude of λ and hence approach Newton's method, which allows the process to increase the convergence to the minimum point of the cost surface.

Due to the speed of the Levenberg–Marquardt algorithm, it is extensively used in neural network training, where sum-of-squared-error functions are utilized to define the objective function. However, the Levenberg–Marquardt algorithm is not made for objective functions that are composed of the root mean square error or the cross-entropy error, and it struggles with regularization terms. Another problem of the Levenberg–Marquardt algorithm is that the Jacobian matrix becomes rather large for large data sets.

A sample implementation of the Levenberg–Marquardt method is given in Table 4.5.

The MATLAB code in Appendix D.7 is a simple implementation of the above optimization algorithm. The update rule for the damping factor is accomplished by changing its magnitude based on the amount of improvement the algorithm yielded. There are much-advanced implementations in use, such as the one

Example 4.11

The MATLAB code detailed in Appendix D.7 uses the Levenberg–Marquardt opti-mization method to solve a specific test function (i.e., the Rosenbrock function).

The results are shown in Figure 4.8. In the figure the adaptation of the damping factor causes a "zig-zag" approach to the optimum.

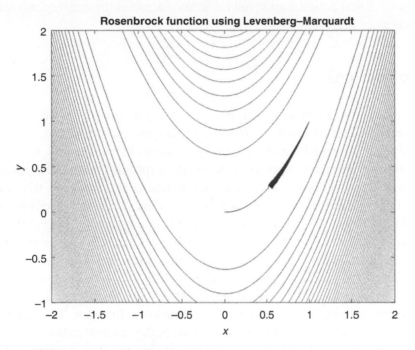

Figure 4.8 Using the simple implementation of the Levenberg–Marquardt method for finding the minimum point of the Rosenbrock test function.

provided by MATLAB (i.e., `lsqcurvefit()`), which is a curve-fitting tool that has the option of using the Levenberg–Marquardt method. For our purposes we will employ the pre-build training algorithm for neural networks with the option of using the Levenberg–Marquardt method.

Modifying the rule for updating the damping factor will result in rather differ-ent solution paths using the provided MATLAB code. Generally, we can distin-guish among the following schemes for the update rules:

- Direct method: Here we increase λ by a fixed factor for uphill steps, and decrease the damping factor by the same fixed factor for downhill steps.
- Delayed gratification – direct method: In this scheme the damping factor λ is increased using a small constant factor, while for uphill moves the damping factor λ is decreased using a large constant factor.

- Indirect method: Utilizing an initial value of λ, we update the damping factor at each step such that the step size is limited by a constant value.

Generally, the direct methods utilizing the same change in λ for either uphill or downhill steps result in moves that are rather large going downhill, which in turn reduces the convergence rate. Using delayed gratification, we effectively impose a small λ that prevents an uphill move and increases the convergence rate in the vicinity of the solution.

4.9 Conclusions

Optimization holds a prominent spot in engineering and technology, as it directly relates to productivity, cost, efficiency, and performance. As we will use optimization with respect to control and intelligent systems, we established in this chapter the mathematical foundation for hard computing optimization algorithms. We looked at the classical optimization approach where the first and second derivatives of the function for which we seek the optimum are of prime interest. We also discovered that for multivariable systems the Hessian matrix and its properties provide insight and guidance in terms of finding optimum values. In preparation for our discussion on neural networks we reviewed the gradient descent method, which is also known as the steepest descent method. It is one of the five iterative methods discussed in this chapter that iteratively seeks to find a local minimum of a function. The search for this optimum location is guided by the negative of the gradient, where we make finite steps proportional and in the direction of this negative gradient until some stopping condition is met.

Newton's method employs the Hessian matrix to provide a more effective guide for finding the optimum. This additional guidance compared to the gradient descent method is due to having access to curvature information which may lead to a better convergence rate.

A method that attempts to utilize the benefits of the gradient descent method and Newton's method is the conjugate gradient method. The conjugate gradient method requires less computation than Newton's method, and has an improved convergence rate compared to the gradient descent method. For this, it aligns its search direction with the conjugate gradient with respect to the Hessian matrix.

The quasi-Newton method does not explicitly compute the Hessian matrix, but attempts to approximate this matrix with a positive definite matrix which is iteratively updated. Just like Newton's method, the quasi-Newton method uses an approximation of the inverse of the Hessian matrix. This allows for a reduction in computational effort compared to Newton's method.

The Levenberg–Marquardt method is often employed in the training of neural networks. Its application is often tied to objective functions that are composed of the sum of squared errors. Just like the quasi-Newton method, it does not employ the Hessian matrix; rather, it utilizes the Jacobian matrix and the gradient vector.

We will encounter hard computing optimization algorithms again in Chapter 6, when we discuss how to train neural networks. In the next chapter we will look at methods that are not based on traditional calculus formulations to find optimum locations in n-dimensional spaces.

SUMMARY

In the following a brief summary is presented of the key topics in hard computing optimization as covered in this chapter.

Objective Function

An objective function, also commonly referred to as a loss function or cost function, serves to guide and assess the progress in finding the optimum during an optimization process. An objective function in optimization is often constructed by a squared error term.

Positive Definite and Positive Semidefinite

A matrix $\mathbf{A}$ is classified as positive definite when $\mathbf{A}$ is symmetric and all of its eigenvalues are positive. The matrix $\mathbf{A}$ is positive semidefinite if one or more of its eigenvalues is zero (nonnegative) and all of the remaining eigenvalues are positive.

Hessian Matrix

A square matrix containing the second-order partial derivates of a scalar-valued function is called a Hessian matrix. A Hessian matrix embeds information about the curvature of the scalar function:

$$\mathbf{H}_f = \begin{bmatrix} \dfrac{\partial^2 f}{\partial x_1^2} & \dfrac{\partial^2 f}{\partial x_1 \partial x_2} & \cdots & \dfrac{\partial^2 f}{\partial x_1 \partial x_2} \\[2ex] \dfrac{\partial^2 f}{\partial x_1 \partial x_2} & \dfrac{\partial^2 f}{\partial x_2^2} & \cdots & \dfrac{\partial^2 f}{\partial x_1 \partial x_2} \\[2ex] \vdots & \vdots & \ddots & \vdots \\[2ex] \dfrac{\partial^2 f}{\partial x_1 \partial x_2} & \dfrac{\partial^2 f}{\partial x_1 \partial x_2} & \cdots & \dfrac{\partial^2 f}{\partial x_2^2} \end{bmatrix}.$$

Gradient Descent Method

The gradient descent method, also referred to the steepest descent method, is a local iterative optimization algorithm involving the first term of the Taylor series expansion of the function to be optimized. Starting with a random initial starting point on the cost surface, the algorithm takes steps toward the minimum by being guided by the gradient information.

Newton's Method

Newton's method is a local iterative optimization algorithm. While the gradient descent method relies on first-order derivatives, Newton's method uses a second-order model, which results in the use of the Hessian matrix. Its search moves along the cost surface by repeatedly taking steps that are derived from the second-order terms of the Taylor series modeling the function to be optimized.

Conjugate Gradient Method

The conjugate gradient method is an iterative local optimization algorithm using a line search. It performs a search aligned with the conjugate directions, as given by the Hessian matrix. Compared to Newton's method, the conjugate gradient method does not involve the inversion of the Hessian matrix.

Quasi-Newton Method

Since computing the Hessian matrix is computationally expensive, an alternative method to Newton's method is given by employing an approximation of the Hessian matrix. This approximation is used in the quasi-Newton optimization method and is given by evaluating a set of linear equations. This set of equations is entailed in each iterative step of the optimization algorithm and allows for updating the Hessian matrix by utilizing the ratio of the change in the gradient to the change in the parameters.

Levenberg–Marquardt Method

The Levenberg–Marquardt optimization is an iterative nonlinear optimization algorithm that is very efficient for medium-sized problems. Similar to the quasi-Newton method, rather than computing and using the inverse of the Hessian matrix, it estimates the Hessian matrix. However, rather than solving a set of linear equations, the Levenberg–Marquardt method utilizes the sum of the outer products of the gradients.

REVIEW QUESTIONS

1. What is meant when referring to a feasible solution?
2. Is a maximization problem fundamentally different to a minimization problem?
3. What is a behavioral constraint?
4. If one of the eigenvalues of a matrix is zero and all other eigenvalues are negative, can we refer to the matrix as being negative definite or negative semidefinite?
5. Can singular matrices be inverted?
6. Is a Hessian matrix symmetric?
7. How is the Hessian matrix related to the Jacobian matrix?
8. Suppose a matrix surface plot exhibits a saddle form, what type of matrix is it most likely associated with?
   a. Positive definite matrix
   b. Negative definite matrix
   c. Positive semidefinite matrix
   d. Indefinite matrix
9. Suppose your candidate optimum point satisfies the necessary condition and yields a positive scalar when used to evaluate the nth derivative of the objective function, where n is an even integer. Is the candidate optimum point an optimum, and if so, a maximum or a minimum?
10. Why do we use a quadratic error function for an objective function in an optimization problem?
11. What is the difference between a multimodal function and a unimodal function?
12. Suppose you have a large learning rate for the gradient descent method; why would you potentially not find the optimum point?
13. What is the difference between Newton's method and the gradient descent method?
14. What is the function of the learning rate when used in Newton's method?
15. What part of Newton's method computation is most taxing in terms of time and computational resources?
16. How does the conjugate gradient method differ from the gradient descent method?
17. Which method is less computational expensive, Newton's method or the conjugate gradient method?
18. How does the quasi-Newton method differ from Newton's method?
19. Does the Levenberg–Marquardt method utilize the Hessian matrix computation?

20. What are the different methods of updating the damping factor for the Levenberg–Marquardt optimization method?

21. For what purpose are optimization algorithms used in neural networks?

PROBLEMS

4.1 Consider a matrix A:

$$A = \begin{bmatrix} 1 & 1 & 1 \\ 2 & 3 & 1 \\ 1e^{-10} & -1 & 1 \end{bmatrix}.$$

Determine if A is singular using MATLAB's functions:

a. `eig()`

b. `rank()`

c. `chol()`

d. `cond()`

Compare the results and comment on the matrix classification in terms of positive, negative, definite, semidefinite, and indefinite. Also find the inverse and comment on the reasons for the resulting magnitudes of the individual terms in the resulting matrix.

4.2 Find the Hessian matrix for the following multivariable cost functions:

a. $f_o(\vec{x}) = -2.5x_1^2 + 2.5x_2^2 + 3x_3^2 - 4x_3 + 12$

b. $f_o(\vec{x}) = \exp(3x_1) * \exp(3x_2) * 6\sin(6x_3)$

c. $f_o(\vec{x}) = 10x_1^2\ln(6x_2)$

Verify your answers using MATLAB's function for computing the Hessian.

4.3 Consider the following multivariable cost functions:

a. $f_o(\vec{x}) = x_1^2 + \sin(x_2)$

b. $f_o(\vec{x}) = x_1^2 - 3x_1x_2 - x_2^2$

c. $f_o(\vec{x}) = \cos^2(x_1)\cos(x_2)$

Find out if $f_o(x_1 = 0, x_2 = 0)$ is a local minimum, maximum, or a saddle point. Verify your answer by plotting the corresponding surfaces in MATLAB.

4.4 Find the Jacobian matrix of the following equations:

a. $x_1 = 4u - 3v^2$

$\quad x_2 = u^2 - 6v$

b. $x_1 = u^2v^3$

$\quad x_2 = 4 - 2u^{\frac{1}{2}}$

c. $f_1(\vec{x}) = x_1x_2 + 2x_2x_3$

$\quad f_2(\vec{x}) = 2x_1x_2^2x_3$

Verify your answers using MATLAB's function for computing the Jacobian.

4.5 Find the local extreme points for the following functions:

a. $f_o(x) = x^3 - 9x^2 + 24x - 7$

b. $f_o(\vec{x}) = \exp(-\frac{1}{3}x_1^3 + x_1 - x_2^2)$

First compute the results by hand and then verify your answer using MATLAB code.

4.6 Consider the equations below and determine if they represent a convex function.

a. $f_o(x) = \alpha_1(x - \beta_1)^2 + \alpha_2(x - \beta_2)^2$ for any $\alpha_i > 0$

b. $f_o(\vec{x}) = \|A\vec{x} - \vec{b}\|^2$

4.7 Consider the equation $f_o(\vec{x}) = 4x_1^2 + x_1 - 4x_1x_2 + 3x_2^2$. Plot the contour plot of this function and use the given MATLAB script in Appendix D.6 and the following main MATLAB function to call this script, and find the minimum value using the quasi-Newton method and MATLAB. Also plot the steps onto the contour plot.

```
clear x y
syms x y
f(x,y) = 4*x.^2+x-4*x*y+3*y.^2;
start= [0,0];
start1= num2cell(start');
f(start1{:})
e = 0.00005;
disp('for bfgs Newton')
localMin2 = QuasiNewton(f,start,e,1) % here opt = 1 for DFP
% use "2" for BFGS
```

4.8 Consider the equation $f_o(\vec{x}) = 4x_1^2 + x_1 - 4x_1x_2 + 3x_2^2$. Find the minimum value of this function as well as the corresponding values for x_1 and x_2 using:

a. gradient descent method and MATLAB;

b. Newton's method and MATLAB.

Compare the results as well as the choices of parameter for the different algorithms.

4.9 Consider

$$f_o(\vec{x}) = \vec{x}^T \begin{bmatrix} 2 & 1 \\ 1 & 1 \end{bmatrix} \vec{x} - \vec{x}^T \begin{Bmatrix} -1 \\ 1 \end{Bmatrix}.$$

Plot the surface plot and determine an approximate minimum value for $\vec{x}^*$. Using an initial starting point of $\vec{x}_o = [0 \quad 0]^T$, find the minimum point and minimum function value using the quasi-Newton method step-by-step procedure. Verify your results by utilizing the MATLAB code provided in Appendix D.6.

4.10 Consider the function $f_o(\vec{x}) = x_1^4 + x_1 x_2 + (1 + x_2)^2$. Utilizing a starting point of $\vec{x}_o = [0 \quad 0]^T$ and Newton's method, find out why this method fails to yield a solution.

4.11 Consider the McCormick function $f(\vec{x}) = \sin(x_1 + x_2) + (x_1 - x_2)^2 - 1.5x_1 + 2.5x_2 + 1$. Using a search area of $x_1 \in [-1.5, 4]$ and $x_2 \in [-3, 4]$, find the global minimum value of $f(\vec{x})$ and the corresponding $\vec{x}$ values:

a. using the computation of the gradient and Hessian information;

b. using Newton's method;

c. using the gradient descent method.

4.12 Consider the McCormick function given in Problem 4.11. Defining a search area of $x_1 \in [-1.5, 4]$ and $x_2 \in [-3, 4]$, find the global minimum value of $f(\vec{x})$ and the corresponding $\vec{x}$ values using the quasi-Newton method with options for the DFP and the BFGS algorithm.

4.13 Consider the McCormick function given in Problem 4.11. Defining a search area of $x_1 \in [-1.5, 4]$ and $x_2 \in [-3, 4]$, find the global minimum value of $f(\vec{x})$ and the corresponding $\vec{x}$ values using the Levenberg–Marquardt method.

4.14 Consider the Trid function:

$$f(\vec{x}) = \sum_{i=1}^{d} (x_i - 1)^2 - \sum_{i=2}^{d} x_i x_{i-1}.$$

Using a search area of $x_i \in [-d^2, d^2]$ for all $i = 1, 2, \ldots, d$, find the global minimum value of $f(\vec{x})$ and the corresponding $\vec{x}$ values for $d = 2, 3,$ and 4.

a. Use the computation of the gradient and Hessian information.

b. Use Newton's method.

c. Use the gradient descent method.

4.15 Consider the Trid function from Problem 4.14. Using a search area of $x_i \in [-d^2, d^2]$ for all $i = 1, 2, \ldots, d$, find the global minimum value of $f(\vec{x})$ and the corresponding $\vec{x}$ values for $d = 2, 3,$ and 4 using the Levenberg–Marquardt method.

4.16 Consider the power sum function:

$$f(\vec{x}) = \sum_{i=1}^{d} \left[\left(\sum_{j=1}^{d} x_j^i \right) - b_i \right]^2.$$

Alter the corresponding MATLAB code to find the maxima instead of the minima. Using $d = 4$ and $\vec{b} = [8 \quad 18 \quad 44 \quad 114]$, find the global maximum value of $f(\vec{x})$ and the corresponding $\vec{x}$ values.

a. Use the computation of the gradient and Hessian information.

b. Use Newton's method.

c. Use the gradient descent method.

d. Use the Levenberg–Marquardt method.

4.17 Consider the two-variable function:

$$f(\vec{x}) = \frac{1}{10}\left(12 + x_1^2 + \frac{1 + x_1^2}{x_1^2} + \frac{100 + x_1^2 x_2^2}{(x_1 x_2)^4}\right),$$

which is commonly known as Wood's function. Defining a search area of $x_i \in [0, 10]$, find the minimum using an initial starting point of $\vec{x}_o = [0.5 \quad 0.5]^T$ and the following methods:

a. steepest descent method;
b. Newton's method;
c. conjugate gradient method;
d. quasi-Newton method – DFP algorithm;
e. quasi-Newton method – BFGS algorithm.

REFERENCES

[1] Keyszig E (2010). *Advanced Engineering Mathematics*, 10th edn. Wiley, New York.
[2] Arora JS (1989). *An Introduction to Optimum Design*. McGraw-Hill, New York.
[3] Polak E, Ribière G (1969). Note sur la convergence de méthodes de directions conjuguées. *Revue Française Informatique Recherche Opérationelle*, 3(1): 35–43.

5 Optimization: Soft Computing

5.1 Introduction

Ever since Euclid of Alexandria in 325 BC, the topic of optimization has received much attention for both theoretical interest and everyday applications. It first addressed shortest-path problems around 325 BC, followed by calculus-based approaches proposed in the seventeenth and eighteenth centuries by Pierre de Fermat and other famous mathematicians, including J.-L. Lagrange. Iterative approaches were introduced during those times by Newton, Gauss, and others. Some of these concepts and methods were reviewed in the previous chapter. This chapter presents a different kind of optimization approach. Much of the material presented in the following is based on concepts found in nature. These nature-inspired notions have received a mathematical explanation and were termed optimization algorithms; they are part of soft computing algorithms. Soft computing contrasts itself from hard computing optimization algorithms by allowing for approximate models, and it is robust, to some degree, to imprecision and uncertainty. Soft computing algorithms entail a number of intelligent computing models such as fuzzy logic (introduced in Chapter 2), neural networks (part of Chapter 6), and the nature-inspired computational algorithms that are part of the focus of this chapter. Soft computing techniques also encompass machine learning, which is discussed at the end of the book. For now, we will treat soft computing techniques for optimization by looking at evolutionary algorithms, in particular genetic algorithms (GA), swarm-based algorithms such as particle swarm optimization (PSO), and tabu search (TS) applied to the optimization problem as introduced in Chapter 4.

In Chapter 4 we were able to formulate crisp system descriptions using exact data and binary logic to find optimality. Such exactness and precision are not always easy to define in applications – for example, nonlinear problems often lack a precise mathematical representation [1]. In addition, the number of feasible solutions tends to increase exponentially with the growth in size of the problem, making hard computing techniques cumbersome or even impossible to apply. Choosing which algorithm to use for a given optimization task may depend on the simplicity of the problem, the existence and availability of a precise mathematical description of the problem, the size of the problem in terms of the parameter space, and whether a closed-form or analytical solution even exists. Simple problems with analytical solutions tend to be solved at less expense using hard computing algorithms, while complex problems with strong parameter inter-dependencies may be addressed by applying soft computing algorithms.

A third category exists by combining the advantages of hard computing and soft computing, resulting in hybrid computing. The objective for hybrid computing algorithms is to obtain the advantages of hard and soft computing while avoiding the disadvantages of both. Just like soft computing, hybrid computing has found its application in control engineering, prediction, and classification problems. However, discussion and further insight on hybrid computing will not be presented in this book.

In the following sections test functions will be employed to measure the performance of the various soft computing optimization algorithms. In particular, we will utilize the spherical, quadratic, Ackley, Bohachevsky, Colville, Easom, Griewank, hyperellipsoid, Rastrigin, Rosenbrock, Schwefel, and the Haupt and Haupt test function as the objective functions in the optimization algorithms. Figure 5.1 depicts some of the different test functions.

5.2   Local Search

A number of soft computing optimization algorithms are based on the concept and short comings of the local search (LS) algorithm. The LS algorithm explores the search space using a single point (or current state), which moves one step at a time to its best-performing neighbor (or successor state). A random set of neighbors is created and evaluated based on an objective function. The path this exploration takes is not utilized in the decision-making process to advance the search. The termination criteria are given by the value of the objective function. This approach is able to find reasonable solutions in large search spaces and has found application in optimization problems. As the path is not tracked in the algorithm, LS has low computational memory requirements. The generic LS algorithm is given in Table 5.1.

Table 5.1 Local search algorithm

Step	Description
Step 1	Choose a random starting point $\vec{w}_o$ from the search space.
	Choose a tolerance value for the objective function value to be used as a stopping criterion ε_f.
Step 2	Generate a set of random n neighbors $\mathbf{W} = \left[\vec{w}_i^1 \ldots \vec{w}_i^n \right]$.
Step 3	Evaluate the set of random neighbors using the cost function $f_o(\mathbf{W})$.
Step 4	Select the best-performing neighbor as the solution of the current step and move in the search space to this new point $\vec{w}^* = \vec{w}_{i+1}$.
Step 5	Test if $|f_o(\vec{w}_{i+1}) - f_o(\vec{w}_i)| > \varepsilon_f$;
	then go to Step 2;
	else go to Step 2.

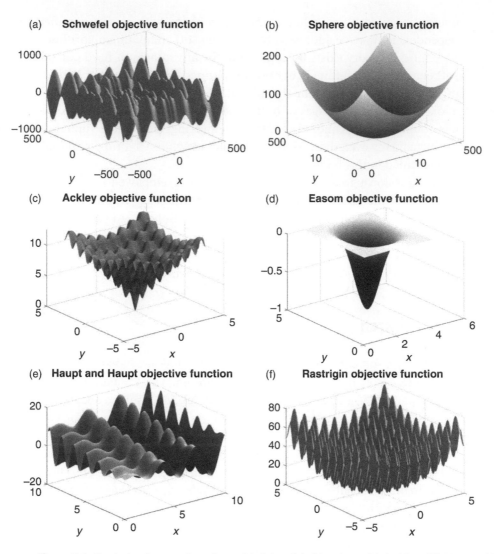

Figure 5.1 Optimization test functions: (a) Schwefel, (b) sphere, (c) Ackley, (d) Eason, (e) Haupt and Haupt, and (f) Rastrigin.

Local search works well for scheduling problems since these types of problems have a natural neighborhood defined. One of the key disadvantages of LS is that it converges to a local minimum when applied to multimodal problems. Generally, we are interested in the global optimum. One of the reasons for the convergence to a local minimum is the cycling phenomena. As we do not keep track of the path the iterative algorithm takes during the exploration of the search space, we have no measure to forbid a move that results in visiting a previous solution. Hence a cycling among old neighbors may ensue without progressing to find new neighbors.

There are algorithms which address exactly this shortcoming, such as the hill climbing (HC) algorithm (for maximization problems), the simulated annealing (SA) algorithm, and the tabu search (TS) algorithm, among others. The HC algorithm is practically identical to the LS algorithm except that during the iteration no neighbor is chosen that results in a lower objective function value. SA allows for moves during the search process that result in a worse value of the objective function with a probability that depends on the change in the objective function and the search time. The TS algorithm keeps a diary – or so-called tabu list – that details a finite set of past neighbors visited during the search and avoids these spaces in the search space for subsequent moves, which eliminates the chance of cycling. Hence, for maximization problems, TS and SA can overcome the problem of the LS by accepting "uphill" moves during the search. None of these algorithms utilize the gradient information – though there have been a number of hybrid computing algorithms proposed that make use of such information. Even though the gradient information is not utilized, the LS, TS, HC, and SA algorithms work best for objective functions that are smooth and gently sloped.

In the following, we expand the discussion of the presented LS algorithm by introducing formally the TS algorithm, while leaving the LS discussion as a base for understanding the TS algorithm without presenting a detailed MATLAB implementation.

5.3   Tabu Search

Suppose we have a multimodal objective function and our goal is to find the global minimum point within a defined search space. Due to the existence of local minima, an optimization algorithm that is based on the LS principles may face problems such as being stuck in a "valley" or local optimum point. An advanced algorithm would need to address how to escape this local minimum, and perhaps how to do this efficiently. Another problem that may have to be addressed is when dealing with an objective function that possesses a relatively "flat" region and/or a disconnected feasibility space. For "flat" sections of the objective function there is no incentive or motivation given by the LS algorithm to move and explore other areas. For fragmented search areas we will need to address how the search can "jump" to another isolated feasibility region. TS incorporates concepts to address some of these issues efficiently. For example, the TS algorithm avoids cycling by keeping a memory of the search history, has a built-in stage to intensify promising areas of the search space, and can potentially learn from the search itself.

TS was proposed as a metaheuristic search method by Fred W. Glover in 1986 [2] by incorporating the above-listed refinements and concepts into the LS algorithm. In particular, the algorithm incorporates a tabu list as well as an aspiration criterion. The tabu list stores the most recent moves of the algorithm within the search space. As the name indicates, the moves in the tabu list are avoided in future moves, hence potentially preventing the phenomena of cycling. The list length may have an influence on the success of avoiding cycling, as the step size as well as the number of items in the list (the list length) have to account for any distance to be traversed by the search in a negative – or uphill – direction when computing a minimization problem. However, there is an option to disregard the entries of a tabu list, which is termed the aspiration criteria. This criterion can be a simple condition stating that if the move yields a better solution than what is currently existing then the move is allowed even though it is in the tabu list. There are essentially two types of aspiration criteria: objective based and search direction based. An objective-based criterion is one that allows a tabu move if it yields a solution that exceeds an aspiration value. An aspiration criterion is guided by the search direction and allows for a tabu move if the direction of the search – regardless of improving or not improving – does not change.

The tabu list can be structured in two different forms: as a static memory or as a dynamic memory. The static tabu list employs a fixed length of items. As the tabu list is updated after each move of the TS algorithm, the oldest record is discarded as the latest move is included. The duration of a move staying in the tabu list is referred as the tabu tenure. For a static tabu list the tabu tenure is fixed for the entire search. For a dynamic tabu list the tabu tenure is not constant during the search operation. The length of the list is adapted based on what the search yields in terms of improvements between two steps and other efficiency measures, as well as the objective function properties. TS is structured into two parts: the diversification stage and the intensification stage. For the diversification stage the algorithm explores the search area as much as possible and keeps a record of areas that may lead to good results (local or even global minima). This record is termed the promising list, which is utilized in the second stage of the algorithm, where the search is intensified by focusing on those regions detailed in the promising list.

A characterization of the TS algorithm and its search space is given in Figure 5.2. Note, the tabu list entries and the promising list entries are not simply the exact coordinates of those listed locations; rather, they prescribe an n-dimensional sphere (for an n-dimensional problem) to encompass a vicinity around those locations listed.

Figure 5.2 depicts a two-dimensional problem defined by some cost function and the two parameters w_1 and w_2 for which we seek the optimum value. The current state of the search is denoted by the black star and its location in Figure 5.2. Around the current position, n-dimensional spheres are created to define a set of neighborhoods. As we deal with a two-dimensional problem, these neighborhoods are simple circles in Figure 5.2. In each neighborhood a set number of random candidate solutions are created (symbolized by the white stars in Figure 5.2). Note that the candidate solutions that fall within a tabu sphere will be discarded unless it satisfies some aspiration criterion. Also, at the exploration stage of the algorithm, candidate solutions that fall within a promising sphere will not be considered. To perform a move in the search space, allowable candidate solutions are evaluated on the given objective function, and the best-performing

Table 5.2 Tabu search algorithm

Step	Description
Step 1	Choose a random starting point $\vec{w}_o$ from the search space. Choose a tolerance value for the objective function value to be used as a stopping criterion ε_f. Define the tabu list, the tabu spheres, the aspiration criteria, and the promising spheres.
Step 2	Choose N neighbors around $\vec{w}_o$ which are not in the tabu list or satisfy the aspiration criteria.
Step 3	Evaluate the neighbors using the objective function $f_o(\mathbf{W})$ and choose the best-performing one.
Step 4	Update the current information in the tabu list and promising list.
Step 5	Test if $|f_o(\vec{w}_{i+1}) - f_o(\vec{w}_i)| > \varepsilon_f$; then go to Step 2; else stop the algorithm.

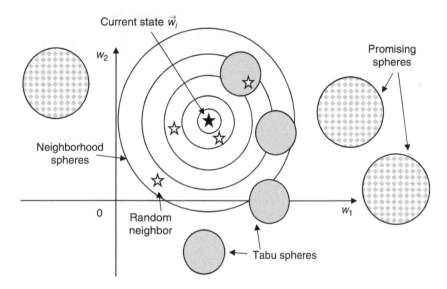

Figure 5.2 Graphical depiction of tabu search constellation in the search field.

location will become the new current location. The previous location will be recorded in the tabu list, while the oldest entry in this list will be dropped – provided a static memory is used. Well-performing locations will also be recorded in a separate list, the promising list. This process repeats itself for a number of steps. After that, the intensification portion of the TS algorithm is used to explore the regions in the promising spheres using the same algorithm, but limited to those spaces defined by the promising list. The detailed steps of the algorithm are given in Table 5.2.

A possible MATLAB implementation of this algorithm is given in Appendix D.8. The MATLAB code represents a TS algorithm that utilizes a

Example 5.1

Consider the Haupt and Haupt [3] test function, as depicted in Figure 5.1(e). This multimodal function has a global optimum point at the location $\vec{w}^T = [9.0389 \quad 8.6674]$ with a minimum objective function value of -18.5547. The function is governed by Equation (5.1):

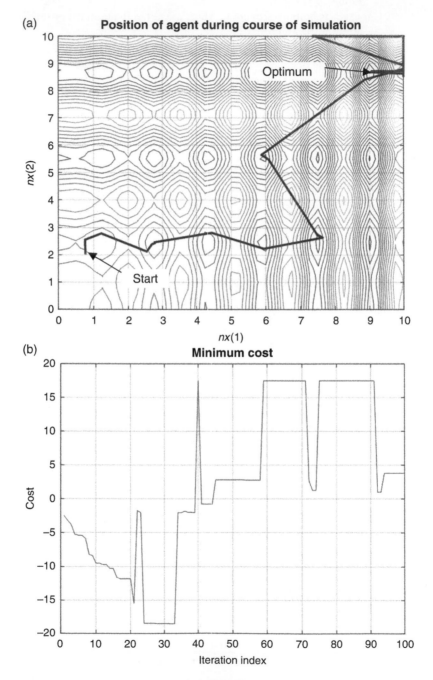

Figure 5.3 (a) Search path during TS optimization of the Haupt and Haupt problem; (b) minimum cost during 100 iterations of the search.

$$f_o(\vec{w}) = w_1 \sin(4w_1) + 1.1w_2 \sin(2w_2). \tag{5.1}$$

Using the TS optimization algorithm as given above and implemented in Appendix D.8, after 100 iterations the algorithm finds an optimum point of $\vec{w}^T = [9.0354 \quad 8.6741]$ and a corresponding objective function value of -18.5531. The path of the search position is depicted in Figure 5.3(a). The path starts at a randomly chosen location within the search area and progresses to the global optimum (minimum) point rather quickly. Figure 5.3(b) shows the corresponding objective function value at each iteration. From these graphs it appears the TS algorithm found the optimum at around iteration number 23. As the TS implementation has a fixed iteration number rather than a quality measure as a stopping criterion, the algorithm searches further in the vicinity of the global optimum. It is worth pointing out that the TS algorithm overcame multiple local minima. The implemented algorithm could be advanced by implementing such stopping criteria and adaptive features.

static tabu list and leaves out the creation of a promising list. The latter can be easily implemented and the program itself can be used for searching these promising areas for the intensification stage of the TS algorithm. The program incorporates a number of traditional unimodal and multimodal testing functions, as listed in Section 5.1.

5.4 Swarm Intelligence-Based Systems: Particle Swarm Optimization

In the previous section we had a single agent explore an n-dimensional search space with the task of finding an optimum location. It is intuitive to assume that, if there were more than one agent exploring the search space, the chance of finding the global optimum point may increase. The question then becomes: How do multiple agents coordinate their exploration such that an efficient and effective search may ensue? It turns out that somewhat similar questions were posed by biologists studying flocks of birds and schools of fish. In particular, Craig Reynolds worked in the late 1980s to capture the behavior of a flock of birds and developed simulated swarm and graphical simulation models, which he later termed "boids." His observation concluded that the behavior of a flock of birds is given by three processes: the separation process where the flock evades crowding of birds within the overall flock; alignment, where individual birds orient themselves to the flock's overall directional tendencies; and cohesion, where subgroups of the flock orient themselves to their average movements [4]. Recognizing that the model Reynolds developed could serve as a solution to finding an optimum point in a search space by multiple agents, James Kennedy and Russell Eberhart established the particle swarm optimization (PSO) algorithm [5]. The adaptation of Reynolds' work to PSO included the alignment and cohesion process without the separation process.

PSO belongs to the larger class of swarm intelligence (SI), where the behavior of swarm systems is studied and modeled. The basis for SI is given by the governing rules of natural self-formation in the absence of a centralized organization (also referred to as collective intelligence). Other optimization algorithms that fall within the SI category are genetic algorithms (introduced later in this chapter), ant colony optimization (ACO), artificial bee colony (ABC), and differential evolution (DE), to name a few.

In the following, we will detail the basic PSO algorithm. Since its introduction in the 1990s there have been propositions of novel nuances and refinements, as well as hybridization with other heuristic algorithms. The objective of this section is to provide the basic understanding and develop the primary formulation of the governing equations of the PSO algorithm. Our premise is as follows: Each particle is tasked with searching for the optimum location within the feasible region of the search area. Each particle has a property of velocity (i.e., they are always moving). Finally, each particle has a memory that records its personal best position (current and past) as measured against the objective function. Those three principles govern individual particles but by themselves do not affect the swarm. Hence, we need to amend a process that coordinates the entire flock of particles. The coordination entails an information exchange where each particle's personal best position is communicated to the entire swarm and the swarm communicates the best position among the individual best positions to all particles in the swarm in return.

The PSO algorithm is executed in steps where each particle moves to a new position, evaluates its position using the objective function, updates its personal best position if the new position is indeed better than any of the previous positions it has visited, and exchanges this position information with all the other particles while receiving the overall swarm's best position it encountered during the exploration so far. The core of the algorithm is given by the individual movement of a particle. The movement (i.e., its velocity) is governed by three different tendencies: (1) the influence of being autonomous and following its own way by using its current velocity vector as the determining factor of where the next location will be; (2) the temptation from the knowledge of where its own personal best location was and the tendency to go back to that location; and (3) the influence by the collective group of all particles who communicate the overall best location and hence the attraction of going to that location as well. PSO manages these three tendencies with the goal to find the optimum location efficiently. Although the tendencies seem to represent some complex dynamics, they can be easily and very efficiently formulated and computed. Equations (5.2) and (5.3) constitute a simple implementation of these three tendencies and formalize one way of computing the position of each particle in the swarm:

$$\vec{v}_i^{(k+1)} = \chi \left[\vec{v}_i^{(k)} + c_1(\vec{b}_i^{(k)} - \vec{w}_i^{(k)}) + c_2(\vec{y}^{(k)} - \vec{w}_i^{(k)}) \right], \tag{5.2}$$

$$\vec{w}_i^{(k+1)} = \vec{w}_i^{(k)} + \vec{v}_i^{(k+1)}, \tag{5.3}$$

where $\vec{w}_i^{(k)}$ is the position of the ith particle at iteration k, $\vec{b}_i^{(k)}$ is the best position encountered by the ith particle so far, and $\vec{y}^{(k)}$ is the best position found by the swarm up to time step k. The coefficients c_i are weighting coefficients to emphasize one of the tendencies more than the others, and χ is the constriction factor to regulate the velocity magnitude. It is interesting to note that PSO algorithms mix velocity and position terms, as $\vec{v}$ is the velocity vector and the other terms represent positions. We can graphically depict this in the search space, as we deal with vectors without making the distinction of position and velocity. Figure 5.4(a) shows the three different tendencies as vectors pulling at the ith particle from its current position. Adding each of the vectors together will result in a new position of this particle, as shown in Figure 5.4(b).

The overall algorithm is given by Table 5.3 and a possible simple MATLAB implementation is provided in Appendix D.9.

Some more advanced formulation of the PSO algorithm includes the creation of different neighborhoods. Here, a particle is associated with one or more neighborhoods and "knows" the best position encountered collectively by that neighborhood. This is in contrast to the above algorithm, where we effectively have just one big neighborhood comprising the entire swarm. These neighborhoods can be established using geographical means, such as the distance to the nearest particles, or social neighborhoods that are assigned at the beginning of the search process and do not change regardless of the positions of its member particles. We leave the exploration of neighborhood creation search performance to the problem section at the end of the chapter.

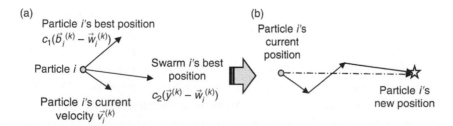

Figure 5.4 (a) Three tendencies as influence vectors applied to the ith particle at iteration k; (b) summation of the individual tendencies – or vectors – yields the new position at iteration $k + 1$.

Table 5.3 Particle swarm optimization algorithm

Step	Description
Step 1	Initialize a set of particles with a random position and velocity, $\mathbf{W}^0, \mathbf{V}^0$. Choose a tolerance value for the objective function value to be used as a stopping criterion ε_f.
Step 2	Evaluate each particle's position using the objective function $f_o(\mathbf{W})$.
Step 3	If the current position is better than the stored past best personal position, substitute $\vec{b}_i^{(k)}$ with the new position.
Step 4	Evaluate if any of the particles' personal best position $\vec{b}_i^{(k)}$ is better than the current swarm global best $\vec{y}^{(k)}$. If so, substitute them for the new swarm global best position.
Step 5	Update each particle's velocity $\mathbf{V}^{k+1}$ and position $\mathbf{W}^{k+1}$ using Equations (5.2) and (5.3).
Step 6	Test if $|f_o(\vec{w}_{i+1}) - f_o(\vec{w}_i)| > \varepsilon_f$; then go to Step 2; else stop the algorithm.

Example 5.2

Utilizing the same objective function from Section 5.3 (i.e., the Haupt and Haupt function) we can find the global optimum point with the MATLAB programs provided in Appendix D.9 entitled `simplePSO.m` and `cost.m`. The nominal choice of the objective function is set to the Haupt and Haupt function. However, many of the other objective functions can be selected along with parameter settings of the algorithm, such as the number of particles and the number of iterations. Using the given algorithm, and recalling that the optimum is at $\vec{w}_{opt}^T = [9.0389 \quad 8.6674]$ with a minimum objective function value of -18.5547, the simulation yields an optimum value of $\vec{w}^T = [9.0669 \quad 8.6853]$ with a minimum objective function value of -18.3116. The results are shown in Figure 5.5 and 5.6. Figure 5.5 displays the distribution of particles during the simulation. For this simulation 40 particles are utilized and initially randomly distributed over the search field. The first subplot depicts the constellation of those 40 particles after two iterations ($k = 2$). The global optimum point is at $\vec{w}^T = [9.0389 \quad 8.6674]$ – that is, in the upper right-hand corners of the diagrams. After five iterations the swarm pools around the upper right corner. Due to the three tendencies, the particles are again drawn away from the upper right corner to explore further the entire search space, as can be seen for $k = 20$ iterations. Over time, and especially after the last iteration at $k = 100$, most of the particles are localized around the global optimum point. The global best cost function during the iteration is depicted in Figure 5.6. Note that after about $k = 70$ iterations there is no improvement in terms of finding a better solution.

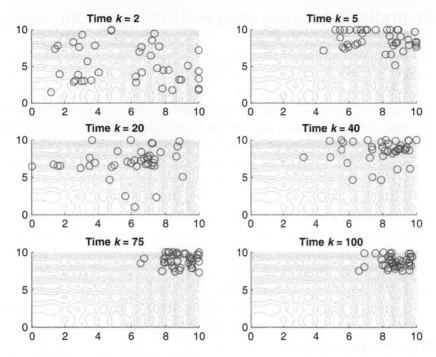

Figure 5.5 Particle movements during PSO simulation for the Haupt and Haupt test function, depicted as a contour plot, using instances of iterations at $k = 2, 5, 20, 40, 75$, and at the end of the simulation at $k = 100$.

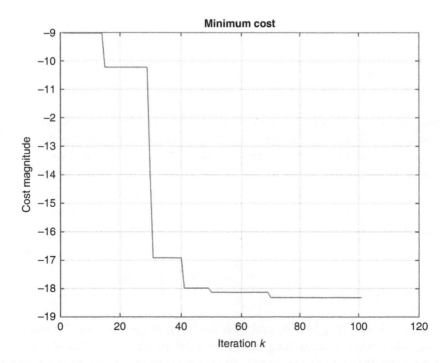

Figure 5.6 Cost of global best-performing location using the Haupt and Haupt objective function and the PSO algorithm.

5.5   Controller Autotuning using Particle Swarm Optimization

In this section our objective is to use PSO to tune a controller. We will utilize a simple but popular controller architecture to illustrate the use of PSO in controller design, the proportional integral derivative (PID) controller. Appendix A provides a brief review of traditional control systems. It includes and discusses the lag, lead, and lag–lead controller designs, which are common topics in a traditional modern control class. Lag–lead controllers are very similar to PID controllers, except that a design methodology is used to arrive at the controller parameters. PID controllers have attained broad usage within the process industry thanks to programmable logic controller (PLC) units and the ease of using these controllers. One of the facilitators that make PID so popular is the abundant literature on how to tune them. Starting with the Ziegler–Nichols tuning rules [6] in the 1940s, a number of other authors have proposed simple experiments that lead to look-up tables in order to find the particular PID controller parameters. We shall take a different route in this section, where we want to find the optimum PID controller parameters using PSO. However, before integrating PSO into PID controllers, let's review the PID controller architecture.

The proportional control component of the PID controller is given by modifying the error signal – which is computed by the difference between the desired output and the actual output of the system we want to control – in a proportional fashion, as given in Equation (5.4):

$$u_p(t) = K_p e(t), \tag{5.4}$$

where

$$e(t) = u_c(t) - y(t) \tag{5.5}$$

and $u_c(t)$ is the reference input or desired output, $y(t)$ is the plant output or actual output, and K_p is the controller gain. The structure of the P-controller is given in Figure 5.7, leaving out the integral and differential block. For the integral portion, the objective is to reduce the steady-state error between $u_c(t)$ and $y(t)$. Usually, the integral portion comes with a proportional term, hence we never see just an I-controller, but a PI controller. The integral part is given by Equation (5.6):

$$u_i(t) = K_i \int_0^t e(\tau)d\tau = \frac{K}{T_i} \int_0^t e(\tau)d\tau, \tag{5.6}$$

where T_i is the integral time and K_i is the integral gain. Figure 5.7 shows the PI controller when disregarding the differential block. Note that Equation (5.6) represents the average error for a given time period. Hence, even though there may be instances where $e(t)$ is zero, the corresponding control action $u_i(t)$ may not be zero. The role of the D part of the PID controller is to anticipate change in the error signal, and is given by Equation (5.7):

$$u_d(t) = K_d \frac{de(t)}{dt} = KT_d \frac{de(t)}{dt}, \tag{5.7}$$

where T_d is the derivative time and K_d is the derivative gain. Usually, the differential part is combined with the proportional part of the control architecture and hence we do not see just a D controller but a PD controller. The corresponding architecture of the control system is given in Figure 5.7 by leaving out the integral block. In real-world applications the differential portion of a PID controller also includes a low-pass filter. This filter prevents false predictions of the change in error due to the noise in signals. We can combine the three individual controller terms to form the PID controller:

$$u_{PID}(t) = u_p(t) + u_i(t) + u_d(t)$$

or

$$u_{PID}(t) = K \left\{ e(t) + \frac{1}{T_i} \int_0^t e(\tau)d\tau + T_d \frac{de(t)}{dt} \right\}. \tag{5.8}$$

Traditionally, modern control systems are treated in the s-domain, or Laplace domain. Taking the Laplace transform of Equation (5.8) with zero initial conditions, we can express the PID controller action as:

$$U_{PID}(s) = K \left\{ 1 + \frac{1}{T_i s} + T_d s \right\} E(s). \tag{5.9}$$

The transfer function of a PID controller block is expressed by taking the ratio of input over output; using Equation (5.9) and individual gains for each portion of the PID controller as specified in Equations (5.4), (5.6), and (5.7), we have

$$G_c(s) = K_p + \frac{K_i}{s} + K_d s. \tag{5.10}$$

When $K_i = 0$ we have a PD controller of the form:

$$G_c(s) = K_p + K_d s.$$

When $K_d = 0$ we deal with a PI controller given by:

$$G_c(s) = K_p + \frac{K_i}{s}.$$

The overall structure with a simple unity feedback loop implementing a PID controller is given in Figure 5.7.

Table 5.4 Effect of increasing PID gains on the step response

PID gain	Percent overshoot	Settling time	Steady-state error
Increase K_p ⬆	Increases ⬆	Small change ⬤	Decreases ⬇
Increase K_i ⬆	Increases ⬆	Increases ⬆	Decreases ⬇
Increase K_d ⬆	Decreases ⬇	Decreases ⬇	No change ⬤

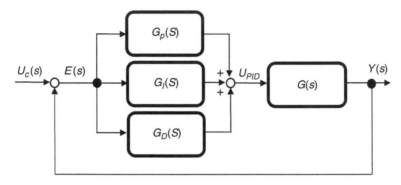

Figure 5.7 PID controller structure with unity feedback and plant dynamics.

With this structure of the PID controller we have three parameters available to tune the system for an acceptable closed-loop performance of the feedback system. As mentioned, there exists a number of tuning rules and associated experiments. They are all based on managing the sometimes-competing effects of the three terms of the PID controller. Table 5.4 illustrates these influences of the three gain values of the PID controller on the performance characteristics of the closed-loop system due to a step input.

Example 5.3

In order to implement the PSO algorithm to find the optimum PID gains, we will utilize Simulink as the plant simulation tool and call it from MATLAB to compute the error signal. The error can be computed in multiple ways. In this example we will utilize the integral of the squared error (ISE). We can utilize the PSO MATLAB script given in Appendix D.9 with a few minor changes. In the original MATLAB script we evaluated the objective function by invoking the following script:

```
S.cost(i)=cost(S.x(i,:),funchoice);
```

We alter this by calling a separate function that connects to a Simulink model which simulates the system with the current values of the parameters entailed in S.x(:,:). We can do this by using a so-called function handle. A function handle is a type of MATLAB value that allows for calling a function indirectly. The latter are functions that exist locally (in the workspace of MATLAB) – for example:

```
fobj = @ObjectiveFunction;
```

Here, @ObjectiveFunction refers to a MATLAB function we defined and stored in the same directory as the PSO MATLAB program:

```
function [iserror] = ObjectiveFunction(x)
% Replaces cost function to simulate a dynamic system
% given by a Simulink program called pso.slx
%
open('pso.slx');sim('pso.slx');
error1=(1-y)*(1-y)';
iserror=abs(sum(error1));
end
```

To call this function, we replace in the MATLAB code of the main PSO program the lines where we call for the cost function with the following script:

```
S.cost(i)=fobj(S.x(i,:));
```

Hence, we submit the current set of parameters contained in S.x(i,:) to the function ObjectiveFunction(). This function utilizes the Simulink model pso.slx by first opening the model and then simulating it. The Simulink model uses the current values from the particle swarm and generates the error values for the duration of the simulation – that is, the integral of the square terms of the error signal. This value is then presented back to the main PSO MATLAB program and assigned to S.cost(i). However, to explicitly outline the current parameter set being tested, we place them in the workspace by making an assignment in the main program prior to assessing the cost:

```
w = S.x(i,:);
```

Here, w carries the PID information which will be used in the Simulink program, where we employ a PID controller block. This PID block takes its values from the workspace by using the settings shown in Figure 5.8.

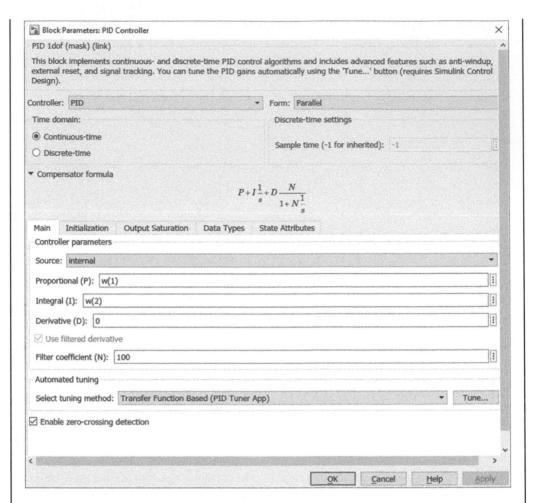

Figure 5.8 PID Simulink block with proportional and derivative gain values $w(1)$ and $w(2)$ set to the current set of PSO particle values. In this example we find the optimum parameters for a PI controller.

The Simulink model is given in Figure 5.9(a). For the example shown in the following, the model plant is essentially a second-order system with a damping ratio of $\xi = 0.25$ and a natural frequency of $\omega_n = 5$ rad/s. The corresponding open-loop step response is also given in Figure 5.9(b).

Using 40 particles and 40 iterations, a search range for K_p and K_i of $[-10 \quad 10]$, the PSO algorithm yielded the optimum PI parameters of $\vec{w}^* = [1.7887 \quad 1.6068]^T$. The resulting step response is shown in Figure 5.10. Note that the overshoot and response time has been reduced compared to the open-loop system. A number of other criteria can be embedded into the algorithm, in particular using the given simple objective function detailed above.

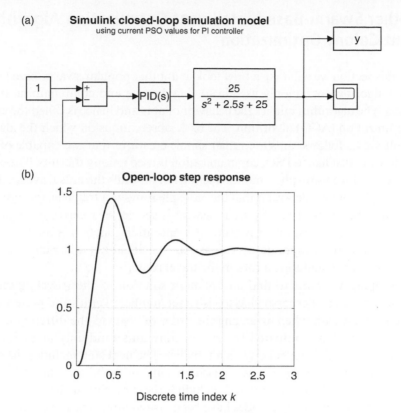

Figure 5.9 (a) Simulink model with simple unity feedback system and an open-loop transfer function of $G(s) = \dfrac{25}{s^2 + 2.5s + 25}$; (b) step response of the open-loop system.

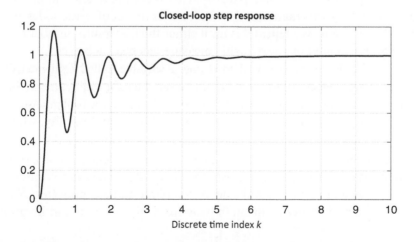

Figure 5.10 Simulation of the closed-loop system using optimum PI parameters.

5.6   Other Swarm-Based Intelligent Optimization Algorithms: Ant Colony Optimization

In this section we will take a brief look at another popular swarm-based optimization algorithm commonly used in solving discrete and combinatorial-type problems. The algorithm mimics the behavior of ants and hence is called the ant colony optimization (ACO) algorithm. The basic observations on which the algorithm is built are as follows: ants are rather simple creatures that are capable of complex behavior. Just like in PSO, communication is used among the ants. The communication is done using pheromones that are left on trails the ants traverse. The effect of the pheromone deposit is that the more pheromone a trail has, the more popular the trail becomes. Also, if a trail segment is shorter, it receives more pheromone when an ant traverses it. However, an evaporation function is built in so that the pheromone will evaporate over time. An individual ant will randomly pick one path when encountering a fork in the travel route.

Suppose we want to find an optimum solution to the traveling salesperson problem (i.e., a salesperson has to visit a set number of cities and prefers to choose the shortest path when arranging the order of visits to the different cities). The ACO algorithm is initiated by first locating ants randomly at each node/city, while placing the current city on a tabu list. The next step includes the determination of which city to visit next based on some probability, followed by actually moving the ant to that city and including the new city in the tabu list. At this point, we test if all cities/nodes have been visited and, if not, we repeat the prior step by determining which city we move to next. However, if we have visited all cities/nodes of the problem we record the length of the tour/overall path and clear the tabu list. Based on the different lengths the ants traveled to visit each city/node, the pheromone amount is distributed for each path and the algorithm starts anew by having ants randomly distributed over the set of cities/nodes and starting their travel again. We repeat this until either the maximum number of iterations have been executed or some quality measure has been achieved.

To make this work we need to define a variable called trail intensity, which indicates the intensity or the amount of pheromone on a trail segment. Suppose the trail segment connects city or node i to city or node j, then the trail intensity is denoted as τ_{ij}. Another variable can now be established: the trail visibility, which is given by $\eta_{ij} = 1/d_{ij}$ where d_{ij} is the distance between city/node i and j. Also, let's define Q as a constant representing the amount of pheromone deposited on a trail segment by one ant traversing it. Another variable we need to introduce is the evaporation rate ρ, determining how fast the deposited pheromone evaporates. Now we can express the probability of an ant moving from city/node i to j at iteration k:

$$p_k(i,j) = \begin{cases} \dfrac{[\tau_{i,j}^k]^\alpha \, [\eta_{i,j}]^\beta}{\sum\limits_{h \in N_i} [\tau_{i,h}^k]^\alpha \, [\eta_{i,h}]^\beta} & \text{if } (i,j) \in N_i \\ 0 & \text{otherwise.} \end{cases} \tag{5.11}$$

In Equation (5.11) we have two parameters, α and β, which represent the importance of the intensity in the probabilistic transition and the importance of the visibility of the trail segment, respectively. Equation (5.11) computes the probability of picking a segment from i to j at an arbitrary iteration time k. The probability is given by the ratio of the current pheromone content in the trail segment multiplied by the visibility of the segment divided by the entire sum of all pheromone portions multiplied by the visibility. Note that the visibility is nothing else than the inverse of the distance of the current path segment.

The trail intensity is also affected by the evaporation rate of the pheromone. We can update the trail intensity as follows:

$$\tau_{ij}^k = \rho\tau_{ij}^{k-1} + \Delta\tau_{ij}. \tag{5.12}$$

In Equation (5.12) we add pheromone proportionally to the path quality. Suppose the length of the path to visit all cities/nodes by ant m is given by L_m, then

$$\Delta\tau_{i,j}^{k\cdot} = \begin{cases} \dfrac{Q}{L_m} & \text{if } (i,j) \in path \\ 0 & \text{otherwise.} \end{cases} \tag{5.13}$$

Hence, using Equations (5.12) and (5.13) allows us to update the pheromone value for each portion of the path. Table 5.5 provides for the step-by-step guide for the ACO algorithm.

The ACO has been widely used for problems that deal with scheduling. A number of implementations in MATLAB can be found on MATLAB's file exchange webpage [7]. However, for our purposes of intelligent control systems, we leave the topic for future control applications without presenting a specific implementation.

Table 5.5 Ant colony optimization algorithm

Step	Description
Step 1	Initialize pheromone amount on all elements.
	Choose a tolerance value for the objective function value to be used as a stopping criterion ε_f.
Step 2	Distribute ants randomly on nodes. Record current node on the tabu list.
Step 3	For each ant, determine the probability of choosing a segment of the path to a next node.
Step 4	Move to the next node and place this node on the tabu list.
	Check if all nodes have been visited. If not, go to Step 3.
Step 5	Record the length of the overall path and clear the tabu list.
	Determine the shortest path and updated the pheromone amount.
Step 6	Test if $\lvert f_o(\vec{w}_{i+1}) - f_o(\vec{w}_i) \rvert > \varepsilon_f$;
	then go to Step 2;
	else stop the algorithm.

5.7   Genetic Algorithms

In Sections 5.2 and 5.3 we discussed heuristic single-agent optimization algorithms which we expanded to multiple-agent optimization algorithms in Sections 5.4 and 5.6. The motivation for expansion of the number of search agents is the potential increase in the probability to find the global optimum point. However, the computational effort associated with multiple-agent optimization algorithms is also enlarged. These multi-agent algorithms are based on the characteristics of swarm behavior (i.e., natural self-formation in the absence of centralized organization). In this section we explore another multiple-agent optimization algorithm. However, this algorithm is not based on those swarm principles. Genetic algorithms (GA) derive their guiding principles from Charles Darwin's theory of evolution (i.e., survival of the fittest). GA belongs to the evolutionary computation (EC) techniques which are used in optimization, learning, design, and control, among other areas. EC techniques do not require rich domain knowledge (e.g., gradient information of the objective function). However, domain knowledge has been incorporated in a number of EC computational algorithms. An EC usually starts out with an initial population that is randomly distributed over the objective function. This initial population of candidate solutions is evaluated by the utilization of the objective function and associated with a corresponding fitness value. Individual agents are selected as parents based on their fitness, and offspring are generated using processes such as cross-over and mutation (these terms will be introduced in detail later in this section). The new generation undergoes the same procedure of evaluation, selection, pairing, mating, mutation, and so on until a stopping criterion is met. We will focus on one of the EC algorithms in this section, the GA. Some of the first formulations of a GA were presented by Holland, who addressed adaptive search techniques [8]. Besides the thesis of survival of the fittest, the algorithm embedded the notion that good parents produce good children.

In GA each agent, or member of the current population, is represented by a string that mimics in function the biological information carrier role of chromosomes. These chromosomes are constructed of genes, where each gene is essentially one of the parameters for which we seek the optimum value. For example, for finding the lowest point or altitude in Yellowstone National Park in Wyoming, USA, the geographical coordinates can be utilized as the two parameters – in this case the longitude and the latitude. Hence, the corresponding chromosome of a candidate location for finding the lowest altitude in Yellowstone National Park would be the assembly of the numbers given by its longitude and latitude.

Let's generalize this concept: Suppose we want to optimize a set of parameters contained in the vector $\vec{w}$ based on an objective function $f_o(\vec{w})$ using a generic GA. The first task is to generate an initial population of n_{ipop} chromosomes. Here, we use a random number generator to find initial values for $\vec{w}$. Hence, a chromosome will be composed as follows:

$$chromosome(i) = \begin{bmatrix} w_1^i & w_2^i & \ldots & w_n^i \end{bmatrix} \text{ for } i = 1, 2, \ldots n_{ipop}. \tag{5.14}$$

Here, the superscript on the parameters denotes the association to which chromosomes belong rather than the mathematical power operator. The next step is to evaluate each chromosome using the objective function to find the chromosome's fitness value or cost:

$$\text{cost}\{chromosome(i)\} = f_o(\vec{w}^i). \tag{5.15}$$

Following this assessment, the next process is given by the utilization of the selection operator. Here, the literature provides for a number of different selection operators to choose from. Some selection methods are rank-based, where the chromosomes with the best cost value (i.e., lowest cost) are selected over chromosomes with higher cost values. Other methods employ tournament selection schemes among other approaches. Research has shown that some amount of randomness embedded in the overall algorithm improves the chances of the GA in finding the global minima. Hence, a common selection method is the weighted cost selection (WCS). The WCS uses the ranking of chromosomes to associate a probability P of being selected, favoring the best-performing chromosome. Therefore, the WCS allows for lower-ranked chromosomes to be picked as well, although with a lower chance:

$$P[chromosome(i)] \propto \text{cost}\{chromosome(i)\}. \tag{5.16}$$

Once the parent chromosomes are selected, a mating operator is responsible for the creation of offspring chromosomes. In particular, after selecting two parent chromosomes, the mating process will yield two offspring chromosomes. As the GA is a well-established method, a number of different mating operators have been proposed, used, and analyzed. We utilize here a popular algorithm called the simulated binary method. The name hints at a particular method used in the original binary implementation of GAs. In the binary GA, each gene is encoded as a binary number and hence a chromosome is represented as a string of 0s and 1s. The mating process for such binary chromosomes is rather simple: Randomly select a point in the string of 0s and 1s, denote this point as the cross-over point λ, and hence establish two parts of the chromosome – the binary numbers to the left and the binary numbers to the right of the cross-over point. The same process is applied to the second parent. However, the location of the cross-over point is the same as for the first parent. Mating is now accomplished by swapping one section of binary numbers between the two parents. This yields a new set of genes, which are for the most part equal to the parent genes, but some of which are exchanged and a few may have been changed by a randomly selected cross-over point.

The simulated binary method follows this approach, though it is applicable to continuous numbers and involves the computation of a cross-over gene w_{new}. Having selected two parents,

$$parent_1 = \begin{bmatrix} w_1^1 & w_2^1 & \ldots & w_\lambda^1 & \ldots & w_n^1 \end{bmatrix}, \tag{5.17}$$

$$parent_2 = \begin{bmatrix} w_1^2 & w_2^2 & \ldots & w_\lambda^2 & \ldots & w_n^2 \end{bmatrix}, \tag{5.18}$$

the w_λ^j cross-over gene is randomly picked at gene number λ. We compute two new cross-over genes using the selected w_λ^j gene from both parents as follows:

$$w_{new_1} = w_\lambda^1 - \mu[w_\lambda^1 - w_\lambda^2] \text{ and} \tag{5.19}$$

$$w_{new_2} = w_\lambda^2 - \mu[w_\lambda^1 - w_\lambda^2], \tag{5.20}$$

where $0 \le \mu \le 1$. Hence the first offspring is composed of the genes from parent 1 that arc to the left of the cross-over gene λ in Equation (5.17), the first new gene given by Equation (5.19), and the genes from parent 2 that are to the right of the cross-over gene λ, as given in Equation (5.21):

$$offspring_1 = \begin{bmatrix} w_1^1 & w_2^1 & \ldots & w_{new_1} & \ldots & w_{n-1}^2 & w_n^2 \end{bmatrix}. \tag{5.21}$$

Analogously, we can create the second offspring. However, here we utilize Equation (5.20) for the new gene at location λ in the second parent chromosome, and change the order of which set of chromosomes are used, as shown in Equation (5.22):

$$offspring_2 = \begin{bmatrix} w_1^2 & w_2^2 & \ldots & w_{new_2} & \ldots & w_{n-1}^1 & w_n^1 \end{bmatrix}. \tag{5.22}$$

The third operator in a traditional GA is the mutation operator. Mutations in biological systems are rare changes to a chromosome. The function of mutation in a GA is providing a means to escape local minima, overcome "flat" spaces in the objective function, and "jump" to isolated feasibility regions within the search space. The mutation operation implanted in the associated MATLAB code given in Appendix D.10 employs a simple substitution of a randomly picked number within the population of chromosomes using a randomly generated number.

Table 5.6 provides the step-by-step guide for the generic GA optimization algorithm.

The corresponding GA implementation is given in Appendix D.10. The provided code entails the Haupt and Haupt objective function, which can be easily substituted with any of the prior objective functions utilized in Chapter 5. The main program, `simpleGA.m`, queries the user for the simulation parameters, such as the initial population size, the maximum number of iterations, the constant population size after the initial population iteration, the number of chromosomes kept for mating, the number of parameters/genes in a chromosome, as well as the mutation rate and the range of parameter values. A trial-and-error approach shows that starting with a large initial population first, then reducing that number to a constant population size, allows for a somewhat faster computation. This is because

Table 5.6 Genetic algorithm

Step	Description
Step 1	Create a random population of n_{ipop} chromosomes using Equation (5.14). Choose a tolerance value for the objective function value to be used as a stopping criterion ε_f or define an iteration number for a fixed number of generations.
Step 2	Evaluate the fitness of each chromosome in the initial population using the objective function $f_o(\vec{w})$, i.e., Equation (5.15).
Step 3	Reduce the population size, i.e., number of chromosomes, to a constant number $n_{pop} \leq n_{ipop}$ using the associated cost as the determining factor.
Step 4	Apply the selection operator (Equation 5.16), the mating operator (Equations 5.17–5.22), and mutation operation.
Step 5	Evaluate the fitness of each chromosome in the new population using the objective function $f_o(\vec{w})$.
Step 6	Replace the old generation with the new population, and update the generation number.
Step 7	Test if $|f_o(\vec{w}_{i+1}) - f_o(\vec{w}_i)| > \varepsilon_f$ or you have reached the maximum number of generations; then go to Step 4; else stop the algorithm.

the initial random distribution of locations in the search field – given by the initial chromosomes – may result in a good starting point for the search and the reduced number of chromosomes henceforth decreases the computational load in the loop going through each generation.

Plots of minimum cost from a GA optimization simulation that employ a retention mechanism of the best-performing chromosome – regardless of the number of generations – always yield a monotonically declining curve. Although this aspect of survival of the fittest principle is not found in nature, some research has found that allowing for a chromosome to survive any number of generations provided it maintains its fitness level compared to all the other

Example 5.4

Consider the Haupt and Haupt test function, using the provided code in Appendix D.10 and an initial population size of 96 chromosomes, a constant population size of 48, where the best 24 chromosomes are chosen for the selection and mating process. It yields after 100 generations an optimum value of $\vec{w}^T = [9.0390 \quad 8.6682]$ and a minimum cost value of -18.5547. Recall that the optimum is at $\vec{w}_{opt}^T = [9.0389 \quad 8.6674]$ with a minimum objective function value of -18.5547. The code provided keeps track of the minimum cost, the mean cost, and the standard deviation of the cost during the simulation. Figure 5.11 depicts the minimum cost for this problem and the 100 generations.

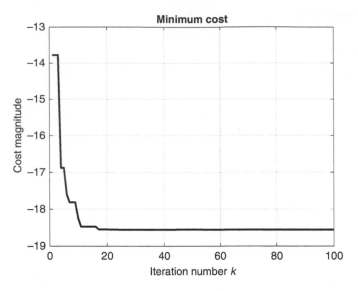

Figure 5.11 Plot of minimum cost value during GA simulation for finding the minimum point of the Haupt and Haupt test function.

chromosomes yields good results. The code provided in Appendix D.10 does indeed keep the best-performing chromosome over each generation, and Figure 5.11 indicates that only about 40 iterations would have been necessary with the given set of parameters to achieve the results listed. In the above simulation, the mutation rate was set to 0.04 or 4%. It is interesting to note that these operational parameter settings are strongly dependent on the landscape of the objective function and are interrelated. Trial-and-error, or a separate optimization routine, could yield good operational parameter values.

5.8   Fuzzy Logic Controller Autotuning using Genetic Algorithms

MATLAB offers the Global Optimization Toolbox that entails GA, PSO, and SA, among other search and optimization tools. In this book so far we have been focusing on the underlying algorithms and principles, rather than ready-made applications featured in specialized toolboxes by MATLAB. The included algorithms in the respective toolboxes are optimized for speed and accuracy, and should be explored if available. In this section we explore some possible ways to utilize GAs for control design. In particular, we revisit fuzzy logic-based controllers and their optimization. The goal of this section is not necessarily to achieve in particular well-performing controllers, but rather to focus on how to utilize GAs to aid in the design of FL controllers and present some alternative ways to code this in MATLAB. In Chapters 2 and 3 we primarily employed the Fuzzy Logic toolbox in conjunction with the Fuzzy Logic Designer application. There are other approaches to design and implement fuzzy inference systems, some of which are presented in Example 5.5.

Example 5.5

To illustrate this alternative approach, we address the problem of creating a fuzzy logic controller to control the output of a dynamic system, similar to Example 5.3. Rather than using a PID controller, a Mamdani fuzzy inference system-based controller is proposed to regulate the output of this second-order dynamical system. To create a fuzzy inference system in MATLAB without the FL Designer app, we use the approach presented in Section 3.3. The system we want to control is a second-order underdamped system, behaving in the fashion given in Figure 5.9(b). Suppose our goal is to regulate the output to a constant reference value – for example, using a step input. Presume we measure the output of the system and are able to compute the error between the output and the reference input, as well as the rate of change of the error. The question is: How do we design the membership functions for the inputs and the output, and how do we come up with the associated rules in some organized fashion? In addition to the question of developing rules, we want to explore some ways to incorporate the optimization of membership functions and the fuzzy logic rule base. The question of designing membership function was already treated in Section 3.3, where we used linguistic variables to express the system's behavior and were able to associate membership functions to it with the corresponding range.

Let's define a Mamdani inference system and two inputs called *error* and *derror*, where the latter represents the change of rate in the error signal of a feedback system. Also, suppose we have three membership functions for each of the two inputs, given in MATLAB as follows:

```
% create Mamdani FIS
fiscon = mamfis('Name', "flcontroller");
% add error input
fiscon = addInput(fiscon, [-2.5 2.5],'Name', "error");
% Add membership functions to "error" input
fiscon =addMF(fiscon, "error", "zmf", [-1.5 0.2],'Name', "Negative");
fiscon = addMF(fiscon, "error", "trimf", [-1.5 0 1.5],'Name', "Zero");
fiscon = addMF(fiscon, "error", "sigmf", [5 0.7],'Name', "Positive");
% add derivative of error input
fiscon = addInput(fiscon, [-2.5 2.5],'Name', "derror");
% Add membership functions to "derror" input
fiscon = addMF(fiscon, "derror", "zmf", [-1 0.2],'Name', "Negative");
fiscon = addMF(fiscon, "derror", "trimf", [-1 0 1.5],'Name', "Zero");
fiscon = addMF(fiscon, "derror", "sigmf", [5 .7],'Name', "Positive");
```

Suppose we utilize one output with five membership functions, as given below:

```
% Add output variable
fiscon = addOutput(fiscon, [-3 3], 'Name', "actuation");
fiscon = addMF(fiscon, "actuation", "trimf", [-100-5-2.5], 'Name', "LN");
fiscon = addMF(fiscon, "actuation", "trimf", [-5-2.5 0], 'Name', "SN");
fiscon = addMF(fiscon, "actuation", "trimf", [-2.4 0 2.4], 'Name', "Z");
fiscon = addMF(fiscon, "actuation", "trimf", [0 2.5 5], 'Name', "SP");
fiscon = addMF(fiscon, "actuation", "trimf", [2.5 5 100], 'Name', "LP");
```

In order to construct the associated rules, let's consider Figure 5.12, which depicts the step response of a generic second-order system, as well as the associated error and derivative of the error signal. Start by considering the first 0.2 seconds of the simulation as depicted in Figure 5.12. We notice that the error, as defined by the difference between desired output and system output, is positive, the derivative of the error is negative, and the system output is positive.

From this characterization, we can conclude that the actuation should counter these influences. For example, we associate the actuation as *LP* corresponding to the *large positive* membership function when these conditions exist. We can go through each combination of the two input signals and establish a logic table by associating the corresponding actuation necessary to mitigate the error signal. A possible inference of these actuation associations is given in Table 5.7.

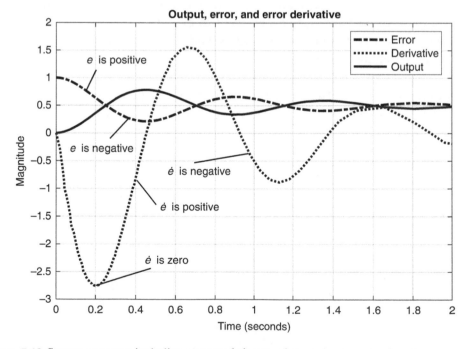

Figure 5.12 System response including error and change of error due to a step input.

Table 5.7 Logic table for FL rules

Actuation MF		Error derivative		
		N	Z	P
Error	**N**	LN	SN	SN
	Z	SN	Z	SP
	P	LP	SP	SP

Having established a logic table, we can proceed with implementing the fuzzy rules as described by the logic table.

In Section 3.3 we implemented the rules using the function `addRule`, where we developed a `rulelist` that was represented by a matrix. This rather abstract formulation is not the only way to incorporate fuzzy rules in an inference system. In the following, we review an alternative way of describing a set of fuzzy rules. The function `addRule` allows for rules to be implemented that have been expressed in a more symbolic fashion rather than the numerical matrix given by the `rulelist`. For example, for the rules developed in Table 5.7 we can utilize the following MATLAB script:

```
% FIS rules
rule1 = "error==Negative & derror==Negative => actuation=LN";
rule2 = "error==Negative & derror==Zero => actuation=SN";
rule3 = "error==Negative & derror==Positive => actuation=SN";
rule4 = "error==Zero & derror==Negative => actuation=SN";
rule5 = "error==Zero & derror==Zero => actuation=Z";
rule6 = "error==Zero & derror==Positive => actuation=SP";
rule7 = "error==Positive & derror==Negative => actuation=LP";
rule8 = "error==Positive & derror==Zero => actuation=SP";
rule9 = "error==Positive & derror==Positive => actuation=SP";
rules = [rule1 rule2 rule3 rule4 rule5 rule6 rule7 rule8 rule9];
fiscon = addRule(fiscon,rules);
```

As convenient and readable as this implementation of the rules is, it turns out that it does not lend itself well to the implementation of an optimization algorithm to find the best composition of the nine rules. For this, the `rulelist` approach is well suited since we deal with a numerical matrix. Each entry of the `rulelist` matrix identifies the membership function of each input; each output allows for adding a weight to each rule and permits the definition of which fuzzy operator is used using numerical values. In particular, only integers are used with the exception of the weight indicator for a rule, which can assume a value between 0 and 1. An optimization algorithm can make use of this description by parameterizing the `rulelist`, where each entry has a certain range given by the number of membership functions and input and outputs.

To optimize the specific membership function for each input and output, another parameterization can be employed. For example, if we utilize the GA to optimize the given membership function for the stated problem above, the implementation into MATLAB can be accomplished as follows:

```
fiscon=addMF(fiscon,"error","sigmf",[w(1) w(2)],'Name',"Positive");
```

Rather than having explicit values for the `sigmf` membership function, we utilize the parameter vector $\vec{w}$ from the optimization algorithm.

Finally, to construct a cost function to assess the fitness value of a set of candidate solutions, or in terms of the GA for a particular chromosome, we used in Section 5.5 a Simulink model that was called from a MATLAB script. Here, we present an alternative method that does not utilize a Simulink model, but rather a simulation constructed with a MATLAB script. The simulation can be done in continuous time or discrete time. The following script is used to define a discrete-time model representing a second-order system with a damping ratio of $\xi = 0.25$ and a natural frequency of $\omega_n = 5$ rad/s. The discretization is done using MATLAB's c2d function with a sampling time of $Ts = 0.1$ seconds:

```
function cost = ObjectiveFunction (CHROMOSOMES)
% Replaces cost function to simulate a dynamic system
% with candidate parameters given in CHROMOSOMES
%_____
[row,~]=size(CHROMOSOMES);
cost=zeros(row,1);
zhi = 0.25;wn = 5;num=wn^2;den=[1 2*zhi*wn wn^2];
[A,B,C,D]=tf2ss(num,den);
% Discrete-time state space model
Ts=0.1;[Ad,Bd]=c2d(A,B,Ts);
% Simulating for computing error
nd = 100;[no,n2]=size(C);

for i=1:row
    w=CHROMOSOMES(i,:);u=0;r=1;e=0;
    x=zeros(n2,1);y=zeros(no,nd);
    for j=1:nd
        y(:,j)=C*x+D*u;b=y(:,j);
        x=Ad*x+Bd*u;eb=e;
        e=r-b;ed=(e-eb)/Ts;
        u = fzcontrolU(w,e,ed);
    end
    error1=(1-y)*(1-y)';
    cost(i,1)=abs(sum(error1));
end
```

In this sample implementation we call a function `fzcontrolU` that utilizes the current parameters in $\vec{w}$ and submits the simulation error e as well as the derivative of the error. The function computes the control input by utilizing a script that is similar to the above description of the two inputs and one output, as well as the nine rules. However, to compute the control output we have to evoke the function `evalfis` in `fzcontrolU`:

```
u = evalfis(fiscon, [e ed]);
```

MATLAB has also incorporated some autotuning tools to work with optimizing fuzzy inference systems. We will encounter these when we discuss adaptive neuro-fuzzy inference systems (ANFIS) in Chapter 7.

5.9   Conclusions

As an alternative to the methods discussed in Chapter 4, we looked at approaches that belong to heuristic algorithms. As the Greek word *heuristic* – to find or to discover – implies, these methods do not employ the standard optimization framework provided by calculus, but rather are derived from observations nature provides, such as movements by swarms or biological processes given by genetics. While there are a wide range of algorithms that have been proposed and even used in control estimation problems, we focused on a few basic algorithms. A common approach to test any such optimization algorithm is to utilize standard test functions that provide specific difficulties – for example, test functions that have large spaces with little gradient while adjacent to sections with a large gradient, or test functions that possess many local minima. In our argument for specific heuristic optimization algorithms we discussed the LS and the HC problem. One of the outcomes of this discussion is the argument for attempting to avoid cycling during a search – that is, where the search goes back and forth to the same location in the search space without advancing. Tabu search optimization is built on this premise – that is, we avoid cycling by engaging and maintaining a tabu list that records the last several movements during the exploration phase of the search and makes those locations unavailable for further visits.

An entirely different class of heuristic optimization algorithms are given by PSO and ACO algorithms. Such algorithms fall under the swarm-based algorithm concept, where we have many agents searching for optimality in the search space. These two particular algorithms are inspired by animal swarm behavior. There are other such algorithms, such as the bat algorithm and the bee algorithm, that utilize some of the swarm concept in their structure but are tailored to the behaviors of bats or bees. Not swarm based but still multi-agent based is the GA. GAs are inspired by Darwin's survival of the fittest principle and use terminology found in genetics, such as *chromosomes* and *genes* to specify a candidate solution and the corresponding parameters, respectively.

One of our objectives in this text is to see how to utilize the different algorithms to solve control problems. In this chapter we used heuristic optimization to formulate some type of optimum control concept. Although we optimized only certain control parameters, the capability of some of the heuristic algorithms to handle large parameter sets and parameter interactions may inspire more complex optimization in control problems, such as including the control architecture itself. One example of hybrid control was also provided, where we used heuristic optimization for fuzzy logic-based controller design. In Chapter 7 we will revisit the idea of hybridization, but before that we need to discuss neural networks, which are treated in the next chapter.

SUMMARY

In the following some brief descriptions of some of the key concepts are presented.

Heuristic Optimization Algorithms

Rather than using exhaustive search techniques heuristic optimization algorithms tend to sacrifice accuracy and precision for speed in finding global optimum solutions. Iterative heuristic algorithms evaluate available information at each step before deciding the exact action to take for the next step. A metaheuristic algorithm is a construct that permits the creation of heuristic algorithms to address optimization problems.

Local Search

An LS algorithm explores the search space in a step-wise fashion by observing local changes until an optimal solution is found or the maximum steps are reached. During the exploration of the search space the path taken is not important. The LS algorithm suffers from the cycling phenomena.

Tabu Search Algorithm

The tabu search algorithm attempts to solve the cycling phenomena observed in the LS algorithm by employing a tabu list that records past instances of the search path. These instances in the search space are taboo for a defined duration of the search algorithm. The tabu search algorithm is often constructed with two phases: an exploration phase and an intensification phase.

Particle Swarm Optimization Algorithm

A PSO algorithm consists of a set number of agents exploring the search space. The coordination among the individual agents – or particles – is given by exchange of information on the topology of the search space encountered by each agent so far during the search process. PSO uses a simple mathematical construct to determine each move of each particle, incorporating three tendencies: personal best performance, swarm best performance, and its own momentum.

Ant Colony Optimization Algorithm

Ant colony optimization embeds probabilistic measures in order to find optimal pathways in graphs. So-called artificial ants are randomly distributed at nodes of the different pathways and use a pheromone deposition mechanism to select the best pathway.

Genetic Algorithm

Genetic algorithms mimic the behavior governed by the theory of natural evolution. The optimization problem is expressed by genes and chromosomes that make out candidate solutions and compete against each other in terms of fitness – as measured by the objective function. Well-performing chromosomes have higher chances to procreate and hence create new solutions that potentially improve in terms of fitness. Generally, GAs employ three operators: selection, mating, and mutation operation.

REVIEW QUESTIONS

1. Why does a local search algorithm get stuck in a local minimum?
2. What type of optimization is solved with a hill climbing algorithm?
3. What is the function of the aspiration criteria used in tabu search algorithms?
4. What is the difference between a static and a dynamic tabu list?
5. When overcoming a local minimum, does the relationship between the length of the tabu list and the shape of a local minima matter? Explain your answer.
6. What are the two stages of a tabu search algorithm and what purpose do they serve (consider the tabu list and the promising list)?
7. What are the three processes embedded in the graphical simulation models called boids, which were copied from nature?
8. What are the three tendencies incorporated in a particle swarm optimization algorithm?
9. What type of neighborhoods can be implemented in a particle swarm optimization algorithm?
10. Can you call a Simulink program within a MATLAB script? If so, explain how to implement such functionality in a MATLAB program.
11. For what purpose are pheromones used in an ant colony optimization algorithm?
12. How does evaporation affect the ant colony optimization algorithm?
13. Suppose you have an optimization problem that is defined by five parameters, where each parameter is an integer. What is the length of a gene and the length of the chromosomes when solving this problem using a genetic algorithm?
14. What is a cross-over point and can there be multiple cross-over points in a chromosome?
15. For a genetic algorithm, why would a selection algorithm work poorly if guided purely by the value of the corresponding objective function?
16. What purpose does the mutation function have in a genetic algorithm?
17. What could a too large mutation rate cause during the optimization using genetic algorithms?

18. Do genetic algorithms work well when there is a large interdependency between the parameters of the optimization problem?

19. When do hard computing methods (as reviewed in Chapter 4) work better than soft computing methods (as presented in this chapter)? What are the conditions for the reverse, using soft computing methods over hard computing?

PROBLEMS

5.1 Generate surface plots and contour plots for Schwefel, Griewank, Ackley, Easom, Bohachevsky, Six-Hump Camel, and Branin objective functions using MATLAB scripts. Note, corresponding functions are defined as follows:

SCHWEFEL: $f(\vec{w}) = 418.9829d - \sum_{i=1}^{d} w_i \sin\left(\sqrt{|w_i|}\right)$, for $w_i \in [-500, 500]$

GRIEWANK: $f(\vec{w}) = \sum_{i=1}^{d} \frac{w_i^2}{4,000} - \prod_{i=1}^{d} \cos\left(\frac{w_i}{\sqrt{i}}\right) + 1$, for $w_i \in [-25, 25]$

ACKLEY: $f(\vec{w}) = -a \exp\left(-b\sqrt{\frac{1}{d}\sum_{i=1}^{d} w_i^2}\right) - \exp\left(\frac{1}{d}\sum_{i=1}^{d} \cos(cw_i)\right) + a + \exp(1)$

   Use $a = 20$, $b = 0.2$, and $c = 2\pi$, $w_i \in [-3.2768, 3.2768]$

EASOM: $f(\vec{w}) = -\cos(w_1)\cos(w_2)\exp\left(-\{w_1 - \pi\}^2 - \{w_2 - \pi\}^2\right)$ for $w_i \in [-100, 100]$

BOHACHEVSKY: $f(\vec{w}) = w_1^2 + 2w_2^2 - 0.3\cos(3\pi w_1) - 0.4\cos(4\pi w_2) + 0.7$ for $w_i \in [-100, 100]$

SIX-HUMP CAMEL: $f(\vec{w}) = \left(4 - 2.1w_1^2 + \frac{w_1^4}{3}\right)w_1^2 + w_1 w_2 + \left(-4 + 4w_2^2\right)w_2^2$ for $w_i \in [-2, 2]$

BRANIN: $f(\vec{w}) = \left(w_2 - bw_1^2 + cw_1 - r\right)^2 + s(1 - t)\cos(w_1) + s$ for $w_1 \in [-5, 10]$, $w_2 \in [0, 15]$, $b = 5.1/\left(4\pi^2\right)$, $c = 5/\pi$, $r = 6$, $s = 10$, and $t = 1/(8\pi)$

   Note, d is the dimension of the problem. Since we want to plot these functions, use $d = 2$.

5.2 Use the provided TS MATLAB script to find optimum locations for the Griewank objective function. Plot the path of the search onto a contour map of the two objective functions and comment on the convergence difference.

5.3 Incorporate a promising list in the TS script provided and modify the MATLAB code to explore those promising areas. Compare the results with the regular TS script in terms of iterations and convergence of the search.

5.4 Using the TS MATLAB code provided in Appendix D.8, modify the script to incorporate Ackley's objective function and find the minimum location as well as minimum cost value.

5.5 Using the TS MATLAB code provided in Appendix D.8, modify the script to find the minimum location and minimum cost function value of the Six-Hump Camel objective function, as defined in Problem 5.1.

5.6 Using the TS MATLAB code provided in Appendix D.8, modify the script to find the minimum location and minimum cost function value of the Branin objective function, as defined in Problem 5.1.

5.7 Using the PSO MATLAB code provided in Appendix D.9, modify the script to produce Figure 5.5 and simulate this for Ackley's objective function.

5.8 Using the PSO MATLAB code provided in Appendix D.9, modify the script to find the minimum location and minimum cost function value of the Six-Hump Camel objective function, as defined in Problem 5.1.

5.9 Using the PSO MATLAB code provided in Appendix D.9, modify the script to find the minimum location and minimum cost function value of the Branin objective function, as defined in Problem 5.1.

5.10 Using the PSO MATLAB code provided in Appendix D.9, incorporate a social neighborhood structure. Using 40 particles, create four different social groups who each have their own social optimum location as one of their three influence factors rather than the overall swarm optimum location. Simulate using the Haupt and Haupt objective function and compare the outcome with the original PSO program.

5.11 Using the PSO MATLAB code provided in Appendix D.9, incorporate a geographical neighborhood structure. Using 40 particles, create four different geographical groups who each have their own optimum location as one of their three influence factors rather than the overall swarm optimum location. Simulate using the Haupt and Haupt objective function and partition the search area into four equal quadrants as the geographical neighborhoods. Compare the outcome with the original PSO program.

5.12 Use the social PSO from Problem 5.10 and the geographical PSO from Problem 5.11 and compare its performance on the Schwefel, Griewank, and Ackley test function.

5.13 Follow the instructions of Section 5.5 to implement the PSO for a PI controller optimization algorithm. Use the following two lines instead of the cost evaluation lines found in the main PSO MATLAB script given in Appendix D.9:

```
w = S.x(i,:,k);
S.cost(i)=fobj(S.x(i,:,k));
```

Incorporate the objective function given in Section 5.5 by utilizing the integral square error (ISE). Replicate the results depicted in Section 5.5.

5.14 Alter the PSO PI controller design (see Problem 5.13) to include a differential term – that is, formulate the problem to find the optimum values for a PID controller and compare the results with the results obtain in Problem 5.13.

5.15 Alter the PSO PI controller design (see Problem 5.13) to use the absolute value error and compare the results with the ones achieved using the ISE computation.

5.16 Alter the optimum PI controller design as given in Section 5.5 to utilize the TS algorithm given in Appendix D.8 rather than the PSO algorithm.

5.17 Alter the optimum PI controller design as given in Section 5.5 to utilize the GA algorithm given in Appendix D.10 rather than the PSO algorithm.

5.18 Use the provided GA MATLAB script to find optimum locations for the Griewank and Ackley objective functions. Using the top chromosome, plot the path of the search onto a contour map of the two objective functions and comment on the convergence difference.

5.19 Using the GA MATLAB code provided in Appendix D.10, modify the script to find the minimum location and minimum cost function value of the Six-Hump Camel objective function, as defined in Problem 5.1.

REFERENCES

[1] Das SK, Kumar A, Das B, Burnwal AP (2013). On soft computing techniques in various areas. *National Conference on Advancement of Computing in Engineering Research*, pp. 59–68.

[2] Glover F (1986). Future paths for integer programming and links to artificial intelligence. *Computers and Operations Research*, 13(5): 533–549

[3] Haupt RL, Haupt SE (1998). *Practical Genetic Algorithms*. Wiley, New York.

[4] Rabin S (2002). *AI Game Programming Wisdom*. Charles River Media, Boston, MA.

[5] Kennedy J, Eberhart R (1995). Particle swarm optimization. *Proceedings of the IEEE International Conference on Neural Networks*, vol. 4, pp. 1942–1948.

[6] Ziegler JG, Nichols NB (1942). Optimum settings for automatic controllers. *Transactions of the ASME*, 64: 759–768.

[7] MathWorks (n.d.). File exchange. www.mathworks.com/matlabcentral/fileexchange.

[8] Holland JH (1973). Genetic algorithms and the optimal allocation of trials. *SIAM Journal on Computing*, 2(2): 88–105.

6   Artificial Neural Networks

6.1   Introduction

In October 2015 a neural network-based program named AlphaGo was able – for the first time – to beat one of the best Go players in the world [1]. The game of Go is played on a board and, unlike chess, it has approximately 2.1×10^{170} different board positions. The incredible number of different positions and moves that the game allows were considered an impossibility for any program to learn and hence for any program ultimately to defeat a human player. At the time of AlphaGo's victory it became apparent that the advancements achieved in neural networks will have great potential for a vast number of everyday applications, well beyond playing games. In this chapter our objective is to develop the basic concepts of neural networks (NNs) as a foundation for developing control algorithms that make use of some of the strategies employed by AlphaGo. We will discuss the basic inner workings of NNs by starting from simplistic network models and discussing how these models work, how "learning" is achieved, and what limitations these models have, and supplement this treatment with the introduction of more sophisticated NNs that are capable of handling more complex tasks.

We will primarily use simple MATLAB scripts to explain these concepts. However, when dealing with more intricate networks we shall make use of MATLAB's Neural Network toolbox to take advantage of the many built-in functions and capabilities that some of these NNs provide. This chapter is an essential building block for the next few chapters, where we combine NNs with fuzzy logic (FL) to develop adaptive neuro-fuzzy inference controllers, explore deep learning, and ultimately work through reinforcement learning-based controllers, which share some of the philosophies employed by AlphaGo.

As stated in some of the previous chapters, the inspiration for many algorithms treated in this book originates from scientists and researchers simply observing nature. This observation is then transcribed as a mathematical construct. The development of NNs followed the same pathway, and is in essence an attempt to mimic how the human brain works. In the case of NNs, this observation and attempt to transcribe it mathematically was done by the psychologist Walter Pitts and the neurophysiologist Warren McCulloch in 1943. Unfortunately, these researchers lacked adequate computational resources to entertain the simple basic construct they concocted, which was termed the first artificial neuron. In the 1960s, with the availability of greater computational resources, the concept of the artificial neuron was extended and training algorithms were developed, leading to the well-established backpropagation method, introduced in 1974 by Werbos [2].

To provide a simplistic view of a basic NN, consider the problem of recognizing numbers or letters. Humans tend to recognize even sloppy handwriting rather quickly without – in some instances – having seen the specific handwriting style prior. The adaptation of NNs to the recognition of letters and numbers can be done by employing the visual cortex of our eyes and comparing it to a digital camera. In particular, the optical nerve serves as a matrix of sensors, each pixel detecting different brightness and color variations. If we simplify this analogy and only consider brightness and exclude any color information, and furthermore exclude shades of gray, we have a set of pixels that each can record a value of black or white. Hence, the sensed data is a matrix of binary recordings. Suppose the image is a 28 × 28 pixel matrix. An NN's approach is to deconstruct the pattern given by the 28 × 28 pixel matrix into basic shapes (for the image processing case) and assigning a probabilistic value to its correlation with a given set of forms or structures. This deconstruction and comparison is done by using so-called neurons that are connected and organized in layers. A symbolic simple NN is shown in Figure 6.1.

The example NN depicted in Figure 6.1 contains one hidden layer, composed of four neurons. The input layer has six neurons, while the output layer is structured with two neurons. For the problem of a 28 × 28 pixel matrix, the input layer would be constructed by 28 × 28 = 784 neurons. The output layer would be constructed by as many neurons as there are different possible outcomes to the problem. For example, if we were to try to recognize a single-digit number,

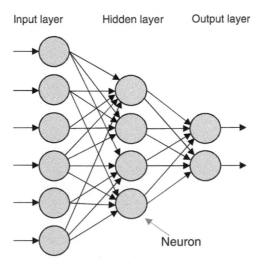

Figure 6.1 Example of a simple feed-forward neural network with one hidden layer.

there would be 10 output neurons, each representing a different digit. We did not define the number of hidden layer neurons for this number recognition problem as this is a design parameter. Also, we could have multiple hidden layers, each with a fixed number of neurons. An NN that is composed of multiple hidden layers is termed a "deep" neural network.

Another observation we can make when considering the directions of arrows shown in Figure 6.1 is that the flow of data/information is from left to right – that is, the input is on the left side and is processed by the input layer neurons, which transmit the data to the hidden layer(s) in sequence, followed by the output layer. Such networks are termed "feed-forward" neural networks. There are other constructs, such as the recurrent neural network (RNN) where cycles and the notion of time are included, the radial basis neural network (RBN) where different functions are embedded into the neurons, to name a few different network topologies. Each of these different NNs is structured differently and hence has different benefits and is suited to certain applications. Before discussing in more detail different network structures and topologies, we need to mention the two primary applications of NNs: classification and regression. The described problem, where we want to identify letters or numbers, is a typical classification problem. In classification problems we intend to predict a label or a class. A classification problem addressed with an NN is often recognized by the number of output neurons, where this number corresponds to the number of different labels or classes. The other primary application of NNs is the regression problem. Regression attempts to predict a quantity at the output of the NN based on some input. A regression problem can be considered as a mapping function from input variables to some continuous output variable.

Before developing the detailed mathematical principles governing the functioning of NNs, let's briefly discuss different NN structures. You may have noticed in the literature that there is an abundant number of different NNs. Each structure or topology will have some properties that allow for certain applications to be more successful than other NN structures. We start with the simplest form of an NN, the perceptron, which is illustrated in Figure 6.2(a). An alternative name to the perceptron is "single-layer neural network." It contains only two layers: the input layer and the output layer. In Figure 6.2(a) the two layers are separated into functional blocks, indicating the simple mathematical operations that are performed in a perceptron network. The perceptron works by taking the input at the input layer, multiplies these with corresponding weights w_i that are associated with the connecting lines of the inputs to the input neurons, and evaluates the resulting sum by an activation function, which corresponds to the output value. Typical applications of the perceptron are simple classification problems. We will provide more detail on this model and discover the severe limitations embedded in this structure. Due to these it is not much used nowadays.

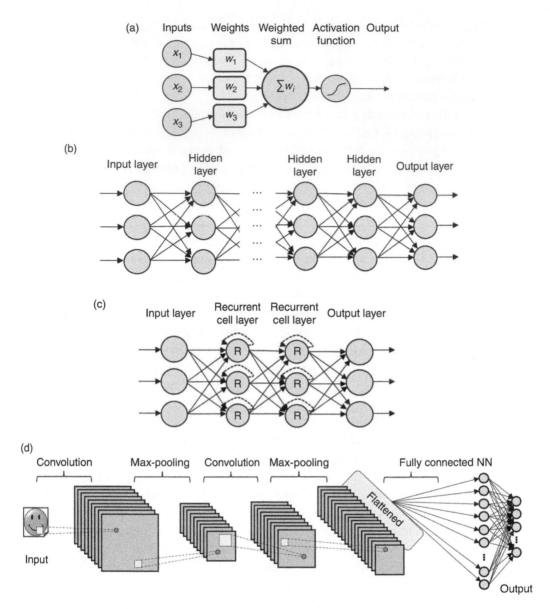

Figure 6.2 Example of (a) perceptron, (b) deep feed-forward neural network (DFFNN), (c) recurrent neural network (RNN), (d) convolutional neural network (CNN).

An extension of the perceptron network is the feed-forward neural network (FFNN), which we described with Figure 6.1, with the exception that every neuron in one layer is fully connected with the next layer. Note that in Figure 6.1 some neurons do not connect to all of the neurons in the next layer. We will discuss the implications and workings of such a network in Section 6.2. The feed-forward descriptor refers to the flow of data in the network, where cycling is structurally impossible. The FFNN has found numerous applications, such as in pattern recognition, speech recognition, and our handwriting problem detailed above, among others.

If the FFNN includes more than one hidden layer it is called a deep feed-forward neural network (DFFNN). One of the potential advantages of a DFFNN is the improved ability to generalize. Generalization refers to the ability of the NN to work on new data, where "new" refers to data that were not used in the NN training process. An example of a DFFNN is shown in Figure 6.2(b).

An interesting NN topology is presented by the RBN. As the name indicates, radial basis functions are used in this network. They are implemented inside the neurons (i.e., they are used as activation functions). Activation functions are essentially a mapping from the input spectrum of values that a neuron receives into a number that is commonly limited to the range between 0 and 1. However, newer networks, including deep neural networks, relax this mapping to a larger scope, which we will discuss when detailing deep learning. RBNs make use of hidden layers and work well for function approximation problems.

With respect to control, quite often we need models that capture the temporal properties of a system. The aforementioned topologies do not lend themselves well to time series and time-stamped data-based systems. There are specific topologies that embed this type of characteristic, such as the RNN and the long short-term memory (LSTM) neural network.

The structure of the RNN is shown in Figure 6.2(c). The RNN extends the FFNN by allowing each of the hidden layer's neurons to receive inputs with a time delay. This time-delay property can be seen as a feedback term similarly to how the infinite impulse response (IRR) filter utilizes feedback. Hence, a single neuron can be used for a number of time-stamped data points. In effect, RNNs can deal with current inputs and past inputs, as measured in time. This allows for entertaining time series data where prior inputs are used in the current output. RNNs are causal system representations (i.e., only current and past inputs can predict current or future outputs). They are used in controls, time series representation and analysis, and speech recognition, among other time-dependent applications.

RNNs have some structural problems which express themselves in difficulties in the learning or training process. These difficulties are termed the vanishing gradient problem and exploding gradient problem. We will discuss these types of problem in the following sections in more detail. However, a solution to this specific problem for time-dependent data and NNs was proposed in 1997 by Hochreiter and Schmidhuber [3] by their presentation of the LSTM network. The LSTM network is a special type of RNN which utilizes so-called gated cells that allow data to be "memorized." Decisions about when to access or modify these memories are taken by these analog gates, which possess an important property when applying the backpropagation learning process. LSTM networks are covered in Chapter 8.

Another popular NN topology is given the convolutional neural network (CNN). CNNs have become rather popular for image classification problems, object recognition, and image identification. CNNs are composed of at least three different layers: the convolutional layer, the pooling layer, and the fully connected layer. We will discuss each layer and functioning of CNNs in detail in Chapter 8. For now, we present the structure in Figure 6.2(d). CNNs have a wide spectrum of applications, such as image recognition for cancer cell detection, drug discovery, image and video analysis, and forecasting problems.

Before we start exploring the details of these networks, their mathematical formulation, and how we can train them, some perspective on computational resources is necessary as much of the progress in NNs has depended on their implementation. Implementation has a strong correlation with available processing power; as mentioned, the inventors of NNs did not have any computational resources at the time of their invention. The central processing unit (CPU) started to be commercially available around 1971 and was used for the backpropagation algorithm later in the 1970s. Around 1999 a new form of computational resource became commercially available, the graphics processing unit (GPU). GPUs have the advantage of being able to increase the amount of data a CPU can process per unit time. More recently, in 2016, the tensor processing unit (TPU) became available specifically for use with NNs and machine learning. These hardware advancements allow for much larger networks to be trained and operated. However, the increase in computational resources also accelerated the need for electrical power. AlphaGo, in 2015, which defeated one of the best Go players in the world, consumed approximately 85 kW of power. If we compare this to the 20 W of power consumed by a human brain for the task of playing the game, we can see that NNs have much need for improvements. Some of the research is focused on this power requirement, and spiking neural networks (SNNs) have been proposed for this reason. The difference between SNN and regular NNs is that rather than using continuous variables, discrete spikes are used in the signal flow within an SNN. This is inspired by how we understand the human brain to signal internally. When using discrete "spikes" rather than continuous number representations, less power is consumed. In addition, SNNs allow for simpler computations. Specific processors are also being developed to accommodate these types of networks, and are termed neuromorphic processing units (NPUs). We will not discuss SNNs any further as they are beyond the scope of this book.

6.2   Single-Layered Neural Networks

We start with a simple NN topology; the single-layered NN. In this section and for this structure, our objective is to generate a specific desired output for

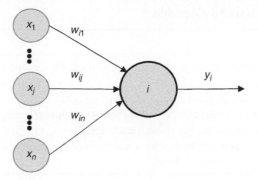

Figure 6.3 Single-layered neural network: one neuron with n input variables and one output.

a specific input to the NN. We term the difference between the desired output d_i and the actual output y_i of the NN as the error e_i:

$$e_i = d_i - y_i. \tag{6.1}$$

To minimize the error we adjust the weights w of the neural network. As an example of the type of network we are working with and the associated weights used, consider Figure 6.3.

In Figure 6.3 the variable x is the input, and for the given example we have n different inputs, hence a vector $\vec{x} \in \mathbb{R}^{n \times 1}$. Each input is multiplied by a weight w_{ij}, which are symbolized by connecting arrows in NN topology graphs. The subscripts of the weights denote the connecting neuron index and the corresponding input, in that order. Suppose we want to minimize the error e_i, we can utilize a simple error-driven adaptation, the so-called delta rule:

$$w_{ij} = w_{ij} + \alpha e_i x_j = w_{ij} + \Delta w_{ij}, \tag{6.2}$$

where the coefficient α is the learning rate. Equation (6.2) is essentially a linear regression. To make this a proper NN we need an activation function. For the given topology, the activation function allows for an effective learning process and scales the correction term Δw_{ij}. Defining the weighted sum of inputs as v_i, where

$$v_i = \sum_{j=1}^{n} x_j w_{ij}, \tag{6.3}$$

and the activation function as φ, the generalized delta rule is given as:

$$w_{ij} = w_{ij} + \alpha \varphi'(v_i) e_i x_j. \tag{6.4}$$

Table 6.1 Single-layer training algorithm

Step	Description
Step 1	Define the number of inputs, the training data set, the desired output, and an initial weight distribution.
Step 2	Define the activation function and the learning rate.
Step 3	For a number of iterations (also called epochs), update the weights using Equation (6.6).
Step 4	Compute the output and the error at each iteration.
Step 5	Test the trained network by feeding it a new set of data and compare the output with the desired output defined in Step 1.

Figure 6.4 Sigmoid activation function.

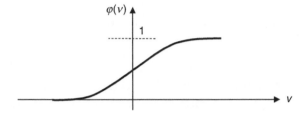

In Equation (6.4) we made use of the derivative of the activation function. The activation function is a mapping of the input into a number that often ranges between 0 and 1. For example, choosing the sigmoid function as the activation function, the graph used for the mapping process of the input is shown in Figure 6.4.

The reason for utilizing the derivative of the activation function is that the adaptation of the weights is an optimization process. Recall from Chapter 4 that one of the methods discussed was the gradient descent algorithm. In the gradient descent algorithm we update the weights based on the gradient of the error surface. The sigmoid function has a convenient property for computing its derivative:

$$\varphi'(v_i) = \varphi(v_i)\{1 - \varphi(v_i)\}. \tag{6.5}$$

Therefore, we can write the adaptation rule from Equation (6.4) as:

$$w_{ij} = w_{ij} + \alpha\varphi(v_i)\{1 - \varphi(v_i)\}e_i x_j. \tag{6.6}$$

For our single-layer NN, the learning algorithm can be given as detailed by the step-by-step procedure shown in Table 6.1.

Example 6.1

To illustrate the training and application of this simple single-layered NN, consider the two classification problems depicted in Figure 6.5. Both problems entail two inputs (i.e., x_1 and x_2) and each of the two inputs can assume a value of 0 or 1. We can plot the complete data set for two variables that comprises four data points. Figure 6.5(a) shows a linearly separable data set, as indicated by the dashed line separating the 1s and the 0s. In Figure 6.5(b), however, we have a linearly nonseparable data set, as can be seen by the territory enclosed by the dashed oval. Table 6.2 provides the truth table associated with both of the examples. Our objective is to see if a single-layer NN is capable of classifying each of the two problems. For this purpose we use a simple MATLAB script for computing the steps, as listed in Table 6.1.

The MATLAB code is broken into several parts: loading the data, training the network, and computing the predicted classification. The code is given in Appendix D.11. Utilizing

Table 6.2 Truth table for classification examples

(a)	Input 1	Input 2	Output
	0	0	0
	0	1	0
	1	0	1
	1	1	1

(b)	Input 1	Input 2	Output
	0	0	0
	0	1	1
	1	0	1
	1	1	0

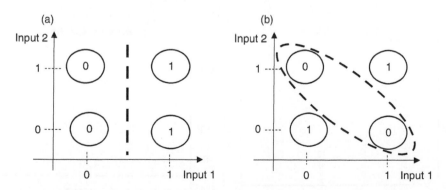

Figure 6.5 Graphical representation of classification examples: (a) linearly separable data set; (b) linearly nonseparable data set.

this MATLAB code for the two stated problems, we obtain the results as presented in Table 6.3. These numbers were generated by utilizing an adaptation rate of 0.9.

If we associate the results as a probability of belonging to a class, any value above a certain threshold value *tr* would mean that the NN classifies the output as a 1, while any value below *tr* indicates a 0. Considering the results tabulated in Table 6.5(b), it seems that the single-layer NN is not able to classify correctly the second data set, given in Figure 6.5(b). However, for the linearly separable data set in Table 6.3(a) the obtained results indicate some success, as the 1s are predicted rather close (0.99) while the 0s are rather off, and certainly not close. This is also reflected when graphing the convergence of the training: Figure 6.6 depicts the error versus the iteration number. The latter is called the *epoch* in the field of neural networks. Figure 6.6(a) is for the linearly separable data set and indicates a convergence to a small error. With a different adaptation rate and a different number of epochs, the error may be reduced further.

Table 6.3 Simulation results for two classification examples using single-layer neural network

(a)	Desired	Simulated
	0	0.5000
	0	0.7148
	1	0.9969
	1	0.9988

(b)	Desired	Simulated
	0	0.5000
	1	0.2832
	1	0.6039
	0	0.3759

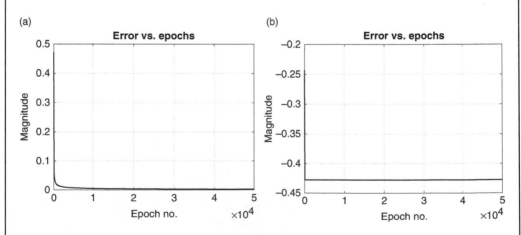

Figure 6.6 Convergence plot of error for (a) example 1 with linearly separable data set and (b) example 2 with linearly nonseparable data set.

The question is: Why was the single-layer neural network not able to classify the second problem? We can investigate the outcome a bit further by looking at what kind of rules the network establishes for making determinations on the classification of data. The output of the single-layer NN is computed by:

$$y = \varphi\left(\sum_{j=1}^{2} x_j w_{ij}\right). \tag{6.7}$$

If we temporarily neglect the mapping that the activation function performs in Equation (6.7), the summation is simply $x_1 w_{11} + x_2 w_{12}$. The activation function scales this summation to be between 0 and 1. Hence, the rules that we impose for classifying the output can be stated as:

$$\text{If } x_1 w_{11} + x_2 w_{12} > tr \rightarrow 1.$$
$$\text{If } x_1 w_{11} + x_2 w_{12} < tr \rightarrow 0.$$

We effectively draw a line tr into the data set plane and categorize the data based on this line. It turns out that single-layer NN are limited to linearly separable data sets. In our example we used two inputs, hence the separation between the classes is given by a simple straight line. If we consider higher dimensions, single-layer NN have the potential to classify correctly the data set if the data set is separable by linear hyperplanes of appropriate dimensions. This shortcoming is our motivation to explore multilayered neural networks, as discussed in the next section.

6.3   Multilayered Neural Networks

In the previous section we learned that entertaining linearly nonseparable data sets using NN requires multilayered NNs. However, adding one or more hidden layers to the network will have an effect on how we train the network. Consider our adaptation rule, which we referred to as the delta rule. This rule found the error at the end of a single-layered NN and used it to change the weights of the connection of the network. The error and the weights were directly related in our learning process. When introducing a hidden layer, the error of the output node is still computed as the difference between the desired output and the output of the NN. However, there is no longer a definition for the desired output of the hidden layers which would allow us to compute some meaningful associated error for the previous set of weights. Hence, to adjust the weights of the hidden layer, we need to formalize some kind of error for that layer as well. This is accomplished with the backpropagation algorithm. Backpropagation refers to "backward propagation of errors." This method computes the gradient of the error with respect to the NN weights and mimics the delta rule. The ensuing optimization of the weights is in essence a gradient descent optimization algorithm. The term "backward" stems

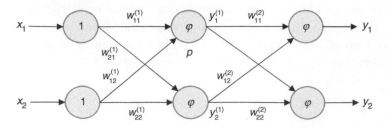

Figure 6.7 Multilayered neural network: two neurons for each layer, with one hidden layer.

from the flow of how we utilize the error at the output of the NN and compute the weights for each layer backward using the gradient of the error information.

In the following, we will derive the backpropagation method using a simple multilayered NN with two input neurons, two neurons in the hidden layer, and two output neurons. The network is depicted in Figure 6.7.

In Figure 6.7 we introduce our notation for identifying which weight belongs to which layer by using superscripts with parentheses for layer information. Also, we use our previous notation for identifying the connections the weights make by a subscript. The first index of this subscript represents the receiving neuron while the second index indicates which is the sending neuron. Considering the network in Figure 6.7 and moving from left to right, we will use this flow to compute the output of each layer. Assuming no activation function for the input layer neurons – that is, only a multiplication with a constant value of 1 – the output of the input layer, which serves as the input to the hidden layer, is computed as:

$$\begin{Bmatrix} v_1^{(1)} \\ v_2^{(1)} \end{Bmatrix} = \begin{bmatrix} w_{11}^{(1)} & w_{12}^{(1)} \\ w_{21}^{(1)} & w_{22}^{(1)} \end{bmatrix} \begin{Bmatrix} x_1 \\ x_2 \end{Bmatrix} = \mathbf{W}^{(1)}\vec{x}. \tag{6.8}$$

Employing an activation function φ for the neurons in the hidden layer, we can calculate the output of the hidden layer based on Equation (6.8) as:

$$\begin{Bmatrix} y_1^{(1)} \\ y_2^{(1)} \end{Bmatrix} = \varphi\left(\begin{bmatrix} w_{11}^{(1)} & w_{12}^{(1)} \\ w_{21}^{(1)} & w_{22}^{(1)} \end{bmatrix} \begin{Bmatrix} x_1 \\ x_2 \end{Bmatrix}\right) = \varphi\left(\mathbf{W}^{(1)}\vec{x}\right) = \begin{Bmatrix} \varphi\left(v_1^{(1)}\right) \\ \varphi\left(v_2^{(1)}\right) \end{Bmatrix}. \tag{6.9}$$

We repeat this computation for the second layer and can compute the output of the network as:

$$\begin{Bmatrix} y_1^{(2)} \\ y_2^{(2)} \end{Bmatrix} = \begin{Bmatrix} y_1 \\ y_2 \end{Bmatrix} = \varphi\left(\begin{bmatrix} w_{11}^{(2)} & w_{12}^{(2)} \\ w_{21}^{(2)} & w_{22}^{(2)} \end{bmatrix} \begin{Bmatrix} y_1^{(1)} \\ y_2^{(1)} \end{Bmatrix}\right) = \varphi\left(\mathbf{W}^{(2)}\vec{y}^{(1)}\right) = \begin{Bmatrix} \varphi\left(v_1^{(2)}\right) \\ \varphi\left(v_2^{(2)}\right) \end{Bmatrix}.$$

$$\tag{6.10}$$

Our objective in training this two-layer network is achieved by adapting the delta rule from the previous section, and propagating the error from the output through the network backward to the input. For the output layer we have

$$\left\{\begin{matrix} e_1 \\ e_2 \end{matrix}\right\} = \left\{\begin{matrix} d_1 \\ d_2 \end{matrix}\right\} - \left\{\begin{matrix} y_1 \\ y_2 \end{matrix}\right\}. \tag{6.11}$$

The activation function plays a central role in the computation of the output as well as for the training in backpropagation. To symbolize the utility of the activation function for both computations, consider Figure 6.8. Figure 6.8(a) shows the forward computation – that is, how we evaluate the output of a neuron – where we have $y = \varphi(v)$. Figure 6.8(b) depicts the backward computation – that is, how we propagate the error through the network starting from the output of the network and going through each layer until we reach the input layer (i.e., $\Delta = \varphi'(v)e$).

Utilizing the activation function, the update information Δ analogous to Equation (6.4) for adjusting the weights at the output of the hidden layer is found as (see also Figure 6.9):

$$\left\{\begin{matrix} \Delta_1^{(2)} \\ \Delta_2^{(2)} \end{matrix}\right\} = \left\{\begin{matrix} \varphi'\left(v_1^{(2)}\right)e_1 \\ \varphi'\left(v_2^{(2)}\right)e_2 \end{matrix}\right\} = \left\{\begin{matrix} \varphi\left(v_1^{(2)}\right)\left(1 - \varphi\left(v_1^{(2)}\right)\right)e_1 \\ \varphi\left(v_2^{(2)}\right)\left(1 - \varphi\left(v_2^{(2)}\right)\right)e_2 \end{matrix}\right\}. \tag{6.12}$$

The error of a node is given by the weighted sum of the backpropagated update information Δ_i:

$$\left\{\begin{matrix} e_1^{(1)} \\ e_2^{(1)} \end{matrix}\right\} = \begin{bmatrix} w_{11}^{(2)} & w_{21}^{(2)} \\ w_{12}^{(2)} & w_{22}^{(2)} \end{bmatrix} \left\{\begin{matrix} \Delta_1^{(2)} \\ \Delta_2^{(2)} \end{matrix}\right\} = \mathbf{W}^{(2)^T}\vec{\Delta}^{(2)}. \tag{6.13}$$

With this, we can compute the new update information (see Figure 6.9):

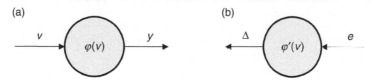

(a)   (b)

Figure 6.8 Neuron action during evaluation and training: (a) computing output of neuron going forward, (b) training of weights going backward.

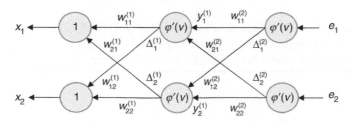

Figure 6.9 Flow of signals during backpropagation.

$$\left\{ \begin{array}{c} \Delta_1^{(1)} \\ \Delta_2^{(1)} \end{array} \right\} = \left\{ \begin{array}{c} \varphi'\left(v_1^{(1)}\right)e_1^{(1)} \\ \varphi'\left(v_2^{(1)}\right)e_2^{(1)} \end{array} \right\}. \tag{6.14}$$

Hence, the adjustments on the weights are accomplished by using the propagated error terms from Equation (6.13) analogous to Equation (6.4).

For example, computing the updated weights for $w_{21}^{(2)}$, we utilize the output of the first layer as well as the "delta" from the output layer $\Delta_2^{(2)}$ along with the learning rate α:

$$w_{21}^{(2)} = w_{21}^{(2)} + \alpha\Delta_2^{(2)}y_1^{(2)}$$

We can "propagate" this to the next layer in reverse order (i.e., the first layer of the network):

$$w_{11}^{(1)} = w_{11}^{(1)} + \alpha\Delta_1^{(1)}x_1.$$

Or, in general terms for the first layer:

$$w_{ij}^{(1)} = w_{ij}^{(1)} + \alpha\Delta_i^{(1)}x_j; \tag{6.15}$$

and for the hidden layer:

$$w_{ij}^{(2)} = w_{ij}^{(2)} + \alpha\Delta_i^{(2)}y_j^{(1)}. \tag{6.16}$$

The backpropagation algorithm is summarized in Table 6.4.

Table 6.4 Backpropagation algorithm

Step	Description
Step 1	Initialize each weight of the network. For example, use random values between 0 and 1. Define an error criterion for the training of the NN.
Step 2	Utilizing the input and desired output, compute the error and the delta using Equations (6.11) and (6.12).
Step 3	Use the computed Δ_i of the output neuron and compute the Δ_i of the immediate prior neurons, using Equations (6.13) and (6.14).
Step 4	Repeat Step 3 for each layer until reaching the layer immediately following the input layer.
Step 5	Update the individual weights using Equations (6.15) and (6.16).
Step 6	For every training data point, repeat Steps 2–5.
Step 7	Repeat Steps 2–6 until the training error criterion is met.

Example 6.2

Addressing the primary problem of the single-layer NN by introducing a hidden layer and the backpropagation algorithm for the training of the NN, we can revisit the two problems that we utilized to test the single-layer NN in Section 6.2. The MATLAB code presented in Appendix D.12 implements this algorithm for the backpropagation algorithm with an NN that has two inputs, one hidden layer, and one output layer. Utilizing an adaptation rate of 0.9 and 10,000 epochs for the training, we obtain the results shown in Table 6.5 and Figure 6.10 for the two examples from Section 6.2.

In Table 6.5 we can recognize that the multilayered NN classifies each data point quite accurately. This is for both examples, for the linearly separable data set and the data set which cannot be separated by a linear hyperplane. The convergence plot shown in Figure 6.10 also indicates that the results are rather quickly obtained, as we used only 10,000 epochs compared to 10^5 epochs for the single-layer NN.

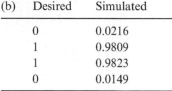

Table 6.5 Simulation results for two classification examples using multilayer neural network

(a)	Desired	Simulated
	0	0.0131
	0	0.0029
	1	0.9918
	1	0.9924

(b)	Desired	Simulated
	0	0.0216
	1	0.9809
	1	0.9823
	0	0.0149

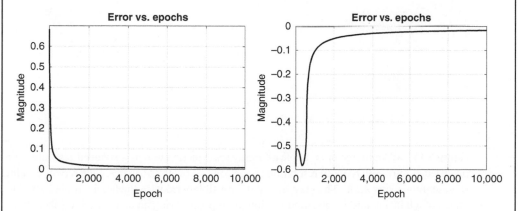

Figure 6.10 Convergence plot of error for (a) example 1 with a linearly separable data set and (b) example 2 with a linearly nonseparable data set.

6.4   Error Functions, Activation Functions, and Learning

Before we proceed with implementation approaches using MATLAB's Neural Network toolbox, we need to make some amendments to the prior discussion, as we have left out some key aspects of NN and its associated data. You may have noticed that the examples presented in the prior sections made use of data that were categorized. Although we used binary values in the examples, there was a clear way to categorize the data. Classified or labeled data comes in all degrees of complexity. For the purpose of our discussion, we will use data with low complexity such as that given by images. An example task for an NN could be to identify and categorize objects in such images. Suppose the objects are different types of fruits, as shown in Figure 6.11.

Suppose that the image itself is composed of thousands of pixels, each having one or more values associated in order to express RGB color values or representing shades of gray. If each of the pictures is labeled with the name of the fruit it is depicting, we have a rather complex data set. However, it is the same type of data – that is, a labeled data set just like our data sets in the previous section. Such

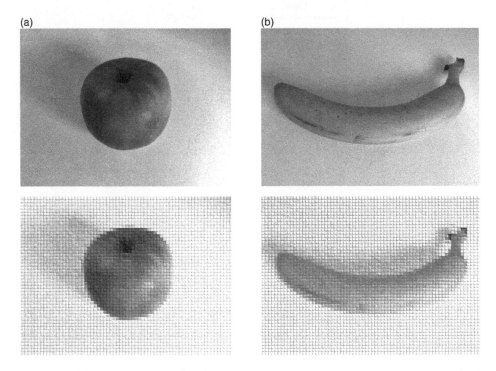

Figure 6.11 (a) An image of an apple as a gray scale picture and as a reduced resolution or pixelized picture. The reduced resolution image consists of 50×66 pixels, each having a value representing a gray scale. The gray scale data are all labeled with the label "apple." (b) An image of a banana, and its pixelized version, each pixel having an associated number that expresses the "grayness" and the entire set of the data categorized as "banana."

data sets permit for supervised NN training (the training we introduced in the prior sections). The supervision allows for the computation of the error, which is then used in the backpropagation algorithm. The data label is used as the supervisor and is equal to our "desired" output of the NN. Supervised learning is used for two types of problem sets: classification or regression. Classification refers to the process of assigning categorical labels to data. The output of the NN is essentially the set of different labels or categories to be assigned. Whereas regression allows for the prediction of numerical values based on a data set, here the relationship between the dependent variable(s) and one or more independent variables is modeled with an NN.

Since there is a supervised learning scheme in NN, it is logical to assume that there is also an unsupervised learning scheme. Unsupervised learning treats data that have not been labeled. One of the utilities of unsupervised learning is to detect hidden patterns in data without supervision. There are three different problem sets that can be addressed with unsupervised learning: clustering, dimensionality reduction, and association. Clustering is the process of grouping the unlabeled data into different clusters based on some similarity conditions or their differences. Dimensionality reduction techniques aim to reduce the number of input variables or features in a given data set, often leading to data compression. The objective of association in unsupervised learning is to detect dependencies and relationships between variables or data items and data sets. Considering the applicability of supervised and unsupervised learning, we can state that generally the goal for supervised learning is to predict, while the goal for unsupervised learning is to gain insight. There exists an in-between type, semi-supervised learning, where the data set is composed of labeled and unlabeled data. Some examples where semi-supervised learning is used is speech analysis or medical image analysis. Reinforcement learning (RL) is another category. In RL, the network attempts to learn how to act based on a feedback and reward system.

For our discussion of supervised learning we used a simple error function (Equation 6.11). With this error function we computed the gradient and used a gradient descent algorithm to perform backpropagation-based training of the weights in an NN. However, we can greatly modify this error function to embed different characteristics into the learning process. In the following, we will discuss some of the more popular error functions used in NN training.

One of the simplest error functions used for regression tasks is the mean square error (MSE). The MSE is the average of the squared difference between the desired and the predicted values of an NN:

$$\text{MSE} = \frac{1}{n} \sum_{i=1}^{n} (d_i - y_i)^2. \tag{6.17}$$

In Equation (6.17), n represents the number of data points considered. In order to reduce the penalty attributed to large differences in the predicted data for regression problems, one can use the mean squared logarithmic error (MSLE):

$$\text{MSLE} = \frac{1}{n} \sum_{i=1}^{n} \left(\log(d_i) - \log(y_i) \right)^2. \tag{6.18}$$

The Huber error function, also applicable to regression problems, has the added benefit of being less sensitive to outliers when compared to the MSE error function. The Huber error function is given in Equation (6.19):

$$H_\delta = \begin{cases} \frac{1}{2}(d-y)^2 \text{ if } |(d-y)| < \delta \\ \\ \delta\left((d-y) - \frac{1}{2}\delta \right) \text{ otherwise} \end{cases}, \tag{6.19}$$

where δ is a design parameter (e.g., $\delta = 1$). For small errors the Huber error function is essentially the quadratic error function. However, for large errors the function becomes linear.

The cross entropy (CE) error function is given by Equation (6.20):

$$\text{CE} = -d \ln(y) - (1-d)\ln(1-y). \tag{6.20}$$

Suppose our desired output, or training output data, is either 1 or 0. Then we can reformulate Equation (6.20) as:

$$\text{CE} = \begin{cases} -\ln(y) & \text{for } d = 1 \\ -\ln(1-y) & \text{for } d = 0 \end{cases}.$$

One of the desired properties of an error function is to be proportional to the output error. If we set $d = 1$, we can plot the corresponding CE, which is shown in Figure 6.12(a). As the output of our NN is given by y and considering the case of zero error, the CE is zero as well. However, if the output y approaches 0 as the error grows, the CE increases exponentially. On the other hand, if $d = 0$ and considering the corresponding CE plot, as given by Figure 6.12(b), we notice that when the output y is 0, the CE is 0 as well, and when the output approaches 1, the error again increases exponentially.

The CE can be adapted to the classification problem – for example, in the case of categorizing data into two sets. This adaptation of the CE to the classification problem yields the binary cross entropy (BCE) function:

$$\text{BCE} = -\{y \log(p) + (1-y)\log(1-p)\}. \tag{6.21}$$

In Equation (6.21) the output y of the NN is binary (i.e., 0 or 1). Hence, p is the probability of predicting this observation. In other words, the probability of a data point taking on a label "1" is given by $p(y)$.

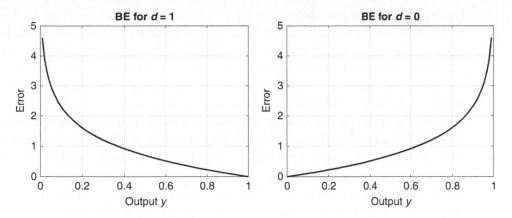

Figure 6.12 Cross entropy error function: (a) when the output is 1 and the label is correct, (b) when the output is 0 and the label is incorrect.

A common problem encountered in the training of NNs, or other data-based models, is the problem of overfitting. This phenomenon occurs when a model – such as an NN – attempts to predict every data point that was used for the training of the model while the data was contaminated with noise or random fluctuations. Hence, the training of the NN will model not only the system characteristics, but also the noise. When using this model with new data, the overfitted model will no longer be able to predict the noise or the random fluctuation, hence affecting the quality of the output of the NN. This leads to a diminished ability of the NN to be generalized. Effectively, an NN that has been overfitted results in a model that is too complex, with too many parameters for the data it is representing. How to detect and avoid overfitting is something we will learn in the next sections. However, we also can address overfitting with modifying the error function. This modification is in essence a regularization (i.e., an addition to our error function). Consider the MSE given in Equation (6.17). A regularization term can be added that involves the weights of the NN:

$$\text{MSE}_R = \frac{1}{n} \sum_{i=1}^{n} (d_i - y_i)^2 + \lambda \frac{1}{2} \|\vec{w}\|^2. \tag{6.22}$$

In Equation (6.22) λ represents a weighting factor, while $\vec{w}$ is the set of weights used in the NN. The same regularization term can also be used for other error functions, such as the CE:

$$\text{CE}_R = \sum_{i=1}^{n} \{-d_i \ln(y_i) - (1 - d_i)\ln(1 - y_i)\} + \lambda \frac{1}{2} \|\vec{w}\|^2.$$

Either error function is part of the optimization process that has the objective of reducing the error and hence finding the corresponding weights of the NN.

Having the weights included in the error function along with the error of the network allows for some of the weights to become zero as the optimization is a minimization problem. A zero weight implies that the associated connection in the NN is not used for the computation of the output of the NN. This effectively eliminates a connection within the NN and is equivalent to reducing the complexity of the NN, which reduces the chance of overfitting.

So far, we utilized one activation function: the sigmoid activation function φ as depicted in Figure 6.4. There is a range of other activation functions used in the construction of NNs. Some have been proposed to solve long-standing problems of deep NN, which we will cover in more detail in Chapter 8. However, for now we introduce formally some of the different activation functions and their general properties. The sigmoid activation function is given by Equation (6.23):

$$\varphi(v) = \frac{1}{1 + e^{-v}}. \tag{6.23}$$

Recall that the sigmoid activation function maps its input into a range between 0 and 1. Hence, a common usage of this activation function for NNs is when we want to predict the probability as an output. We made use of the simple form of the derivative of the sigmoid activation function for the backpropagation algorithm. It turns out that the sigmoid activation function is not very well suited for the gradient descent method, despite its simple derivative. NNs with sigmoid activation functions may "kill" the gradients during the training process and hence the neuron's activation will be 0 or 1 (i.e., saturated). This means that the gradient at those places is almost zero and there will be no more signal activity flowing to its weights.

Another activation function that looks similar to the sigmoid activation function is the Tanh function, or hyperbolic tangent function, as shown in Figure 6.13(a). This function is rather similar to the sigmoid function; however, the output range is different: the Tanh function has an output range from -1 to $+1$ and is zero-centered. The Tanh function is defined by Equation (6.24):

$$\varphi_{\text{Tanh}}(v) = \frac{(e^v - e^{-v})}{(e^v + e^{-v})} \tag{6.24}$$

Outputs that relate to processes with negative and positive results are easily mapped with Tanh. When used in hidden layers, the zero-centering property of the Tanh function helps with the learning for the next layer. One of the disadvantages of the Tanh function is that, just like the sigmoid function, it may cause the vanishing gradient problem in the learning process of the NN.

The rectified linear unit (ReLU) activation function is shown in Figure 6.13(b). Note, compared with the prior two activation functions, the upper limit of the ReLU function is not given. Due to the zero output for a negative v, the ReLU function may cause that not all of the neurons are activated at the same time. The ReLU function is given by Equation (6.25):

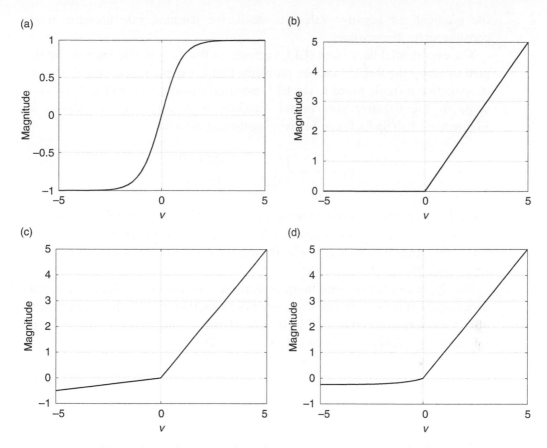

Figure 6.13 Different activation functions: (a) hyperbolic tangent function (Tanh); (b) rectified linear unit (ReLU); (c) leaky ReLU function; and (d) exponential linear unit (ELU) function.

$$\varphi_{\text{ReLU}}(v) = \max(0, v). \tag{6.25}$$

Because only a certain number of neurons are activated when ReLU is used, the NN becomes more computationally efficient. Also, the learning process using the gradient descent with the ReLU function as the activation function converges faster. However, since negative values for v result in a zero output, the gradient value is also zero. Hence, during backpropagation, some of the weights will not update and the process will create so-called "dead" neurons.

The "dead" neuron problem for the ReLU activation function inspired the leaky ReLU activation function, which is shown in Figure 6.13(c) and defined by Equation (6.26):

$$\varphi_{\text{Leaky ReLU}}(v) = \max(0.1v, v). \tag{6.26}$$

With this simple modification of the ReLU function, backpropagation for negative v values is possible and the "dead" neuron problem disappears. However, as the gradient for negative values is small, the learning rate becomes time-consuming for those values.

The exponential linear unit (ELU) activation function is also inspired by the goal of fixing the ReLU function problems that it experiences during the back-propagation learning process. The ELU essentially changes the ReLU so that the slope of the negative part of the function is a log curve, as shown in Figure 6.13(d). The ELU can be given mathematically as:

$$\varphi_{\text{ELU}}(v) = \begin{cases} v & \text{for } v \geq 0 \\ \alpha(e^v - 1) & \text{for } v < 0 \end{cases}. \tag{6.27}$$

The ELU function does not suffer from the "dead" neuron problem the ReLU function possesses, but has a new problem: the exploding gradient problem. The exploding gradient problem results from large error gradients that cause large updates on the weights of the NN.

Finally, we introduce one more activation function, the softmax function. Graphically it is very similar to the sigmoid function and the mapping results also in an output that is limited by 0 and 1. The mathematical representation of this activation function is given in Equation (6.28):

$$\varphi_{\text{softmax}}(v_i) = \frac{e^{v_i}}{\sum_{k=1}^{p} e^{v_k}}. \tag{6.28}$$

In Equation (6.28) the input vector $\vec{v} \in \mathbb{R}^n$ has n elements. As the sigmoid function has been used in NN to describe a probability as output, the softmax function has a very similar usage. However, the softmax function computes the relative probabilities, which makes it very useful for multi-class classification problems.

Recall Equation (6.7), the computation of the output of a neuron. It involves the weighted sum of all its inputs applied to the activation function. The activation function introduces the nonlinear complexity that allows us to model complex relationships. The weights are responsible for the trigger-ing level of the activation function. If we want to shift the output by some constant amount, we need to introduce a new term: the bias term. The bias term, as shown in Figure 6.14, can move the activation function curvature. In Figure 6.14 we use the sigmoid activation function. A positive bias term to the neuron will move the sigmoid activation function to the right, while a negative bias term will move the curve to the left. The bias term is just a constant, but applied to an activation function we can effectively better fit the data to our model as we include an additional model parameter. The updated equation for computing the output of a neuron with a bias term is given in Equation (6.29):

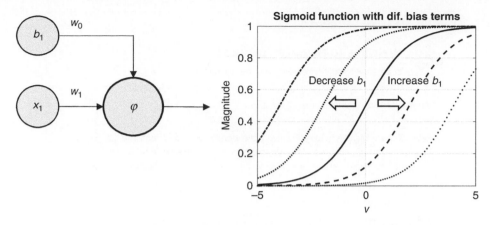

Figure 6.14 Bias term addition to a neuron and its effect on the activation function output.

$$y = \varphi\left(\sum_{j=1}^{2} x_j w_{ij} + b_0 w_0\right). \tag{6.29}$$

6.5   Neural Network MATLAB Implementation

As NNs can become rather complex structures and the associated calculations computationally demanding, from here on we will make use of MATLAB's Neural Network toolbox. There are different ways to employ the functionality of this tool set. We will explore two different approaches to how to construct, train, visualize, and simulate NNs with MATLAB's Neural Network toolbox.

The first approach involves MATLAB's `nnstart` GUI (for older Neural Network toolbox versions, this is the `nntool` function). This app guides the user graphically through the steps to develop NNs. It includes the functionality to do NN fitting, NN pattern recognition, NN clustering, and NN time series modeling. Conveniently, MATLAB includes built-in data sets and examples. To start the GUI, either type into the command prompt `nnstart` or select the corresponding app under the App tab.

Suppose you want to train an NN for the purpose of pattern recognition and classification: Select the appropriate button as shown in Figure 6.15, which will lead you to the next screen that provides you with some introduction and the graphics MATLAB uses to represent the network. The next window will allow you to define inputs and targets, and to select the corresponding data. Note that targets refer to our desired output information d. To practice creating and training an NN, lets chose an example data set provided by MATLAB: the breast cancer data set for pattern recognition, as shown in Figure 6.16.

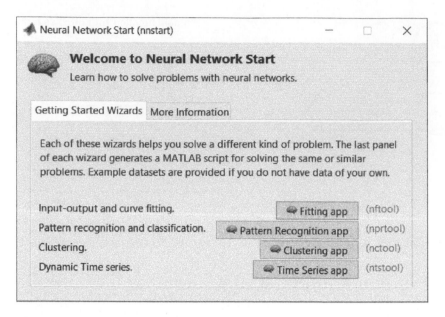

Figure 6.15 nnstart GUI welcome window.

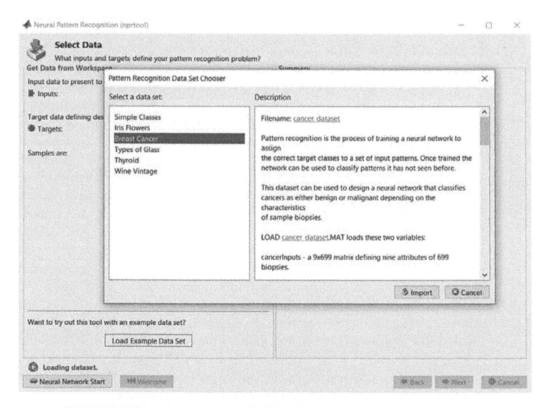

Figure 6.16 Data section window with data description.

The cancer data set that MATLAB is providing utilizes a *.mat file (i.e., `cancer_dataset.mat`), which holds two variables. The two variables represent data matrices – that is, the *cancerInputs* which is a 9 × 699 matrix and the *cancerTargets* which is a 2 × 699 data matrix. The rows of the input data variable *cancerInputs* contains nine attributes from 699 biopsies. The target variable, *cancerInputs*, has two rows, each representing a category: benign or malignant. When using your own data, make sure you use the same format for your input and target data sets. One important practice we have yet to discuss is how to utilize data for training, testing, and validation. Validation is the process used to evaluate the quality of a model during training. Testing is the process of quantifying the quality of the model after training. To properly assess the quality of the network, we should not use the same data that it has already seen during the training process. Hence, validation and testing data are always data that have not been used to train the network. For the given example, the data provided by MATLAB is just one input and one target set. To divide the data set into three sets for three different tasks, the GUI allows us to decide how much of the data entailed in the two matrices is utilized for the training and the testing, and how much of it is used for validation. Figure 6.17 illustrates the prompting of MATLAB for this information.

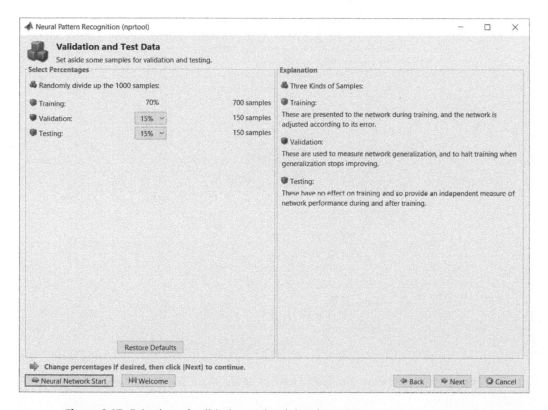

Figure 6.17 Selection of validation and training data sets.

Now we are ready to define the actual structure of the NN. The window "Network Architecture" allows us to define how many neurons the hidden layer will utilize. It also provides for a graph of the current network. The network we are creating has nine input neurons, a hidden layer with weights w and bias terms b, and an output layer with two neurons. The input layer accounts for each separate input and is automatically adjusted to the number of rows your training data has. The output layer functions in a similar way; as the targets include two classes, MATLAB determines that we need two output neurons.

Proceeding to the next window, we are informed about which training method is being used. For example, `trainscg` implies that MATLAB will use the scaled conjugate gradient backpropagation method. After executing the training of the network, the window updates with key information on the training results. A second window provides information on the process itself and allows for plotting the function of the results. Figure 6.18 depicts the window for the NN training process and results plot functions. The *Algorithm* section indicates the different methods used for the training process; the *Process* section provides information on

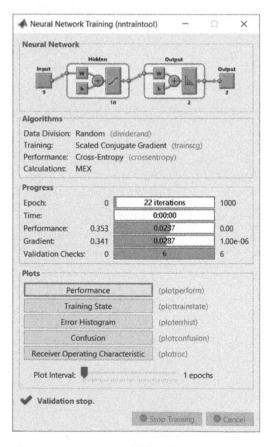

Figure 6.18 Neural network training window with algorithm, process, and results information.

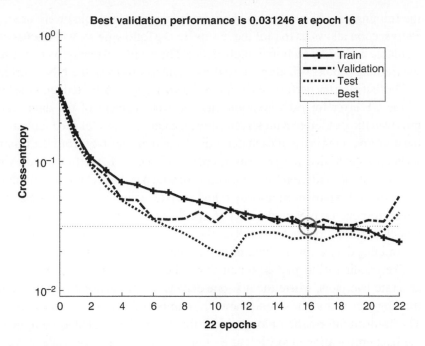

Figure 6.19 Performance plot of network during training using three different data set segments.

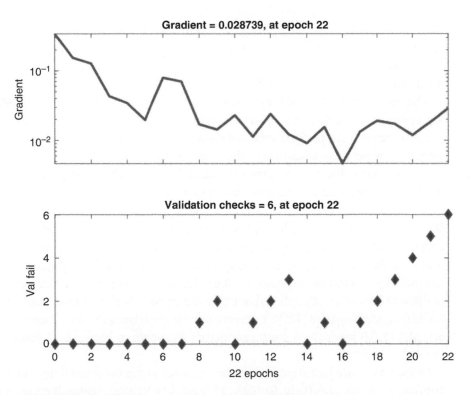

Figure 6.20 Training state plots for a trained neural network.

the training parameters, such as how many iterations (epochs) were used; and the *Plots* section allows us to plot the results. In the following we will discuss each plot available in the *Plot* section in more detail. The graphical depiction of the results of the training process will allow us to draw conclusions from the NN's performance.

The first plot, the performance plot, is shown in Figure 6.19. It details the logarithm of the CE error for each iteration and for each segment of the data sets. As we provided three different data set segments, the figure shows for each separate data set the corresponding error at each iteration. Note that the error from the training data set is consistently declining with more epochs. However, the error from the validation set is the guiding criterion for the selection of which is the best model. In this case, the validation error plot indicates that its lowest error occurs at epoch number 16. Hence, the corresponding NN model after 16 epochs of training will be used as the trained model. The selection that is based on the error using the validation data rather than the training data allows for minimizing the chance of overfitting.

The training state plot selection from the menu depicted in Figure 6.18 will generate two plots, where the first one depicts the gradient information during the training process and the second shows the number of validation checks. (Figure 6.20) The gradient information plot provides for the gradient value as computed during the backpropagation algorithm at each iteration using the log scale. Note, for our example, the gradient is the smallest at epoch 16, indicating we reached the bottom of the error surface plot – that is, the local minima of the error function.

The second plot provided by the training state plot function refers to validation failures during the iterations. Usually, a failure implies that overfitting is occurring. MATLAB stops the training automatically if six failures occur for one epoch, which happened in our example at epoch 22. The error histogram plot option will generate a histogram detailing the history of the errors between the target values and the trained NN's predicted values.

The confusion plot option will yield a confusion matrix for training, testing, validation, and the three sets of data combined. A confusion plot – or table – is a way to describe the performance of a classification model. Consider Figure 6.21, where we depict a classification problem to label fruit items.

In such plots, the x-axis carries the actual values, while the y-axis lists the predicted values. For the example shown in Figure 6.21 we trained a network to predict if an image depicts an apple or not an apple. Hence, we have two classes. We can relate now the number of times the classification algorithm predicted the correct label for the item "apple", which is referred to as true positive (TP), how often it assigned to an apple the wrong label (i.e., false negative [FN]), how often the model predicted the item not to depict an apple correctly (i.e., false positive [FP]), and how often it predicted not to be an apple when the image depicted no apple (i.e., true negative [TN]). For our example classifying the breast cancer data set, MATLAB generates a confusion matrix as shown in Figure 6.22 using class labels "1" and "2."

Good results are indicated by high percentages along the main diagonal of the confusion matrix, referring to high TP and TN values. Using the confusion

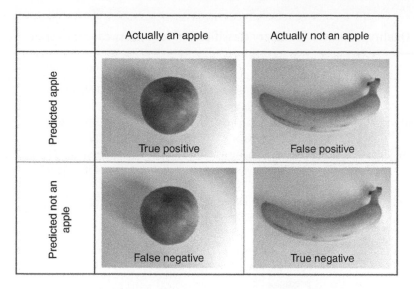

Figure 6.21 Confusion matrix for classification of fruit items.

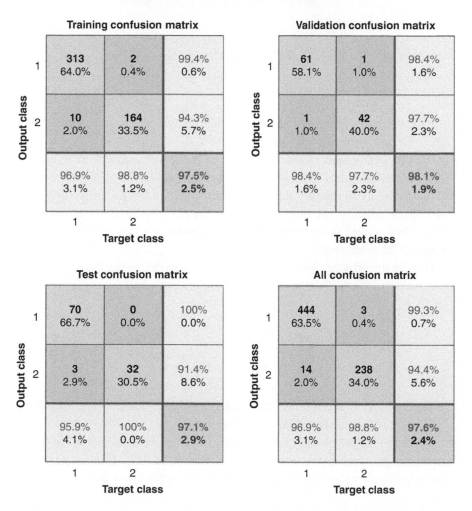

Figure 6:22 Confusion matrix generated by MATLAB for the breast cancer classification problem.

Table 6.6 Quality measures for binary classification for breast cancer overall data

Name	Description	Equation/example
Accuracy	Measures overall how often the classifier is correct	(TP + TN)/total = (444 + 238)/699 = 97.57%
Misclassification rate	Measures overall how often the classifier is wrong	(FP + FN)/total = (3 + 14)/699 = 2.43%
Recall (true positive rate)	When it is actually a "1," compute how often does it predict a "1"	TP/actual "1" TP/(TP + FN) = 444/458 = 96.94%
False positive rate	When it is actually a "2," compute how often does it predict a "1"	FP/actual "2" = 3/241 = 1.24%
Specificity (true negative rate)	When it is actually a "2," compute how often does it predict a "2"	TN/actual "2" = 238/241 = 98.76%
Precision	When it predicts a "1," compute how often is it correct	TP/predicted "1" = TP/(TP + FP) = 444/447 = 99.33%
Prevalence	How often does the "1" category actually occur in the sample	Actual "1"/total = 458/699 = 65.52%

matrix, other quality measures can be computed for binary classification problems. We list them in Table 6.6.

A Type I error refers to the items in FP, hence indicating if the model predicts something that is not there. A Type II error is defined by FN and indicates the model is failing to identify something that it should have identified. In Table 6.6 we defined recall as the ratio of the correctly classified items over the number of items that should have been classified. In essence, recall indicates whether the model is missing labels.

Having introduced precision and recall, we can define the F-score. The F-score is a measure of a test's accuracy. There are multiple F-score versions:

$$F_1 \text{ score: } F_1 = 2\frac{P \times R}{P + R}, \tag{6.30}$$

$$F_2 \text{ score: } F_2 = 5\frac{P \times R}{4P + R}. \tag{6.31}$$

In general, the F_1 score favors both measures (i.e., precision and recall) equally. The F_2 score has some bias toward precision.

The last plot that MATLAB provides in the plots section (Figure 6.18) is the receiver operating characteristic (ROC) plot. The ROC plot is constructed by graphing recall vs. specificity. However, in this application we use a probabilistic classifier – that is, a classifier that yields a probability of belonging to a class. By varying the threshold value for this probability, we can create a curve that indicates the trade-off between recall and specificity. Recall is also sometimes referred to as sensitivity (TPR) and specificity is computed by (1 – false positive rate [1 – FPR]). Hence, a purely random classifier should result in a curve that is a straight diagonal line (i.e., FPR = TPR). A classifier that yields a curve that comes close to the top-left corner of the diagram indicates good performance. Figure 6.23 shows the results for our breast cancer MATLAB data set example. For each data set as well as for the combined data set, the ROC indicates a near-ideal discriminator characteristic, as all curves are close to the top-left corner.

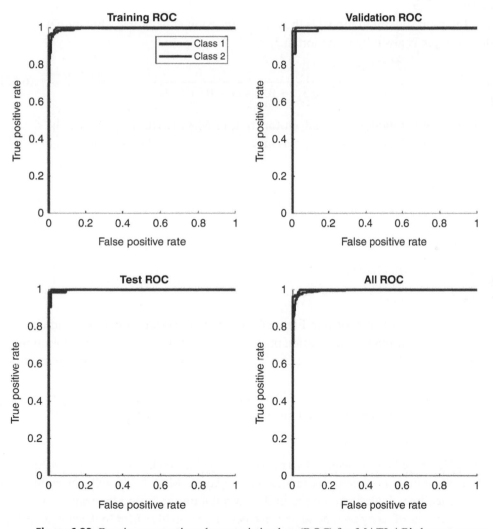

Figure 6.23 Receiver operating characteristic plots (ROC) for MATLAB's breast cancer data set for all three data sets as well as for the combined data set.

Going back to the main MATLAB Neural Network toolbox window, MATLAB allows for modification to the NN model by changing the process and model parameters under the "Retrain" button option. Once satisfied with the trained NN model performance, MATLAB provides for three deployment options: as a MATLAB function, using code generation, or Simulink block diagram. We will explore more of these deployment options at the end of this chapter.

The second approach in designing and training an NN using MATLAB is the script option without the explicit help of the NN GUI. We will demonstrate this by an example where we try to match a function output using an NN.

Example 6.3

The function we aim to model is the Humps function. The Humps function we use in this example is given by Equation (6.32):

$$y = \frac{1}{(x - 0.3)^2 + 0.1} + \frac{1}{(x - 0.9)^2 + 0.4} - 6. \tag{6.32}$$

The Humps function is a built-in function in MATLAB. To generate the desired data set for the NN training using this function, we can use the following commands in MATLAB:

```
% 1. Generate input vector
x=0:0.025:2;
% 2. Generate desired output vector
d=humps(x);
% 3. Plot input and output
plot(x,d,'k.-');
title('Humps Function');
xlabel('Time Index k')
ylabel('Magnitude')
```

The resulting plot is shown in Figure 6.24. The Humps function has found applications in optimization and integration problems as a test function – for example, finding the roots of this function. Our objective is to find an NN that generates the same output as the Humps function.

Our next task is to create the NN structure. For this purpose, we can use the command `feedforwardnet()` to establish a feed-forward NN:

```
net = feedforwardnet(10);
```

The NN called `net` consists of 10 neurons in the hidden layer. Note that we did not define the number of neurons in the output layer. MATLAB will determine this automatically by looking at the size of the output data set during the training process. The advantage of using command line instruction or *.m file script to develop an NN is the access we have to each of the hyperparameters of the training process. For example,

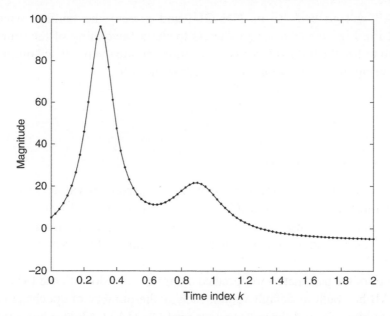

Figure 6.24 Data generated using the Humps function for developing an NN using the command window in MATLAB.

we can define the error function, the training algorithm, and the activation function for the first and second layer, as shown in the MATLAB script:

```
net.performFcn = 'mse';
net.trainFcn = 'trainlm';
net.layers{1}.transferFcn = 'tansig';
net.layers{2}.transferFcn = 'tansig';
```

In each of the lines of code we access the properties of the NN called net by using the dot operator. The error function is called `performFcn` and we chose the MSE function or `mse`. The training algorithm we selected in the code above is the Levenberg–Marquardt backpropagation algorithm (i.e., `trainlm`). For the activation function in both layers, we assigned the sigmoid function. To control the training parameters, we can assign the following values to the number of epochs used, the training goal or stopping requirement, the minimum gradient and the maximum failure number:

```
net.trainParam.epochs = 1000;
net.trainParam.goal = 10^-10;
net.trainParam.min_grad = 10^-10;
net.trainParam.max_fail = 100;
```

These are starting values. Once we obtain results and measure the quality of the trained network we can alter these to improve the NN.

The last item we need to specify before proceeding with the actual training is the training data set, and whether we want to utilize a validation data and testing training set. To illustrate that MATLAB's `train` function can be operated also by only using all of the data for training, we use the following line to train our NN called net:

```
net=.train(net,x,d);
```

MATLAB will go back to its NN GUI and present the familiar window we encountered in Figure 6.18 to give us access to check the quality of the trained NN. We can also utilize the trained NN called net and generate the predicted output values of the NN using the input vector x by simulating the NN:

```
y=sim(net,x);
figure
plot(x,y,'k.-');
hold
plot(x,d,'k:');
legend('Predicted','Actual');
```

The generated output comparison between the predicted values of the Humps function y and the actual output d is shown in Figure 6.25. If improvements are necessary, one has access to any of the training parameters as well as the selected functions and limits, which makes command line NN training quite convenient.

Note, some of the parameters we specified before training do not need to be specified as MATLAB has built-in default values, such as the number of epochs (1,000), the number of failures (6), and the performance goal (0). MATLAB also has a rich set of different training algorithms, including the ones we discussed:

```
traingd    gradient descent backpropagation
trainlm    Levenberg–Marquardt backpropagation.
```

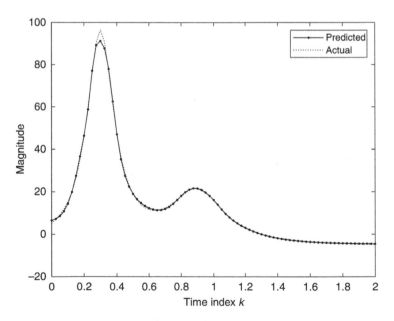

Figure 6.25 Training "actual" and simulated "predicted" data using a command line MATLAB script to generate and train a feed-forward neural network.

There are a number of training algorithms we left out of our prior discussion, but are documented in MATLAB's help documentations, such as:

`traincgb`   conjugate gradient backpropagation with F–R updates
`traingdm`   gradient descent with momentum
`trainb`   batch training with weight and bias learning rules
`trainbu`   unsupervised batch training with weight and bias learning rules.

This is just a sample of the available training algorithms. Use the MATLAB documentation to explore other training methods, including different activation functions.

Considering the design of a controller as one of our objectives, NN can serve in a number of ways to accomplish good control laws by facilitating the modeling aspects. In Chapter 7 we explore NNs in combination with FL to form adaptive controllers. However, in this chapter we want to utilize the Neural Network toolbox to infer an NN model that behaves dynamically equivalent to the actual system without using first principles to model the actual system. Such a model can be quite useful for the process of controller design. MATLAB provides for a number of built-in models and their associated input/output data sets. One such set is based on a magnetic levitation system. The data is generated by the use of a Simulink model, which is accessible by typing into the command window the following name:

`narmamaglev`

This command will cause MATLAB to open Simulink and load the corresponding model that bears this name. The model entails a feedback controller. The data we are interested in are generated by the block "Plant" and are also stored in MATLAB by default. To access these data, use the following command:

`[x,t] = maglev_dataset;`

The variable x holds the electrical current values for the actuation of the magnetic levitation system, while the variable t represents the system's vertical position in response to the applied electrical current. A plot of the applied current and the resulting position using the `narmamaglev` system is shown in Figure 6.26.

The NN structure we aim to extract for this system needs to be able to accommodate the characteristics of the dynamics of the system. Assuming that the magnetic levitation system is nonlinear, we chose to employ a nonlinear autoregressive network with exogenous input (NARX) model structure. MATLAB provides for three different model structure types for time series-based modeling:

- NARX
- nonlinear autoregressive (NAR)
- nonlinear input–output.

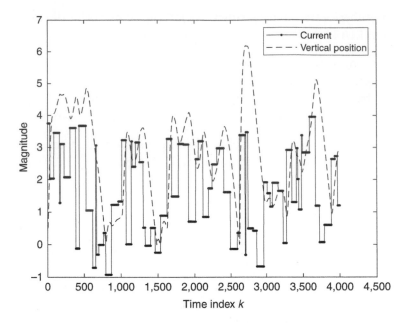

Figure 6.26 Training data for the magnetic levitation system, depicting applied current to the coils of the electromagnet and the response in terms of vertical position of the levitated specimen.

The NARX model structure is a recurrent dynamic NN. A simple illustration of this structure is given in Figure 6.27(a). Note that there is a feedback term. This feedback term allows for utilizing past data in terms of time. The NAR model structure is shown in Figure 6.27(b). In comparison to the NARX model, the NAR model does not have an input. It can represent time series data where we do not have any information on the input. The third model structure option MATLAB offers is the nonlinear input–output structure, which resembles a feed-forward NN.

The NARX model equation is of the form:

$$y_k = F\big(y_{k-1}, y_{k-2}, y_{k-3}, \ldots, y_{k-p}, x_k, x_{k-1}, x_{k-2}, \ldots, x_{k-p}\big). \tag{6.33}$$

Here, the subscript k indicates the discrete time step, y is the output, and x is the input to the system. In Equation (6.33), p represents the number of delay elements. This number indicates how far back we go in time to model the time series.

To create the NARX NN structure in the command window environment, use the following command:

```
net = narxnet(10);
```

This will set aside a NARX structure with 10 neurons in the hidden layer and 2 delay elements (i.e., $p = 2$). The structure is shown in Figure 6.28.

(a)

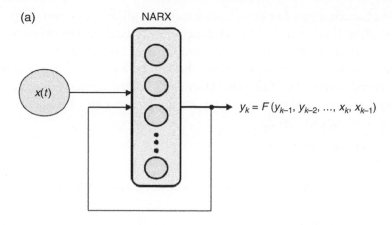

(b)

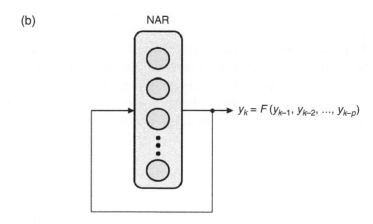

Figure 6.27 (a) Nonlinear autoregressive with exogenous input (NARX); (b) nonlinear autoregressive (NAR).

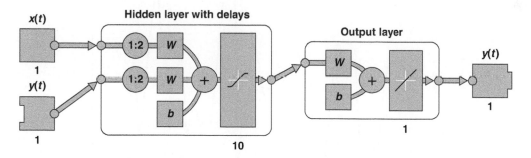

Figure 6.28 Command line-generated NARX structure with 10 neurons in the hidden layer and 2 delay elements.

Our next task is to prepare the data for training. This is accomplished with the command preparets(.):

```
[xo,xi,~,to] = preparets(net,x,{},t);
```

The function `preparets()` reformats the given input and output data set by shifting the input and target data points based on the number of delays. The input arguments for this function are `net`, which is the NN we plan to train, `x` which is a cell representing the inputs, and `t` which is a cell of target or desired values. The function returns the shifted inputs `xo`, the initial input delay states `xi`, and the shifted targets `to`. To train and simulate the NN, use the following commands:

```
net = train(net,xo,to,xi);
y = net(xo,xi)
```

The corresponding simulated output of the trained NN in comparison to the training data is shown in Figure 6.29.

The same results can also be obtained with the `nnstart` NN app, as detailed earlier. Either training approach will result in MATLAB generating the Neural Network Training (`nntraintool`) window, which allows for inspection of the results. For time series and dynamic system modeling some of the quality check tools are different than what we found for the classification problem. For example, having a time series as a data set and generating a model that attempts to capture the underlying dynamics, we can compare the response of the trained NN model with the given time series data set. For the NARX model trained in the above example using the magnetic suspension system data set, we can

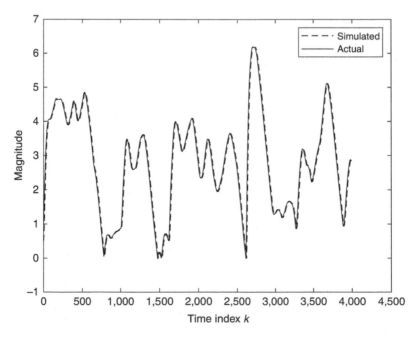

Figure 6.29 Simulated and true output for the levitation system using a trained NARX model and comparing it to the "desired" training data.

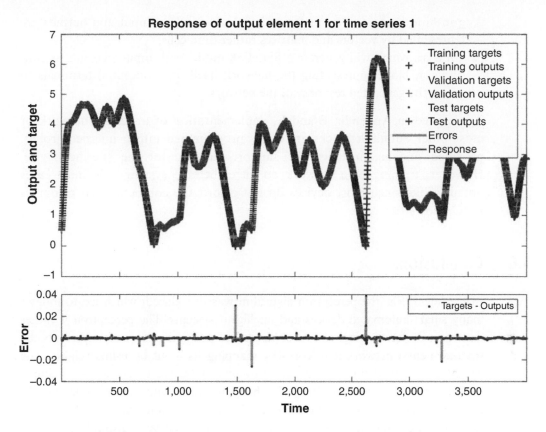

Figure 6.30 Time series data generated using the trained NARX model and given target data set, as well as error plot.

compare the response of the NN model with the response data, as shown in Figure 6.30.

The last functionality we want to include in our discussion is the deployment of trained networks. Either using the MATLAB command window or the `nnstart` app, we will arrive at a trained model that hopefully meets our quality expectations and hence is ready to be implemented in a simulation or on an embedded system. MATLAB provides three pathways to do so:

1. `genFunction`: this will generate a MATLAB function as a *.m file using a cell array, for example:

```
function [Y,Xf,Af] = myNeuralNetworkFunction(X,Xi,~)
```

   where `myNeuralNetworkFunction` is the function name and takes on the input quantities as matrix and cell arrays, generating an output cell with the output data spread over the simulation time provided by the input cells.

2. `genFunction`: used as matrix-only function, where input and outputs are represented by vectors and matrices rather than cells.
3. `genSim`: which will generate a Simulink model with input elements as constants, a block representing the network itself, and output(s) terminals to access the generated response of the network.

As detailed in Appendix E and F, implementation of a trained network or most other Simulink blocks are easily programmed into a microcontroller, provided one has installed the support package for the specific device. However, training of a network on an embedded system is in most cases not possible, since most devices do not afford the computational resources a PC allows for.

6.6   Conclusions

In the late 1950s, the perceptron algorithm was introduced, which led to today's many smart automated devices and intelligent systems. The perceptron is rather limited in its functioning, as it is composed of just one neuron. It represents the smallest neural network and works by mapping its input by using weights. The weights are learned through the use of training data. We noted that such simple neural networks are limited in their ability to work with certain functions. In particular, such networks can handle only linearly separable functions. To accommodate more complex systems we need next to an input layer also a hidden layer prior to the output layer. With such networks we can entertain all kinds of complexity. However, to achieve image recognition as needed for current applications, deep neural networks are being used. The "deep" in deep neural network refers to having more than one hidden layer. Neural networks and deep neural networks are trained on data by the use of backpropagation methods. Backpropagation refers to the process where we compare the output of the network to the desired output (supervised learning) and feed the difference back through the network in reverse. The adaptation of the network weights is accomplished by the use of the activation function. This function plays a critical role in the learning process, especially for deep neural networks, as the error signal – or rather the gradient information – gets lost during the backpropagation algorithm when using certain activation functions. In addition, new concepts such as drop-out have been introduced to counter the chance of the vanishing and exploding gradient problems during the training stage of the network.

In our discussion we also included some consideration of quality measures, such as the accuracy, recall, specificity, precision, prevalence, and some of the derived quantities such as the F-score and the ROC plot. We also noted that overfitting can be a substantial problem in supervised learning algorithms. To have some utility for detecting such occurrences during training, we make use of validation data sets.

However, overfitting is a delicate problem, as we want to balance accuracy and generalizability. This is something we will encounter again in the following chapters, in particular in Chapters 8 and 9. However, before we discuss deep learning and machine learning in those chapters we will look at hybridization of some of the algorithms we have so far encountered.

SUMMARY

In the following, key concepts of neural networks are presented.

Supervised and Unsupervised Learning

When applying supervised learning the network has access to labeled data sets. Labeled data sets are used during the training of the network and the network objective is to make predictions when presented with new data sets. Unsupervised learning does not utilize labeled data sets. Unsupervised learning allows for the analysis of uncategorized data to find hidden patterns and structures, to perform clustering or dimensionality reduction.

Perceptron

The perceptron is a simple neural network used for supervised learning of binary classification problems. Perceptron networks are limited to classifying linearly separable data sets.

Multilayered Feed-Forward Neural Networks

A multilayered FFNN has at least one hidden layer, and the data flow is restricted from the input of the network to the output of the network. This is contrary to an RNN, where the data can move in different directions within the network.

Backpropagation

Backpropagation is used in the training of neural networks. Backpropagation is responsible for computing the gradient of the loss function with respect to the network's weights. The network's output error is backpropagated through the network using the gradient information to update its weights.

Activation Function

An activation function is entailed in a neuron of a neural network and allows for modeling complex patterns. Essentially, the activation function computes the output of the neuron based on the input and input weights to the neuron.

Confusion Matrix

A confusion matrix allows assessing the performance of a classification model. The matrix entails predicted categories versus the actual categories by computing the true positive, the false positive, the false negative, and the true negative.

F-Score

The ratio between true positive and the sum of the true positive and false positive is called the precision $P = \text{TP}/(\text{TP} + \text{FP})$. The recall R is the true positive rate, which is computed by taking the ratio of the true positive over the sum of true positive and false negative, $R = \text{TP}/(\text{TP} + \text{FN})$. Then, the F-score can be computed by

$$F_1 \text{ score: } F_1 = 2\frac{P \times R}{P + R},$$

$$F_2 \text{ score: } F_2 = 5\frac{P \times R}{4P + R}.$$

The F-score is a measure of a test's accuracy.

REVIEW QUESTIONS

1. What are the limitations of single-layer neural networks?
2. What is an application of unsupervised learning?
3. Why use a sigmoid activation function in a neural network for classification problems?
4. How many output neurons does a neural network have when classifying Greek letters?
5. How many input neurons does a neural network have for classifying 25×25 pixel images (black and white pictures)?
6. What is a TPU?
7. What is an epoch?
8. What primary difference is there between a recurrent neural network and a feed-forward neural network?
9. What does the delta rule compute?
10. How are bias terms included in a neural network?
11. What is the "dead" neuron problem?
12. What is achieved when a neural network is generalized?
13. What is the purpose of the learning rate in a neural network?
14. Why use the MSLE as an error function over the MSE error function?
15. What is the benefit of using the BCE error function in a neural network training algorithm?

16. What is the benefit of using a leaky ReLU over a ReLU activation function?
17. What purpose does the validation data serve in the training process of a neural network?
18. How is accuracy defined when classifying binary data?
19. Define specificity for a binary classification problem.
20. What type of neural network can model time series data?

PROBLEMS

6.1 Research the AlphaGo project and summarize in layperson terms the basic working principle behind the AlphaGo program, the approach, the competition, and the outcome of the competition.

6.2 a. How many input neurons will a simple FFNN have if it has to classify a 2 megapixel image?
  b. How many output neurons will a letter classification NN have?
  c. If provided a target data set, what kind of training are you most likely to perform?
  d. Why is an NN called deep? Provide an example.
  e. What is the function of the bias term?
  f. For what is the term "generalization" used in NN training?
  g. What is overfitting? Provide an example.

6.3 Suppose you want to train a simple neural network using two inputs for nine different values. The values can be found in Figures P6.1 and P6.2. Alter the MATLAB program called `SingleLayerMain.m` and `example1data.m` so that they can train each data set. Comment on the outcome of the training and indicate why or why not the training was successful.

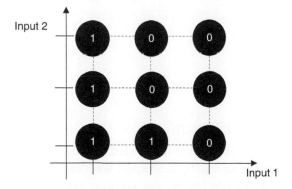

Figure P6.1 Input–output relationship for Problem 6.3 part (a).

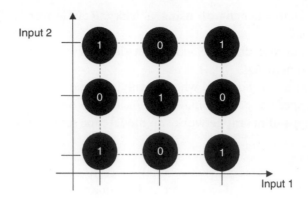

Figure P6.2 Input–output relationship for Problem 6.3 part (b).

6.4 Given the following target data set: $D = [1\ 1;\ 0\ 0]$, will a single-layered NN be able to successfully classify such a data set? Explain your answer.

6.5 Use the two data sets from Problem 6.3 and modify the two-layer NN MATLAB code provided in the text to train an NN for each data set. Compare the results with the results obtained from Problem 6.3 based on accuracy and number of epochs.

6.6 Show the effect of a bias term by varying its magnitude to the following activation functions (plot to show the effect):

   a. linear activation function (use $\varphi(w) = mx + c$ for the linear activation function and chose the constant m and c)

   b. hyperbolic tangent function (Tanh)

   c. rectified linear unit function (ReLU)

   d. leaky ReLU function

   e. exponential linear unit (ELU) function.

6.7 Consider a two-dimensional neural input $x(t) = [x_1\ x_2]$. Design and program in MATLAB the appropriate neural networks for the following logic operations:

   a. OR

   b. AND

   c. NOT

   d. NOR

   e. NAND

   f. EXCLUSIVE-OR (XOR).

6.8 Consider a mass–spring–damper system having a natural frequency of 2 rad/s and a damping ration of 0.25. Suppose we want to model this system using an NN by conducting an experiment to collect time series data for the input and the output. Instead of the real experiment, we will utilize a simulation. The MATLAB code below simulates this mass–spring–damper system with additive process and measurement noise.

a. Use the given MATLAB code and generate an adequate number of input and output data points with the objective to use the data set for NN training. Provide some reasoning on the number of data points you chose.

b. Use a NARX NN model structure and train this network with the data generated in part (a). Compare the step response of the trained network with a step response of the simulation model. *Hint: You can use the variables "num" and "den" to create a transfer function and you can generate the step response of the simulation model without additive noise by using* `step(num,den);`

c. Modify the number of data points to investigate the effect on your results.

d. Modify the network training hyperparameters and comment on the sensitivity of these parameters with regard to the quality of the trained network.

```
% Problem_6_8.m
% Generate data for mass-spring-damper system
%_____
clear;clc
wn = 2; % natural frequency
zhi = 0.25; % damping ratio
num = [wn^2];den = [1 2*zhi*wn wn^2];
[A,B,C,D]=tf2ss(num,den);
TS=0.01; % discrete sampling time
[Ad,Bd]=c2d(A,B,TS); % discrete states-space system
% Simulation parameters: input by user
input('How many data points ');nd=ans;
[~,ni]=size(Bd);[no,n2]=size(C);
% Quantify noise level
X=dlyap(Ad,Bd*Bd');M=sqrt(diag(C*X*C'+D*D'));X=sqrt(diag(X));
Noise=0.1; % Noise level for process and measurement noise
x=zeros(n2,1);u=zeros(ni,nd);y=zeros(no,nd);% initialize variables
nd2=nd/2;zero=zeros(ni,1);
pn=Noise*X;mn=Noise*M;
% Simulate discrete stochastic system
for i=1:nd
    u(:,i)=randn(ni,1);
    y(:,i)=C*x+D*u(:,i)+randn(no,1).*mn;
    x=Ad*x+Bd*u(:,i)+randn(n2,1).*pn;
end;
figure(1);
mar=linspace(1,nd,nd);
subplot(2,1,1);plot(mar,y);grid;xlabel('Discrete Time Index
k');ylabel('y magnitude');title('Output');
subplot(2,1,2);plot(mar,u);grid;xlabel('Discrete Time Index
k');ylabel('u magnitude');title('Input')
```

6.9 Use the NN model developed in Problem 6.8 and create in Simulink a simulation model for the following situations:

a. Simulation of the open-loop system with a square wave as input. Use an amplitude of 1.0 units and frequency of 2 rad/s.

    b. Design a PID controller and simulate the closed-loop system with a square wave that has a frequency of 2 rad/s and amplitude of 1.0 units.

    c. Design an FL controller and simulate the closed-loop system with the same input as given in part (b). Compare your results with the PID controller-based closed-loop simulation. Modify your FL controller to achieve a minimal overshoot and short settling time.

6.10 Momentum is a concept used in NN training to modify the update magnitude of the weights. For the delta rule with momentum, the updated weights can be computed as:

$$\Delta w_{ij} = \alpha \varphi'(v_i) e_i x_j,$$

$$m = \Delta w_{ij} + \beta m,$$

$$w_{ij} = w_{ij} + m.$$

Modify the MATLAB program called `SingleLayerMain.m` and use `example1data.m` as well as `example2data.m` to include momentum into the learning algorithm. Use $\beta = 0.9$ and $\alpha = 0.9$ for the update parameters. Compare the performance of a single-layer NN with momentum and without momentum. Comment on the effects of including momentum into the learning algorithm.

6.11 The MATLAB examples listed in Section 6.5 use the stochastic gradient descent (SGD) algorithm during the training of the network. We introduced the gradient descent (GD) algorithm in Chapter 5. The difference between the SGD and the GD is given by how many data points we use in the evaluation of the error function. While the GD uses all the data points, the SGD uses a single data point for the computation of the error function. Hence, the term "stochastic" refers to the behavior of the learning process. Consider the SGD algorithm, which computes the error for each training data point and adjusts the weights before computing the next update. If there were 1,000 training data points in the training data set, the SGD would update the weights 1,000 times during the training process. A modification can be made to the SGD by processing a "batch" of data points. This approach is called the batch mode. Suppose a batch is composed of n_b data points, then the batch mode of the SGD update algorithm computes the change in weights as:

$$\Delta w_{ij} = \frac{1}{n_b} \sum_{p=1}^{n_b} \Delta w_{ij}(p).$$

Here, p is the index of the training data point. Hence, $\Delta w_{ij}(p)$ is the weight update for the pth training data point. The mini-batch mode refers to an update process that is between the SGD and the GD batch mode algorithm, where $n_b < L$ and L is the total number of data points available.

Using the MATLAB file `Backpropagation()` from Section 6.3, modify the update rule to include the batch mode and mini-batch mode. Use the examples from Problem 6.3(a) to compare its performance for different batch sizes.

6.12 Using the propeller–pendulum system described in Appendix B.3, and using a random input for the propeller action, record the response of the system using the Arduino microcontroller and MATLAB setup for data collection. Train an appropriate NN using the experimentally collected data. Simulate the trained NN using a sine wave with different frequencies and amplitudes and compare the response to the physical system receiving the same input.

6.13 Using the trained NN model of the propeller–pendulum experimental setup:

a. Design an FL controller with the objective to achieve a fast steady-state response of the pendulum.

b. Simulate the response using the FL controller and NN in Simulink.

c. Deploy your FL controller onto the Arduino microcontroller and apply the same input as for part (b) to compare the performance of the simulated closed-loop system with the physical setup operating in closed-loop.

6.14 MATLAB maintains a webpage with a number of different data sets, categorized by image, time series, video, text, audio, and point cloud data types. You can find this repository at: www.mathworks.com/help/deeplearning/ug/data-sets-for-deep-learning.html.

The "Concrete Crack Images for Classification" data set consists of 40,000 images using two classes: 20,000 images for the class "Negative" and 20,000 images for the class "Positive." "Negative" refers to an image depicting no cracks in the concrete, while "Positive" indicates an image with cracks. Download this data set (instructions are given in the data set description on the website) and construct an NN for categorization. Modify the hyperparameters and evaluate the following quantities:

a. accuracy

b. misclassification rate

c. recall

d. false positive rate

e. precision

f. prevalence

g. F_1 score

h. F_2 score.

6.15 For the NN model structure utilized in Problem 6.14 for classifying images with and without cracks in concrete, investigate the ROC plot for different training algorithms. Research each training algorithm and comment on its characteristics based on your NN and the "Concrete Crack Images for Classification":

a. Levenberg–Marquardt `trainlm`
b. BFGS quasi-Newton `trainbfg`
c. scaled conjugate gradient `trainscg`
d. Fletcher–Powell conjugate gradient `traincgf`
e. one step scant `trainoss`
f. variable learning rate backpropagation `traingdx`.

REFERENCES

[1] Silver D, Huang A, Maddison C (2016). Mastering the game of Go with deep neural networks and tree search. *Nature*, 529: 484–489
[2] Werbos P (1974). Beyond regression: new tools for prediction and analysis in the behavioral sciences. PhD dissertation, Harvard University, Cambridge, MA.
[3] Hochreiter S, Schmidhuber J (1997). Long short-term memory. *Neural Computing*, 9(8): 1735–1780.

7 Hybrid Intelligent Systems

7.1 Introduction

In this chapter we will explore how one intelligent algorithm can work with another intelligent algorithm. The objective of such a collaboration is to solve a specific problem, most commonly an optimization, classification, or a control problem. This discussion could be extended to allow for a multitude of different intelligent algorithms to help and cooperate with each other. Each algorithm would have a different task assigned, and combining all the different tasks will solve a specific problem. However, for simplicity, we will limit our exploration to the interaction of two algorithms. Structurally, we can imagine that one algorithm works in support of another algorithm, effectively functioning as a sub-algorithm. An example for such a hybrid structure is when one algorithm finds the optimum operating parameters of another algorithm. Here, the latter algorithm is responsible for solving the stated problem and the former algorithm accounts for efficiency and accuracy. This could be a particle swarm optimization (PSO) algorithm finding the optimum hyperparameters of a genetic algorithm (GA), such as mutation rate, initial population size, etc. Hence, the PSO functions in support of the GA as a sub-algorithm. Another structure used in the composition of hybrid intelligent systems is when all sub-algorithms are working to solve the main problem directly but are employed at different stages of the course of solving the problem. An example of such a structure is when one algorithm is responsible for finding the vicinity of the global optimum while another algorithm utilizes the results of this search and refines the search in this vicinity to find a precise location of the global optimum.

Unfortunately, hybridization can also result into degradation of the performance of the resulting intelligent system. When we compose a hybridized system, we hope to utilize the advantages of one algorithm to mitigate the disadvantages of the other, and *vice versa*. However, the effects of individual intelligent systems are not always mutually exclusive when combined and hence may not have the desired outcome. An example of a potential degradation in effectiveness is found when we combine two evolutionary algorithms to form a new algorithm. Such evolutionary hybrid systems are used in solving optimization and estimation problems but also could perform controls tasks. In general, objectives for such applications can be stated as a global optimization problem. If the problem is highly nonconvex we may likely encounter two issues: premature convergence and slow convergence of the hybrid evolutionary optimization algorithm. Premature convergence could result in a solution that has low accuracy but whose location in the search space is in the vicinity of the global optimum point. Slow convergence may cause excessive computational time, or nonoptimum solutions. Both of these issues can be related to population diversity of the hybrid evolutionary system.

Population diversity allows for the evolutionary search algorithm to explore different spaces of the search area. However, when premature convergence occurs, the population reproduction mechanism has the effect of producing a set of similar genotypes during the first few generations of the algorithm. This leads to the case where the algorithm has not yet had the time to fully search and explore the entire search space. In effect, the population of the hybrid evolutionary algorithm becomes redundant and may behave similarly to a local search algorithm. However, an evolutionary algorithm that maintains a diverse population by including dissimilar traits and elements in the population allows for the algorithm to properly explore the search space. Unfortunately, having an evolutionary algorithm with a highly diverse population also leads to slow convergence. The reason for slow convergence is the lack of a mechanism that is part of the algorithm and is responsible for migrating the focus of the search to specific areas in the search space after the exploration phase has completed. This process is referred to as intensification. The opposite is true too: a hybrid evolutionary algorithm possessing a low population diversity may lead to faster convergence. However, this comes with an increased risk of finding a local optimum rather than a global optimum solution.

Example 7.1

To illustrate hybridization, let's consider the PSO and tabu search (TS) algorithms as applied to a minimization problem. In addition, we assume a sequential structure where TS is the sub-algorithm in support of the PSO algorithm. PSO is governed by Equations (5.2) and (5.3). These equations make use of the weighting coefficients c_i and the constriction factor χ, whose responsibility is to regulate the velocity magnitude of the particles. The weighting coefficients have the function of balancing the exploration and intensification part of the PSO and have been reported to be problem-dependent [1]. The literature [2] suggests that for most problems one chooses $c_1 = c_2$, in essence weighting equally the influence of the particle's best solution and the influence of the swarm's best solution. However, if more information about the cost surface is available, then recommendations are such that for unimodal optimization problems the literature suggests choosing $c_2 > c_1$ and for multimodal problems $c_1 > c_2$. Since we are discussing hybridization, we can consider the case when the weighting coefficients of the PSO are made adaptive to the problem during the search. For this, we employ a second optimization algorithm – the TS algorithm – to be responsible for finding the instantaneous optimum values of the weighting coefficients during the PSO process. The hybrid algorithm is graphically represented by Figure 7.1.

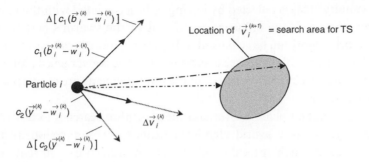

Figure 7.1 Hybridization of a PSO algorithm with a TS algorithm to find the optimum influence coefficients for the PSO algorithm.

The three influence vectors of the PSO for a single particle are shown with the current location of the individual particle under consideration. The PSO will compute the updated particle position using these influence vectors. However, before moving the particle to this new location, each of these influence vectors – representing the influence coefficients – are optimized using the routine provided in Table 5.2. The sub-algorithm is employed at each iteration and for each particle, searching each time an n-dimensional space in the vicinity of the solution the PSO algorithm implied. The TS algorithm searches this vicinity by allowing the weighting factors c_1, c_2 and the magnitude of $\vec{v}_i^{(k)}$ to have some predefined range in terms of variation. This range is searched by the TS algorithm. The new particle position $\vec{v}_i^{(k+1)}$ is found by utilizing the optimum values of these weighting factors.

To test our proposed hybrid optimization algorithm, we choose a multimodal test function – that is, the hyperellipsoid function – which can be given as:

$$f_o(\mathbf{w}) = \sum_{j=1}^{n} j^2 w_j^2, \tag{7.1}$$

where $w_j \in [-1, 1]$ and the optimum location is $f_o(\mathbf{w}^*) = [0, 0]$ for $n = 2$. As the PSO–TS hybrid algorithm is an iterative algorithm, we have the opportunity to observe the changes occurring within the algorithm by investigating the magnitude of the influence coefficients during the simulation.

Recall from our earlier discussion that for such a multimodal cost function the coefficients should start out with a relationship such that $c_2 > c_1$. By plotting the influence coefficients over the course of the optimization, we can recognize that this is the case, as shown in Figure 7.2(a). Note that the magnitude of each of the influence coefficients started with the same value. However, again considering Figure 7.2(a), we notice that after a few iterations of engaging the hybrid algorithm, the magnitude of these coefficients changes and their relationship inverts such that we have $c_1 > c_2$.

A possible interpretation of these results is that our algorithm has completed the exploration part of the search early during the simulation and is switching to the intensification portion of the algorithm to find a more precise optimum location within the current vicinity. This is reflected by having c_1 be associated with the particle's best past position while c_2 is the influence coefficient of the swarm's best past position. Hence, our hybrid algorithm has changed to focus on a more local search after starting out on a global search. This inference is supported by the convergence plot, as shown in Figure 7.2(b), where we notice that most of the improvements are accomplished within the first few iterations.

Hence, the exploration phase is over and the algorithm searches locally for the optimum. Effectively, our new hybrid algorithm treats the cost function as a unimodal function once it starts searching locally and switches the influence coefficient's magnitudes accordingly.

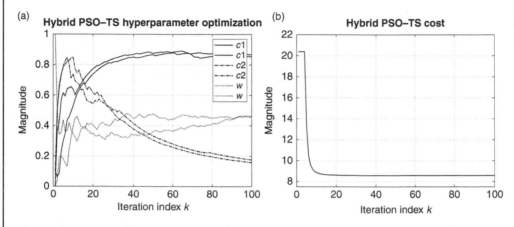

Figure 7.2 Hybridization of PSO and TS: (a) influence coefficients during optimization iteration; (b) convergence plot of cost function value.

Although Example 7.1 is based on the material presented in Chapter 5, this chapter will cover material that is mostly independent of those topics. In the present chapter we will select a few hybrid algorithms that have some potential to contribute to the field of modeling and control. In particular, we will look at neural expert systems, neuro-fuzzy systems, and adaptive neuro-fuzzy inference systems. Much of the material covered in this chapter requires knowledge of the material presented in Chapters 2 and most of Chapter 3, as well as Chapter 6.

7.2   Neural Expert Systems

Neural expert systems combine logical inferences of expert systems with data processing done using neural networks. In contrast to what we discussed in Chapter 6 regarding the inspiration for neural networks by the study of the human brain, an expert system is guided by the modeling of human reasoning. The expert system maps observations into a knowledge set described by

IF–THEN rules. And unlike a neural network, an expert system is not capable of learning through experience without human interaction. However, because the embedded knowledge – as given by IF–THEN rules – can be organized into individual rules or set of rules, interpretability of the results is possible. We know from our discussion in Chapter 6 that such interpretability is not accessible when working with neural networks, as knowledge is integrated into the entire network rather than in some individual parts of the network.

The objective of the hybridization between neural networks and expert systems is to create a neural expert system (NES) that allows for learning without human interaction, and which can present results that can be interpreted. The combination of the expert system's and neural network's advantages seems to be complementary, and a suitable structure is needed to ensure these properties are preserved.

Figure 7.3 depicts the general structure of an expert system, a neural network system, and – through combination of the two systems – the resulting NES. For an NES, the knowledge base resides in a trained neural network. However, the rule extraction mechanism is still utilized by examining the neural network output. The inference engine is responsible for applying logical rules to the knowledge

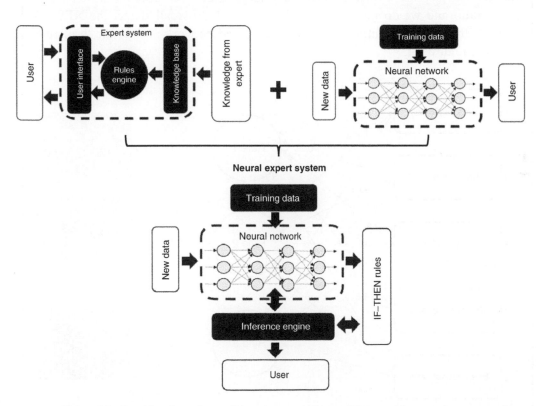

Figure 7.3 Combination of a traditional expert system with a neural network system to form a neural expert system.

base contained in the neural network in order to infer the new information. By employing a neural network, the NES is capable of using imprecise or noise-contaminated data to infer imprecise conclusions, similar to fuzzy logic. This imprecision is often referred to as approximate reasoning. This is different from an expert system, where precise matching is required to provide an inference.

Example 7.2

To illustrate the functioning of an NES, let us consider a simple classification example. Suppose you are presented a leaf from a tree. The goal of our NES is to be able to tell on which species of tree a random leaf grew. Suppose we limit our leaf catalog to a European beech, a red oak, and a Norway maple. For simplicity, we limit the input to a few characteristics. Hence, the input to our neural knowledge base is given by the recognition and determination of the following properties: "sharp spikes," "round shape," "large spikes," "elongated shape," and "even aspect ratio." We will use numerical values for the association of the given properties as (+1) for "true," (−1) for "false," and (0) for "unknown."

Also assume we trained a neural network based on a large data set that resulted in a set of weights, which are shown in Figure 7.4. If the input is characterized as a leaf that has sharp spikes, elongated shape, and large spikes, but does not have an even aspect ratio and we do not know if it has a rounded overall shape, the NES will compute the following outputs based on the trained network and on the three rules shown in Figure 7.4:

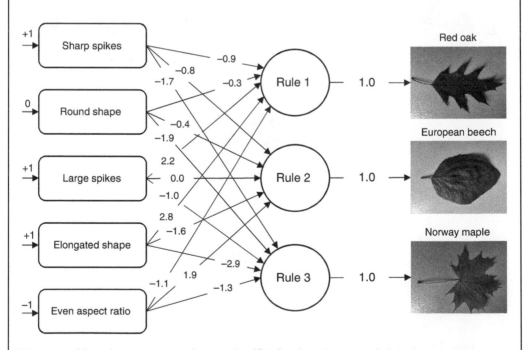

Figure 7.4   Neural expert system for tree classification based on sample leaves.

Rule 1: $y_1 = 1 \times (-0.9) + 0 \times (-0.3) + 1 \times (2.2) + 1 \times (2.8) - 1 \times (-1.1) = 5.2,$

Rule 2: $y_2 = 1 \times (-0.8) + 0 \times (-0.4) + 1 \times (0.0) + 1 \times (-1.6) - 1 \times (1.9) = -4.3,$

Rule 3: $y_3 = 1 \times (-1.7) + 0 \times (-1.9) + 1 \times (-1.0) + 1 \times (-2.9) - 1 \times (-1.3) = -4.3.$

Our inference engine can be constructed in a rather simple fashion to accommodate the stated problem. For example, we compute the weighted output from each rule and assign "TRUE" to any value greater than 0 and "FALSE" to any outcome with a value smaller than 0. Hence, looking at the above computations and resulting rule values, our example classified the given leaf as belonging to a red oak.

Example 7.2 made use of a simple neural network. We could greatly expand on this by making use of the material covered in Chapter 6 for multilayered neural networks, and extend the range of tree types. However, the presented mechanism is the same. Note also that the input neurons are associated with a questionnaire where the user is asked about the properties of the leaf to be classified. As it is not necessary to actually have a leaf in front of the user, but just the characteristics, not all of the information is always needed, and hence we use an approximation of the characterization for the classification task. In our example not all of the information was available as we did not identify if the presented leaf has a rounded shape. This lack of precision makes NES more effective than compared to regular expert systems. Also, the questionnaire employed provides for interpretability of the results. These are two beneficial properties we gained by combining two algorithms into one hybrid algorithm.

7.3 Neuro-Fuzzy Systems

In the previous section we combined a simple expert system with a neural network. A more sophisticated hybrid system can be constructed by utilizing fuzzy logic with neural networks as fuzzy logic and neural networks are naturally complementary intelligent systems. The neural network portion provides for means to deal with raw data while fuzzy logic contributes a method for reasoning. The reasoning is accomplished by using linguistic information that originates from some domain knowledge or experts. Constructing solely the fuzzy logic part would result in a static model, as no learning is built into the algorithm. However, if combined with a neural network the hybrid system gains the ability to learn from new data and new environments. Considering a reverse supporting role by utilizing fuzzy logic for a neural network, the neural network gains transparency. In general, a neuro-fuzzy (NF) system can be trained to define the different IF–THEN rules of a fuzzy inference system. In addition,

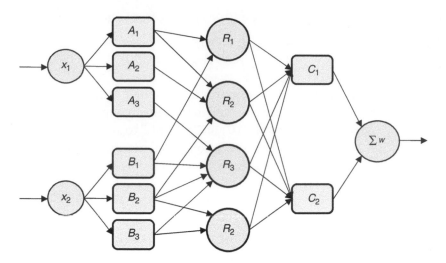

Figure 7.5 Structure of a typical neuro-fuzzy system.

such training can shape the membership functions for the input and output variables. The resulting structure of an NF system resembles a multilayered neural network. Typical NF systems have an input and an output layer and three hidden layers. The hidden layers function as fuzzy rule implementations. An example of such a typical NF system is shown in Figure 7.5.

The first layer of an NF system is responsible for distributing the crisp input to the second layer. The distribution may include a weighting function, but commonly the weight is set to unity. The second layer is composed of fuzzy sets which takes the crisp inputs from the first layer and evaluates the degree of membership by applying the crisp input to a membership function. The output of this evaluation will generate a value in the range $0 \leq \mu \leq 1$.

Just as we used different membership functions in our discussion of fuzzy logic systems, layer 2 can employ any type of membership function. In our example, as shown in Figure 7.5, we have a set A of membership functions and a set B of membership functions. The membership function is equivalent to an activation function in a traditional neural network. The output of the second layer of the neural network is transmitted to the third layer without weighting – that is, the degree of membership as a value μ is transmitted. Layer 3 incorporates the different fuzzy rules of the fuzzy logic system. Each rule is embodied as a single neuron in the NF system. Figure 7.5 depicts an NF system with four rules, hence layer 4 has four neurons. However, any number of rules can be implemented. Unlike the weights, the connectivity between layers 2 and 3 is part of the neural network training. As depicted in Figure 7.5, we imply that the trained neural network allows rule R_2 to compute its output based on membership functions A_1, A_2, and B_2. The output of R_2 is computed by using the product operator, given by:

$$y_{R_2}^{(3)} = \mu_{A_1} \times \mu_{A_2} \times \mu_{B_2}. \tag{7.2}$$

Layer 4 represents the output membership layer of the hybrid NF system. Layer 4 computes the consequence of the fuzzy rules. The output membership function C – as depicted in our NF schematic in Figure 7.3 – utilizes the UNION operator – that is, the probabilistic OR operator. For example, the output of layer 4 is computed as:

$$y_{C_i}^{(4)} = \mu_{R_1} \cup \mu_{R_2} \cup \mu_{R_3} \cup \mu_{R_4} = \max\{\mu_{R_1}, \mu_{R_2}, \mu_{R_3}, \mu_{R_4}\}. \tag{7.3}$$

The output from layer 4 is essentially computed by determining the integrated firing strength of the respective connected rule neurons. Hence, the last task we need to accomplish to obtain a useful output of the NF system is to generate a crisp output. Layer 5 acts as the defuzzification layer. Any of the detailed defuzzification computation methods introduced in the corresponding sections of fuzzy logic – found in Chapters 2 and 3 – can be utilized to accomplish the task of generating a crisp output.

Having an understanding of the basic structure of an NF system, the question arises of how to train such a hybrid system. Since the NF is in essence a neural network system with fuzzy logic components, standard neural network learning algorithms can be utilized for the training of the hybrid NF system. A popular choice for training the hybrid system is the backpropagation algorithm, as discussed in Chapter 6. For this, we make use of the domain knowledge by incorporating IF–THEN rules. These rules are derived from the knowledge we have of the system we intend to model. Since such prior knowledge can be biased or faulty, the training of the NF system must be able to identify rules associated with this biased or inaccurate domain knowledge. Also, the parameters of the membership functions employed by the NF system can be part of the training process.

MATLAB provides for an easy-to-use NF toolbox. However, this toolbox utilizes a Sugeno fuzzy model. The literature provides for some implementation of the discussed hybrid NF system using a Mamdani fuzzy inference model [3]. Since the MATLAB GUI for NF incorporates some adaptive features, we will discuss such systems first in the next section before we look at the implementation of NF systems in MATLAB.

7.4   ANFIS: Adaptive Neuro-Fuzzy Inference System

Recall the Sugeno fuzzy inference system from Section 3.2, where the rule consequent is accomplished by utilizing a function that yields a crisp value. We distinguished between zero-order Sugeno fuzzy models, where singletons were employed, and higher-order Sugeno fuzzy models, where the order referred to the order of the polynomial equation $f(x)$ of the consequent. The format of a Sugeno fuzzy inference system was given as:

Rule 1: IF x_1 is A_1 and x_2 is B_1 THEN y is C_1
    IF x_1 is A
    AND x_2 is B
    THEN y is $f(x)$

where x_1, x_2, and y are the linguistic variables, A and B are the fuzzy sets used for modeling the system, and $f(x)$ is a function of x serving as the equation of the consequent. To incorporate this into an NF system we can alter the structure given in Figure 7.5 by changing the interaction between layers 4 and 5. This alteration is shown in Figure 7.6, where we included the Sugeno-type consequent evaluation.

Note that the ANFIS structure has six layers. The additional layer – as shown in Figure 7.6 – contains a normalization process. Let's introduce each layer in more detail, including the new normalization layer:

Layer 1 The first layer functions in the same fashion as an NF system; it is the input layer and its responsibility is to transport the crisp input to layer 2.

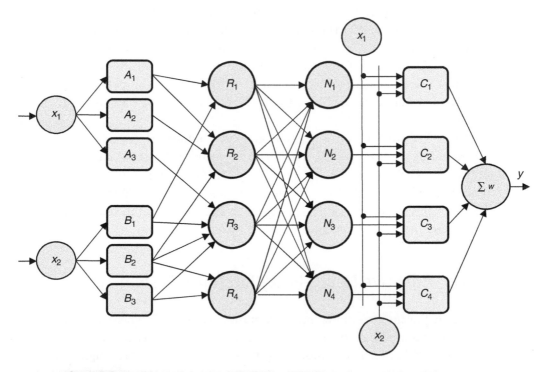

Figure 7.6 Structure of a typical neuro-fuzzy system.

Layer 2 The second layer acts as the fuzzification layer, where each neuron houses a membership function and evaluates the degree of membership, which is set as the output of layer 2. Note that the membership function is in place of an activation function to perform the computation of the output of these neurons.

Layer 3 This layer incorporates the Sugeno-type fuzzy rule set. Here, each neuron holds a rule that receives a fuzzified input from the prior layer and computes the firing strength of the given rule.

Layer 4 Since layer 4 receives an input from each of the neurons of the prior layer, some normalization is needed. Layer 4 computes a normalized firing strength for a given rule. The magnitude of this firing strength is computed by evaluating the ratio of the firing strength of a particular rule to the firing strengths of all rules combined:

$$y_{N_i}^{(4)} = \frac{\mu_i}{\sum\limits_{j=1}^{p} \mu_j}, \tag{7.4}$$

where p is the number of rules.

Layer 5 The objective of layer 5 is to compute a crisp output, hence it is called the defuzzification layer. Here, each neuron is connected to the corresponding normalization neuron. It also has access to the polynomial coefficients contained in vector $\vec{x}$. The output is computed as follows:

$$y_i^{(5)} = y_i^{(4)}(a_{i_0} + a_{i_1}x_1 + a_{i_2}x_2). \tag{7.5}$$

Here, a_{i_j} are the coefficients of the consequent given by rule i.

Layer 6 In layer 6 we compute the sum of all outputs from layer 5, which yields the overall output of the ANFIS system:

$$y = \sum_{i=1}^{p} y_i^{(4)}(a_{i_0} + a_{i_1}x_1 + a_{i_2}x_2). \tag{7.6}$$

An ANFIS system can be a powerful tool for modeling and control tasks. In order to make use of this system, an effective and efficient implementation is necessary. Rather than developing code on our own, for this modeling MATLAB provides a tool as both a command line function and as an interactive application. We will explore both implementations in the next section along with some nuances and details so far not covered.

7.5   ANFIS in MATLAB

We start with a modeling task, where a given set of input and output data is available from some experiment. Our goal is to find a model that captures the underlying characteristics given by the data set. We can make a distinction of whether we have some insight into the behavior of the system we want to model or not. This inside information is generally referred to as domain knowledge. If such knowledge exists we may be able to find a suitable model structure and have sufficient information to form the rule base for the fuzzy inference system. For our discussion purposes, let's assume we do not have such insight and we want to use MATLAB's ANFIS tool to develop a model for the given data set [3]. Composing IF–THEN rules for a system we do not have domain knowledge for is generally difficult. The same challenge presents itself when working with a complex system where we have some or only partial knowledge of the system interactions. Researchers have investigated this problem and proposed a number of solutions. One of the more popular methods to deal with this unknown is to use a partitioning of the system inputs [4]. Considering a fuzzy rule, the antecedent of such a rule defines a narrow fuzzy region, while the consequent models the behavior within that region. The consequent modeling is accomplished by using membership functions for the Mamdani and Tsukamoto fuzzy system, and a zero- or first-order linear equation for the Sugeno fuzzy system. There are three popular partitioning methods: grid partition, tree partition, and scatter partition. A symbolic representation of these partitioning approaches is given in Figure 7.7.

For the grid partitioning method, square areas – or hypercubes for high-dimensional input spaces – are constructed. One disadvantage of the square partitioning method is that the number of hypercubes and therefore the number of rules become very large if the system has many inputs. For example, if there are n_o inputs and n_r rules, using the grid partitioning method we will effectively have $n_r^{n_i}$ IF–THEN rules. Suppose your system has four inputs and five membership functions for each input, the grid partitioning approach will yield $5^4 = 625$ rules.

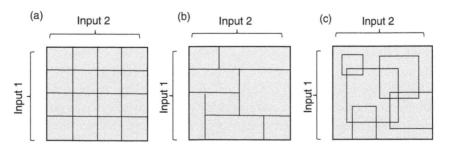

Figure 7.7 Popular partitioning methods for fuzzy system input space modeling: (a) grid partitioning, (b) tree partitioning, and (c) scatter partitioning.

Using the tree partitioning approach will reduce the number of rules but it will require more membership functions for each input. The disadvantage of the tree partitioning approach is that we lose some of the linguistic meaning that membership functions bring to the system and hence the resulting fuzzy system is less descriptive. Tree partitioning is accomplished by making random cuts across the subspace to be partitioned. These cuts are applied in sequence, so effectively at the $(n - 1)$ iteration the space is cut into n regions. The next cut is executed on these regions to create $(n + 1)$ regions, and so on.

For the scatter partitioning approach, shown in Figure 7.7(c), each region is defined by layering a subset of the entire input space, called a hyperbox. This is accomplished by taking the input space and dividing it into a set of intervals (temporally). The hyperboxes are defined by the input region associated with the corresponding output interval and by determining the extremum values of the input data for each output interval. Overlapping regions are called inhibition hyperboxes, which are resolved by defining new hyperboxes.

We will make use of MATLAB's built-in option for creating a partition through MATLAB's ANFIS tool and create a Sugeno-based system that has a single output. As shown in Figure 7.6, for an ANFIS structure the number of rules is equal to the number of output membership functions. MATLAB's ANFIS structure employs the same architecture, where each rule is associated with an output neuron, while using unity weights for each rule. To achieve such a model using MATLAB we can use the GUI MATLAB provides, or a command line approach in MATLAB. We will start with the command line approach first in order to introduce a few concepts we have yet to discuss.

Example 7.3

We start with a simple data set consisting of 100 data points to model an ANFIS system. The data set is generated by the following MATLAB code:

```
u = (0:0.1:10)';
y = cos(3*u)./exp(u/4);
```

For this purpose, we will set up an initial structure of the ANFIS system. This structure will utilize grid partition and five Gaussian membership functions. The MATLAB commands to set these options are:

```
genOpt = genfisOptions('GridPartition');
genOpt.NumMembershipFunctions = 5;
genOpt.InputMembershipFunctionType = 'gaussmf';
```

Next, we create the fuzzy inference system using the MATLAB routine called genfis:

```
inFIS = genfis(u,y,genOpt);
```

The data set for the training is contained in the input data vector u and the output data vector y generated with the above MATLAB code. Note, both vectors are row vectors to ensure that genfis() is able to recognize the correct number of inputs (there is only one output allowed).

Next, we need to establish the training options. As this is the initial training, we can suppress most of the output options ANFIS has built in automatically. The following MATLAB code establishes the initial training options:

```
opt = anfisOptions('InitialFIS',inFIS);
opt.DisplayANFISInformation = 0;
opt.DisplayErrorValues = 0;
opt.DisplayStepSize = 0;
opt.DisplayFinalResults = 0;
```

The first option pairs the initial fuzzy inference system we generated with genfis(), called inFIS, to the InitialFIS for training. The next four lines are display settings such that no response appears on the screen after executing the training. There are other parameters that one can specify – for example, the number of epochs for the training. Additionally, some of the options can be bundled into one command. As an example, if we want to specify an ANFIS system with five membership functions and 50 epochs, the command would read as:

```
opt = anfisOptions('InitialFIS',5,'EpochNumber',50);
```

The training using the specified options is performed by the following command:

```
outFIS = anfis([u y],opt);
```

We can assess the outcome of this training by comparing the output of the trained fuzzy inference system model with the given data set:

```
y_sim=evalfis(outFIS,u);
errory=y-y_sim;
figure
plot(y_sim,'k-');hold
plot(y,'k-');
plot(errory,'k:');
legend('ANFIS output','Original output data','Error')
xlabel('Index k')
ylabel('Magnitude')
title('ANFIS simulation vs. data')
```

The results are shown in Figure 7.8(a). The ANFIS system captures the basic characteristics, but fails to be accurate. Improvements can be made by tuning the system. For example, if we increase the number of membership functions from 5 to 20, we can model the output of this system rather accurately. This is shown in Figure 7.8(b).

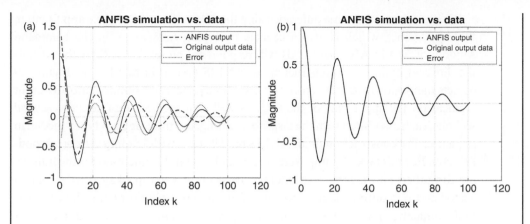

Figure 7.8 Trained system output and original output data as well as error between predicted and actual output: (a) using five membership function and (b) using 20 membership functions.

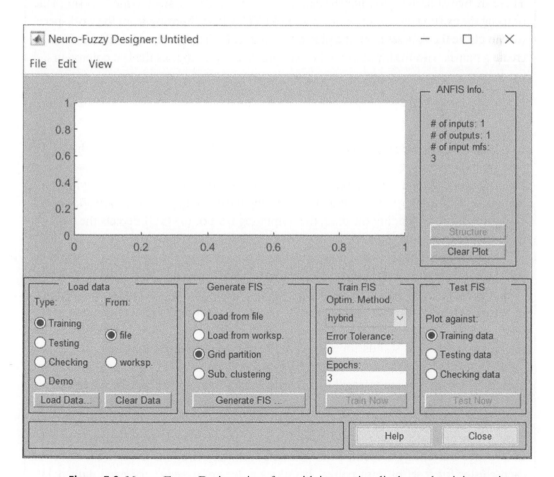

Figure 7.9 Neuro-Fuzzy Designer interface with interactive display and training options.

We can achieve the same outcome using the Neuro-Fuzzy Designer app found in the Apps panel, which is also accessible by typing `anfisedit` in the command window of MATLAB. MATLAB will respond by presenting a GUI that details the different options for constructing an ANFIS system in MATLAB. Figure 7.9 depicts the Neuro-Fuzzy Designer app GUI. The panel is partitioned into an output segment on the upper part of the GUI as well as a segment that contains several sections, each associated with the tasks for creating and training an ANFIS model, in sequence. The construction of an ANFIS model starts with loading the corresponding data. For this task, it is important to remember that the data used for the training are formatted into a table that contains each input value packaged as a column vector and one output data vector placed in the last column of this data matrix. Unfortunately, MATLAB's ANFIS models can only handle one output, regardless of whether it is created using the command window or the GUI environment.

Example 7.4

There are two options for loading the data we want the ANFIS system to model: using a file or using the existing data in the workspace in MATLAB. If choosing from the workspace, we can utilize the data set from the previous example. In order to comply with the format, create a matrix with the input as the first column and the output as the last column in the workspace:

```
traindata=[u y];
```

With this matrix we can load the data using the radial button for *worksp.* on the GUI's first input panel and designate them as the training data. If a second set of data stemming from the same system is available, use these data as the checking data. Checking data helps with determining if overfitting occurs during the training part of the ANFIS model construction. After loading the data, the output segment of the GUI depicts the data we have available for training and possibly for checking an ANFIS system.

To generate the fuzzy inference system and the corresponding rule base, we have options to use a previously designed fuzzy logic system – perhaps by following the steps outlined in Chapter 3 – or to create automatically a set of rules by using partitioning. One of the partitioning methods offered is the *grid partition* method, which we utilized in the prior example. Another option the GUI provides for this task is the *sub. clustering* method for partitioning. This stands for a process that uses subtractive clustering. Generally, clustering refers to the process where we assign a set of observations – or data – to different groups. These groups are referred to as clusters. The criterion behind the assignment is that the data in one cluster have some trait in common. Most clustering methods either use a distance measure or a conceptual criterion as the trait and hence for assigning an observation to a cluster. Clustering is a major topic in data mining and machine learning. Subtractive clustering is a method that estimates the number of clusters and the center of each cluster for a set of data points. It utilizes a density measure for each data point. Suppose we are looking at a single data point x_i; the density measure for this point is computed as:

$$D_i = \sum_{j=1}^{n} \exp\left(-\frac{\|x_i - x_j\|^2}{\left(\frac{1}{2}r_a\right)^2}\right), \tag{7.7}$$

where n is the number of data points and r_a geometrically defines the neighborhood radius for the cluster. Equation (7.7) implies that we are using a distance measure approach and D_i represents the potential of x_i being the center of the cluster. The subtractive clustering algorithm starts by assuming the first cluster center to be the point $x_1^{(c)}$ that has the largest density value $D_{c1}(x_1) \geq D_i(x_i)$.

The next step is to update the density measure for each data point by utilizing $D_{c1}(x_1)$ as follows:

$$D_i = D_i - D_{c1}\exp\left(-\frac{\|x_i - x_1^{(c)}\|^2}{\left(\frac{1}{2}r_b\right)^2}\right). \tag{7.8}$$

In Equation (7.8) we changed the original radius r_a to a new radius r_b whose associated density measure is smaller. Sequentially, we select the next cluster center by choosing the data point with the largest density value. We continue this selection process until a sufficient number of clusters is obtained.

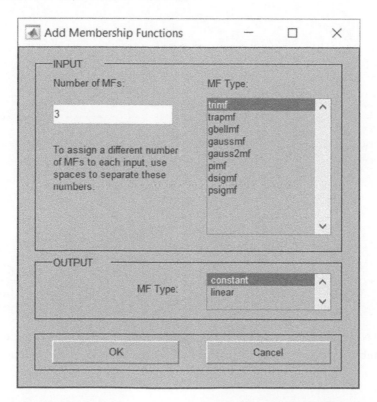

Figure 7.10 ANFIS GUI: FIS editor to allow for altering the number and type of membership functions for the input.

For our ANFIS example using the GUI, let's choose either the *grid partition* or *sub. clustering* method and proceed by pressing the button *Generate FIS …* to construct the initial ANFIS model. The GUI will respond by presenting a new window that allows us to tailor the fuzzy inference system. The window is depicted in Figure 7.10. Notice that the *Add Membership Functions* window's OUTPUT panel provides an option to distinguish between zero- and first-order outputs. This corresponds to the Sugeno-type consequent evaluation, occurring in layer 6 of the ANFIS structure. For our example, choose five Gaussian membership functions and a linear MF-type output. At this point we have created the initial structure of the ANFIS system. Although the system is not trained yet, we can visualize this model by invoking the *Structure* button in the ANFIS Info section. The resulting model is shown in Figure 7.11.

We can investigate the generated rules as well as plot the different membership functions by utilizing the *Edit* function in the Neuro-Fuzzy Designer GUI. For example, to inspect the automatically generated rules, go to *Edit* and select *rules*. The rules for the above-listed parameter settings are shown in Figure 7.12. Note, that by using the rule viewer we not only have access to see the rules, but we can also modify, add, or delete them.

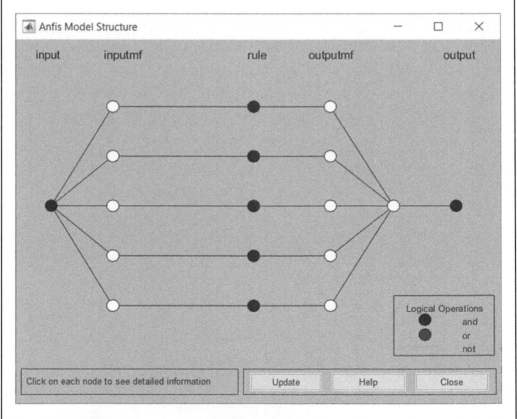

Figure 7.11 ANFIS GUI: generated ANFIS structure with one input and one output, five Gaussian membership functions for the input, and a first-order equation for the output evaluation.

Our next step is to train the created ANFIS model structure using the given data set. For this purpose we have different options offered by the third segment in the lower Neuro-Fuzzy Designer GUI toolbox window. The optimization of the neural network as constructed with membership functions for activation functions can be done by using backpropagation learning, which is detailed in Chapter 6. An alternative method is the *hybrid* method. The hybrid method also uses backpropagation for adaptation of the parameters describing the input membership function. However, since the output is either a constant (zero-order) or a linear (first-order) function, the associated coefficient(s) can be estimated using a least-squares estimation approach. The hybrid option uses the backpropagation method for the input membership function, while for the output equation a least-squares regression method is used.

Additionally, we can specify the training goal by setting the *error tolerance* to some numerical value. The number of epochs is also defined in this step. If provided with a second set of data in the first panel under *Load Data* and selecting *Checking*, the ANFIS tool will train the fuzzy inference system and plot the training error denoted by little stars, as well as the checking error, given by dots at each epoch during the

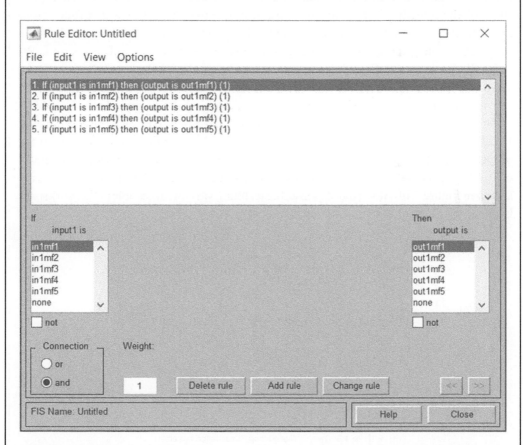

Figure 7.12 ANFIS GUI: generated rule set using the rule viewer from the *Edit* function of the GUI.

training. This feature is useful in detecting overfitting. Overfitting will manifest itself in this plot when the checking error plot first decreases and at one point starts to increase. The instance – in terms of epochs – when the error plot magnitude increases is considered the point when overfitting starts taking place. Fortunately, if provided this additional data set, the ANFIS app will pick the model associated with the best fit prior to overfitting.

As we used fuzzy logic to design controllers, the question arises on how we can use an ANFIS model, which incorporates fuzzy logic, to solve a control problem. To illustrate this case, we will utilize Simulink to showcase the third environment for ANFIS to operate in. However, we will make use of the ANFIS GUI to construct and train an ANFIS block prior to deploying it in the Simulink environment.

Example 7.5

For this example we will consider a simple second-order system possessing a natural frequency of $\omega_n = 1.581$ rad/s, a damping ratio of $\xi = 0.3162$, and a static sensitivity of $K_s = 0.4$. The system is underdamped and when experiencing a unit step input the steady-state error is rather large (i.e., about 60%). Hence, a controller with an integral action on the error would help with the performance. However, ANFIS does not have explicitly built in a structure that allows us to tune a component that is solely responsible for the integration. Fortunately, we can utilize the data from such a function to train an ANFIS model. Consider Figure 7.13, where we have the Simulink block diagram for the given system operating as an open-loop system as well as a simple PI-controlled system. The Simulink program is constructed such that we collect the three separate signal sequences: the error signal, the integral error sequence, and the PI (proportional integral) controller action signal actuating the system plant.

This data are sent to the workspace and made available for us to use as a training data set. Recall that the last column of the training data for ANFIS training corresponds to the output signal. Since ANFIS will work as a controller, the last column should be the PI controller output data. The open-loop response and the simple PI closed-loop response are shown in Figure 7.14. Along the response, the resulting variable accessible in the workspace is shown symbolically and as a variable that was renamed as `trainingdata`.

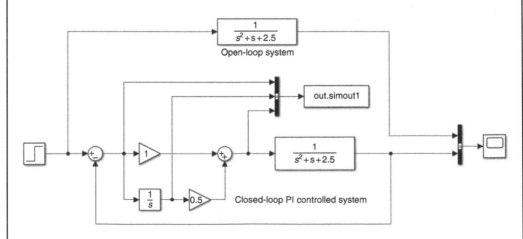

Figure 7.13 Simulink program for simulating the system in an open loop and in conjunction with a simple PI controller.

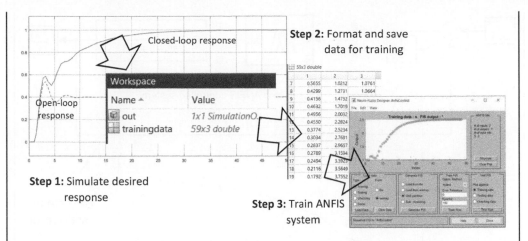

Figure 7.14 Generating data for ANFIS training: using the closed-loop response data from the workspace and ANFIS GUI to train a controller that acts like a PI controller.

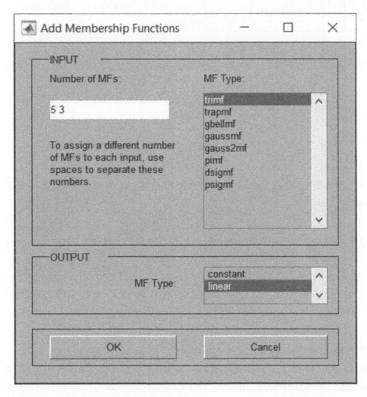

Figure 7.15 ANFIS GUI interaction for selecting membership function. Note, the input has to accommodate two different signals: the error signal and the integration of the error signal.

When proceeding with the generation of the ANFIS structure, the user will notice that the number of membership functions is presented as two separate numbers. This is shown in Figure 7.15, where the numbers 5 and 3 appear separated by some space.

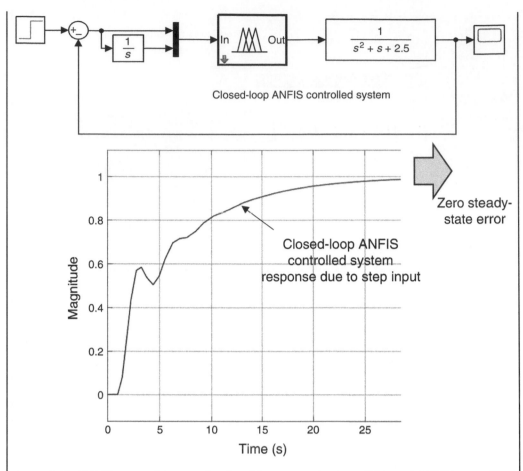

Figure 7.16 ANFIS controller example: closed-loop control using a PI-based ANFIS controller, resulting in zero steady-state error.

Along with our specification of the number of membership functions, we also need to select the type of membership functions and the type of output, as shown in Figure 7.15. Upon defining the training parameters, an ANFIS model is created by the training process. It is good practice to inspect the rules by using the rule viewer and the resulting membership function distribution. For our example, using 100 epochs and exporting the resulting ANFIS system to the workspace, we can construct a closed-loop system in Simulink by implementing a *fuzzy logic controller* block and specifying the exported ANFIS system in the dialog panel of the block parameters. Figure 7.16 depicts the implemented ANFIS controller and the resulting closed-loop performance. The latter matches the desired response provided during the training process of ANFIS and shown in Figure 7.15.

7.6   Conclusions

In heuristic optimization we often attempt to balance the exploration and the intensification phases during the search process. As we found in Chapter 5, some of the heuristic algorithms have embedded mechanisms to address this balance. However, an alternative approach is to combine other heuristic methods to partition the responsibilities of finding optimum locations in the feasible search area and finding the optimum hyperparameters of the search algorithm itself. This dual optimization can be done adaptively, as we have shown in our example of a hybrid particle swarm–tabu search algorithm. Such hybrid algorithms are just as useful beyond solving optimization problems, where we essentially combine two or more intelligent algorithms to solve a specific problem. For example, neural networks may find a prominent role in providing an automated learning mechanism that can be incorporated with another intelligent structure to form an expert system. The neural network's functionality is paired with a logical inference system represented by IF–THEN rules when we work with neural expert systems. Generally, the relationship between two or more intelligent systems working as a hybrid system can be of primary–subordinate type, where one algorithm is supporting the other, or equal-partner type, where each algorithm works in sequence with the other.

In controls, optimization and adaptation along with system modeling are key concepts that sometimes necessitate being addressed in circumstances that do not allow for the use of conventional hard-computing methods. Our discussion of neuro-fuzzy systems led to adaptive neuro-fuzzy inference systems (ANFIS), which possess inherently those capabilities of optimizing and adapting. Hence, ANFIS systems have been rather popular control structures. We arrived at the ANFIS structure by expanding the neuro-fuzzy system from a five-layer network to a six-layer network. ANFIS incorporates a Sugeno fuzzy inference system, which acts as its logical inference system, with a neural network that functions as the data processing unit.

However, until now we only utilized regular neural networks (i.e., networks that had one hidden layer). As more capabilities are achieved by adding multiple hidden layers to a neural network, and hence constructing a deep neural network, we will explore such systems in the next chapter. Adding more layers to a neural network generates a new set of problems. However, if addressed, deep neural networks can also be hybridized in a similar fashion as presented in the current chapter.

SUMMARY

In the following, some brief descriptions of some of the key concepts in hybrid intelligent systems are presented.

Neural Expert System

A neural expert system is constructed by utilizing a logical inference system (expert) with a data processing system (neural). The expert system uses IF–THEN rules to model observed knowledge. The neural part is given by a neural network that is trained with data representing the system we are trying to model.

Neuro-Fuzzy System

A neuro-fuzzy system is constructed by combining a fuzzy logic inference system with a neural network. The neural network is responsible for raw data, while the fuzzy logic inference system provides for the reasoning function of the neuro-fuzzy system. Typical neuro-fuzzy systems have at least five layers: one input layer, three hidden layers, and one output layer. The hidden layers accommodate the fuzzy rule implementations.

Adaptive Neuro-Fuzzy Inference System

The adaptive neuro-fuzzy inference system ANFIS is a special version of a neuro-fuzzy inference system, where the Sugeno fuzzy inference system is utilized for the rule-base implementation. The ANFIS system's architecture consists of six layers. The first layer is the input layer, the second layer is the fuzzification layer, and the third layer holds the Sugeno-type fuzzy rule set. Each neuron in this layer represents one rule. Layer 4 normalizes the output from the rule layer, while layer 5 computes the crisp output value for each rule. Finally, layer 6 combines all of the crisp outputs to compute the overall output of the ANFIS system. ANFIS systems can be trained just like neural networks.

REVIEW QUESTIONS

1. For what purpose are different intelligent systems combined?
2. When combining two different evolutionary algorithms to solve a highly nonconvex problem, what are the possible issues that may cause poor performance of the hybrid system?
3. What effect does the diversity of a population have in an evolutionary algorithm?
4. For multimodal optimization problems using particle swarm optimization, which of the two influence factors (individual particle best or swarm best) is more weighted during the search process?
5. In a neural expert system where does the expert knowledge reside?
6. How is the expert knowledge incorporated in a neural expert system?
7. What is the reason for considering a neural expert system as a system that can process imprecise data?

8. What is approximate reasoning?

9. Can a neuro-fuzzy system be constructed using Mamdani fuzzy inference systems?

10. In MATLAB, using the ANFIS model, how many outputs of the system to be modeled can be accommodated?

11. Which of the two fuzzy inference systems is the ANFIS model from MATLAB using, Mamdani or Sugeno?

12. What is the difference between grid partitioning and tree partitioning?

13. What does the subtractive clustering algorithm compute?

14. When training an ANFIS model in MATLAB, how should one avoid the problem of overfitting?

PROBLEMS

7.1 Using the MATLAB script for the PSO and GA algorithms in Appendix C, develop a hybrid algorithm that optimizes the GA parameters (number of chromosomes, mutation rate, and number of generations) using the PSO algorithm. Test your hybrid system using the Griewank test function:

$$f(\vec{w}) = \sum_{i=1}^{d} \frac{w_i^2}{4,000} - \prod_{i=1}^{d} \cos\left(\frac{w_i}{\sqrt{i}}\right) + 1, \text{ for } w_i \in [-25, 25].$$

7.2 Use the resulting GA system from Problem 7.1 with the optimized GA parameters based on the Griewank test function and apply it to the optimization problem for the Six-Hump Camel test function, which is given by:

$$f(\vec{w}) = \left(4 - 2.1w_1^2 + \frac{w_1^4}{3}\right)w_1^2 + w_1 w_2 + (-4 + 4w_2^2)w_2^2 \text{ for } w_i \in [-2, 2].$$

7.3 Follow this by tuning your GA with the hybrid system from Problem 7.1's PSO algorithm. However, optimize the parameters based on the Six-Hump Camel test function and compare the results with the one obtained in Problem 7.2.

7.4 Develop a neural expert system for distinguishing between bicycles and cars. Develop this system on paper without MATLAB or any training using backpropagation; rather, use trial-and-error to find the approximate weights for this neural expert system. Test your neural expert system by utilizing random images from the Internet and feeding your system with the information gleaned from these images. Comment on the results and on what kind of improvements could be made to enhance the results.

7.5 Develop an ANFIS system using a MATLAB script to model a simple step response function. Play with the different grid partitioning options while keeping all other ANFIS options constant and compare the results. Use a minimal number of epochs for the training.

7.6 Use the Neuro-Fuzzy Designer interface to create an ANFIS system that models the following transfer function:

$$G(s) = \frac{3.125}{(s^2 + 0.625s + 1.5625)}.$$

7.7 Following direction from Appendix G and information from Appendices E and F, use the propellor–pendulum system to conduct an experiment that yields input and output data for use in training an ANFIS model. For this experiment, chose the operating point as the equilibrium point and generate inputs that cause the system to randomly oscillate about the equilibrium point. Make sure your random input allows for the system to react (the change in the input magnitude should not occur before the system has time to properly respond to the previous input). Using the collected data, generate an ANFIS model and compare the ANFIS response due to a step input with the response given by the physical system. Comment on the comparison and what changes would yield a better outcome.

7.8 Repeat the design of an ANFIS system modeling the propellor–pendulum system as described in Problem 7.7; however, use an operating point away from the equilibrium point. Compare the ANFIS system response with the physical system response due to a step input.

7.9 Develop the governing equation of motion of your propellor–pendulum system and extract a transfer function describing this system operating around a selected operating point away from the equilibrium. Use linearization about this operating point to compute the transfer function.

7.10 Design a PID (proportional integral derivative) controller to maintain the angle of the propellor–pendulum system's operating point as selected in Problem 7.8. Test your PID controller by simulating the system in Simulink and apply random perturbation of the angle.

7.11 Use the PID controller designed in Problem 7.10 to maintain the angle of the propellor–pendulum system's operating point. Test your PID controller on the physical system by perturbing the angle repeatedly (by hand) and noticing the reaction of the system. Make sure the controller is capable of moving the pendulum back to the operating point after each perturbation.

REFERENCES

[1] Shi Y, Eberhart RC (1998). A modified particle swarm optimizer. *Proceedings of the IEEE Congress on Evolutionary Computation*, 69–73.

[2] Engelbrecht AP (2006). *Fundamentals of Computational Swarm Intelligence*, Wiley, New York.

[3] Pham Viet (2021). MANFIS_S. www.mathworks.com/matlabcentral/fileexchange/54686-manfis_s.

[4] Serge Guillaume S (2001). Designing fuzzy inference systems from data: an interpretability-oriented review. *IEEE Transactions on Fuzzy Systems*, 9(3): 426–443.

8 Deep Learning

8.1 Introduction

Artificial intelligence (AI) is a discipline dedicated to the creation of systems that reason and act like humans. A subfield of AI is machine learning (ML), where we specifically let computers solve and accomplish tasks without the explicit programming but by the utilization of models that are inferred from sample data. If an ML system employs an artificial neural network (NN) and a mechanism that allows for automatic feature learning, we refer to this type of system as a deep learning (DL) system. This distinction between ML and DL is shown in Figure 8.1.

We will cover a broad range of ML systems in Chapter 9 and focus in this chapter solely on DL systems. We already encountered neural networks (NNs) in Chapter 6, and expanded their capabilities in Chapter 7, where we were able to create neural expert systems (NES) and adaptive neuro-fuzzy inference systems (ANFIS) by combining NNs with other methods. For creating NNs and NES we employed user-defined features. As an example, for the NES discussed in Section 7.2, we crafted simple features by describing the shape of specific leaves. Those features were used for the classification algorithm to allow for the classification among different tree types. It may have been intuitive and rather easy to create our set of features for the leaf classification example using NNs and NES. However, generally this is not a trivial task. It usually requires good domain knowledge of the problem that we are attempting to solve. Features are very influential in the resulting NN's performance. If there is little or no domain knowledge available, we need a different approach to create effective NN-based systems.

A popular DL system that automates the feature generation – to some extent – is the convolutional neural network (CNN). The automation of feature creation allows for the resulting CNN to be more universal. This process of automated feature generation and classification is shown in Figure 8.1(b), where we distinguish from the manual task of creating features – as shown in Figure 8.1(a), indicating the ML approach – with the inherent feature generation of a CNN.

CNNs are networks that have many layers. Such networks with more than one hidden layer give rise to the "deep" in deep learning systems. As there are many layers for a deep NN, the number of parameters can become quite large. Some of the NNs we will encounter in this chapter have millions of parameters. Tuning and

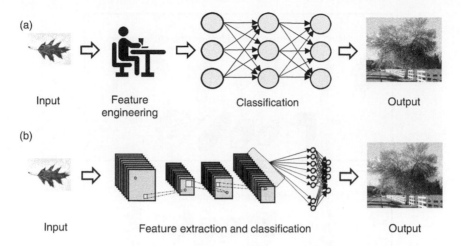

Figure 8.1 Difference between (a) machine learning and (b) deep learning.

training such networks require substantial computational resources and time. However, adapting a trained DL system for a given problem can substantially reduce the amount of work, as the training process is greatly reduced. In addition, by adapting a trained network that employs generalized features, we are less dependent on domain knowledge. The process of adapting a trained deep NN to a specific problem is called transfer learning. In this chapter, we will use trained deep NNs for solving problems, where we adapt a trained NN to a given problem. This will save computational resources and allow for easier implementations.

However, before we adapt deep NNs for particular problem sets, let's discuss the general working mechanism of a DL system and its associated workflow for generating it. Figure 8.2 depicts the general workflow of a DL system employed for a classification problem.

Suppose we start off with an untrained DL system. The first task for such a DL classification system is to train its network. Training requires a large data set to ensure that the network acquires all the necessary traits and characteristics of the system or classes we want to be able to categorize. Once the DL system is trained, we can employ it to classify the type of objects it was trained on. In Figure 8.2, the objects are leaves and the goal of the DL system is to categorize a leaf based on that training. To start a categorization task, we present the DL system with a data point that we want to classify, in this case a test picture of a leaf. The first layer of the DL uses this test picture and looks for small and simple features such as basic shapes. The further the image is processed in terms of the layers of the DL system, the more complex features are being categorized, until the last layers, where the DL system looks for recognizable patterns and complex shapes. The output layer of the DL picks from the most likely categories and assigns the corresponding label to the test picture.

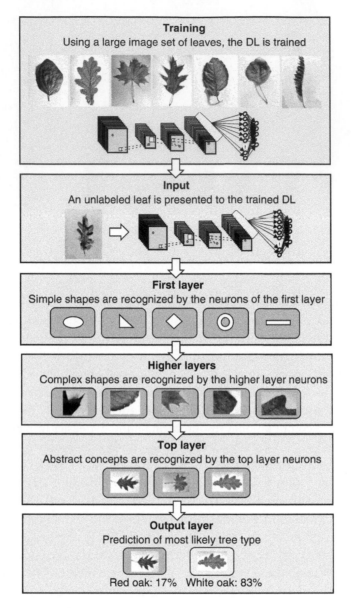

Figure 8.2 DL system solving a classification of tree leaves task.

In the above discussion we oversimplified the DL creation part, as we left out the formation of the DL structure, any discussion on generating a suitable and rich data set, the data exploration and preprocessing, and any algorithm development that may be needed for generating such a DL system.

Generally, there are four basic steps involved in the workflow of a DL system: (1) data collection or generation; (2) data analysis; (3) DL structure selection, modeling, and simulation; and (4) deployment. The first step, data collection, can be achieved in different ways, such as accessing an online data repository for a specific object, acquiring your own data set through measurements, generating data through simulation, or extend an existing database by using measured or simulated data. Once the data are available, we may need to process them in order to conform them to the DL structure. Processing may also include filtering and transforming. For example, if our data set is an image set, we may use the original images and crop these to focus on the objects we are interested in; we may pad the images; and we may perform edge detection. Another process that is commonly used in image classification problems using DL systems is the reduction of dimensions. For this, we can imagine that the number of pixels an image is composed of may cause the image to have too many details that are not of importance to the classification problem. Hence, we could reduce the dimension of the image by reducing the number of pixels – that is, effectively increase the size of a pixel to compose the image with fewer pixels. However, what if we do not know the type of feature we need for the specific classification problem due to a lack of adequate domain knowledge? One could argue that the more dimensions we include in the data set and the more features and variations we will accommodate in our sample, the higher the likelihood of proper classification.

However, the more features, which means more dimensions, the more training data we will need. This is commonly referred to as the "curse of dimensionality," and is something that we will discuss in Chapter 9 when treating ML algorithms that are particularly susceptible to this phenomenon. The size of the data set for training a DL system and the amount of domain knowledge are closely related and can be associated with the type of ML system to be used. Graphically, this relationship is generically depicted in Figure 8.3.

Another difficulty in designing a proper DL system is the selection of a suitable DL network architecture. The task is made in some sense more difficult, as there are constantly new DL systems proposed and published. In the following we will

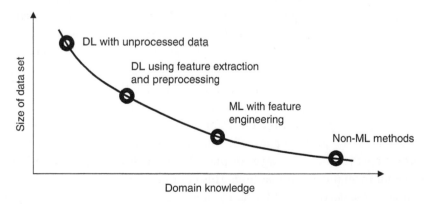

Figure 8.3 Relationship between the required data set size and the amount of domain knowledge in classification schemes.

mention a few popular DL structures with some characterization. A few of these algorithms we will see later in this chapter in more detail.

- Convolutional neural networks: CNNs employ convolution and pooling layers for tasks such as detection and classification. For applications in image processing, the convolution and pooling layers effectively generate n-grams and perform groupings with its results before utilizing a fully connected NN. Applications of CNNs are found in image recognition, image classification, image segmentation, and image analysis, but also in natural language processing, brain–computer interfaces, and time series analysis. Some popular CNN architectures are VGG16 and VGG19 [1], ResNet [2], and SqueezeNet. We will introduce the concepts of convolution and pooling as well as the specific architecture in the CNN section of this chapter.
- Recurrent neural networks (RNNs): Unlike classification tasks, a control system is often dependent on solving problems that have a time dependency. RNNs introduce a recursive component into their architecture to accommodate the notion of time. With the ability to entertain time dependencies, RNNs are used for speech recognition, language translation, prediction, and controls.
- Long short-term memory networks (LSTMs): An LSTM is a special type of RNN, where the recursive component is designed with a gate. This gate decides if information is to be passed on to the next layer or should be forgotten.

Deploying a DL system is done as a trained system on a minimal computational platform, while untrained DL systems need the capacity to acquire data, process the data, and train the DL system. Usually, the latter approach requires large computational resources such as computers or cloud support.

8.2   Issues in Deep Learning

As the name indicates, DL employs NNs that have more than one hidden layer. For some time, training NNs with multiple hidden layers was considered problematic due to the vanishing and exploding gradient problems, as well as the possibility of overfitting. For DL algorithms this problem was solved by introducing specific activation functions and a training that involves a concept called dropout. Before going into detail of these remedies, let's consider the vanishing and exploding gradient we face when training a network with multiple hidden layers.

During backpropagation the information flow is such that we utilize the network's error and propagate it from the NN's output through the network until it reaches the input layer of the NN. At each layer the algorithm imposes some mathematical computation. This computation includes the error, which is scaled by the derivative of the activation function contained in each node and in each layer. Recall from Chapter 6 that our update computations were dependent on a quantity we called Δ, which combined with the learning rate and the signal from the prior layer allowed us to compute the updated weights of the current layer. We can see this when we generalize Equation (6.16):

$$w_{ij}^{(n)} = w_{ij}^{(n)} + \alpha\Delta_i^{(n)}y_j^{(n-1)} = w_{ij}^{(n)} + \alpha\left\{\varphi'\left(v_i^{(n)}\right)e_i^{(n)}\right\}y_j^{(n-1)} = w_{ij}^{(n)} + w_{update}. \quad (8.1)$$

We can characterize the changes in the update quantity w_{update} as it progresses through the network during backpropagation. There are two possible scenarios that may be concerning to the success of the backpropagation algorithm. These two scenarios are depicted in Figure 8.4. The first instance is when the gradient becomes monotonically smaller with the progression through the network. The other scenario is when the gradient becomes exponentially larger with each progression through the layers of the network during training. When the gradient consistently increases during backpropagation, we call this the exploding gradient problem. The opposite, when the gradient becomes consistently smaller and smaller during the training, is called the vanishing gradient problem.

For the exploding gradient scenario, the updates, and hence the weights of the network, can become so large that it results in numbers that are no longer manageable for computations and the network is deemed unstable. When dealing with a vanishing gradient problem, the weights no longer update since the updates are close to zero, and learning ceases to exist. Deep NNs employing

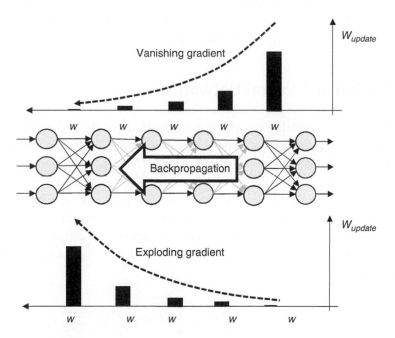

Figure 8.4 Vanishing gradient and exploding gradient depiction during backpropagation training of a deep neural network.

sigmoid or tanh functions are especially prone to the vanishing gradient problem. Resolving the vanishing or exploding gradient problem for a given network may often require the elimination of excess layers. An alternative approach for countering the exploding gradient problem is the gradient clipping method, where the gradient is limited to some finite value. Alternatively, the NN weights can be taxed with a penalty within the error function for large weight values. In effect, this is another regularization approach. When using absolute values for weights it is referred to as L1 regularization; if square weights are used we label those regularization terms as L2. To mitigate the vanishing gradient problem, one should choose the ReLU (rectified linear unit) activation function over the sigmoid or tanh activation functions.

Another common problem encountered in ML is the problem of overfitting. We discussed this problem in Section 6.4 and also introduced specific remedies, such as the introduction of regularization terms with the error function. For DL-based networks another process has been used to avoid overfitting. This process, termed *dropout*, is a simple approach to temporarily mimic a smaller network in terms of complexity, nodes, and weights during the training process. Using dropout during the backpropagation implies that we train only a set of randomly selected nodes of the original network. This set of nodes is smaller than the overall number of nodes of the network. Hence, we are leaving out some nodes which leads to the term dropout. Mechanically, dropout can be established by setting the randomly selected dropout nodes to have a zero output. Overfitting is reduced by constantly selecting randomly different nodes to dropout during the learning process. Some network designers suggest having a 50% dropout rate for hidden layers and 25% dropout rate for input layers.

8.3   Convolutional Neural Networks

Convolutional neural networks are often used for data that has a grid-like structure, such as images where the corresponding two-dimensional grid comprises the boundaries of the individual pixels contained in the image. However, time series data can also be thought of as grid-like structures when considering the signal at discrete time instances as members of a one-dimensional grid. CNNs are named after the convolution operation they seem to employ at some of their network layers. If we think of convolution of two functions $f(t)$ and $k(t)$ in regular mathematical terms we may recall that this can be expressed as

$$c(t) = f * k = \int_{-\infty}^{\infty} f(\tau)k(t - \tau)d\tau. \tag{8.2}$$

In controls we prefer to compute such integrals in the Laplace domain, where the convolution of these two functions becomes a simple multiplication $C(s) = F(s)K(s)$. For CNNs we will often deal with grid-like structures for the input, and hence may prefer the discrete convolution of the function $f_d(q)$, which will serve as the input to our NN, and $k_d(q)$, which we will label as the kernel. The kernel is basically a filter; however, in DL it is referred to as a kernel. Also, note q is the discrete time index and for discrete time convolution we can compute:

$$c_d(q) = f_d * k_d = \sum_{a=-\infty}^{\infty} f(a)k(q-a). \tag{8.3}$$

Equation (8.3) is for one-dimensional input data. However, when dealing with images we need to define the discrete convolution for two-dimensional data (i.e., matrices). Suppose our image is given by the matrix $\mathbf{I}$, then $\mathbf{C}$ can be computed as follows:

$$\mathbf{C}_{i,j} = \mathbf{I} * \mathbf{K} = \sum_m \sum_n \mathbf{I}_{m,n}\mathbf{K}_{i-m,j-n}. \tag{8.4}$$

In Equation (8.4) we identify the pixel by its position in the image matrix $\mathbf{I}$ and compute the convolution by the discrete summation using the two-dimensional kernel function $\mathbf{K}$. Since the commutative law is applicable to the convolution operation, Equation (8.4) can be rewritten as:

$$\mathbf{C}_{i,j} = \mathbf{K} * \mathbf{I} = \sum_m \sum_n \mathbf{I}_{i-m,j-n}\mathbf{K}_{m,n}. \tag{8.5}$$

In Equation (8.5), note that when m is increased the index in $\mathbf{I}$ decreases while the index in the kernel $\mathbf{K}$ increases. Effectively, the kernel $\mathbf{K}$ is flipped relative to the input. It turns out that many CNN implementations use the cross-correlation function instead of the convolution operation. The two are related, as the cross-correlation function is the same as the convolution function; however, without having the kernel function flipped:

$$\widetilde{\mathbf{C}}_{i,j} = \sum_m \sum_n \mathbf{I}_{i+m,j+n}\mathbf{K}_{m,n}. \tag{8.6}$$

Having the convolution operator introduced in Equation (8.2), we can utilize the correlation operator as follows:

$$\widetilde{\mathbf{C}}_{i,j} = \mathbf{I} \otimes \mathbf{K} = \sum_m \sum_n \mathbf{I}_{i+m,j+n}\mathbf{K}_{m,n}. \tag{8.7}$$

Example 8.1

We can demonstrate both operations and compare the outcome using an example. Suppose our image matrix $\mathbf{I}$ is composed of integers representing the intensity of gray scale:

$$\mathbf{I} = \begin{bmatrix} 1 & 2 & 3 & 4 & \cdots \\ 7 & 8 & 9 & 10 & \cdots \\ 4 & 3 & 2 & 1 & \cdots \\ 5 & 2 & 7 & 8 & \cdots \\ \vdots & \vdots & \vdots & \vdots & \ddots \end{bmatrix}, \mathbf{K}_1 = \begin{bmatrix} 0.3 & 0.2 \\ 0.2 & 0.3 \end{bmatrix}, \text{ and } \mathbf{K}_2 = \begin{bmatrix} 0.2 & 0.2 \\ 0.2 & 0.3 \end{bmatrix}.$$

Also, we have defined two kernel matrices, $\mathbf{K}_1$ and $\mathbf{K}_2$. $\mathbf{K}_1$ is a symmetric matrix and is normalized so that the sum of its elements add up to 1.0. $\mathbf{K}_2$ is not symmetric and also not normalized. Kernels are often normalized; however, for our purposes, the fact that $\mathbf{K}_2$ is not normalized has no implication on our discussion in the following.

Let's compute the correlation:

$$\widetilde{\mathbf{C}}_{i,j}^{(1)} = \mathbf{I} \otimes \mathbf{K}_1.$$

For this, we can use Equation (8.7) as our guide on how we progress through the individual elements and compute the products. Graphically, we can imagine the kernel matrix overlaying the image matrix and computing a dot product for each element matching the overlay. As we progress with the indices, we move the kernel matrix incrementally over the image matrix for each change in index. For example, if we want to compute $\widetilde{\mathbf{C}}_{2,2}$, we have

$$\widetilde{\mathbf{C}}_{2,2} = \begin{bmatrix} 1 & 2 & 3 & 4 & \cdots \\ 7 & 8 & 9 & 10 & \cdots \\ 4 & 3 & 2 & 1 & \cdots \\ 5 & 2 & 7 & 8 & \cdots \\ \vdots & \vdots & \vdots & \vdots & \ddots \end{bmatrix} \circ \begin{bmatrix} 0.3 & 0.2 \\ 0.2 & 0.3 \end{bmatrix}$$

$$= (1 \times 0.3 + 2 \times 0.2 + 7 \times 0.2 + 8 \times 0.3) = 4.5.$$

If we attempt to compute the (1,1) element of the correlation matrix, we notice that the overlay of the kernel extends past the image matrix. To remedy this situation we can use padding. There are different padding methods in use. For example, one can use the adjacent value and copy it into the space for the padding. We will demonstrate padding by using the zero-padding method, where we fill the missing spaces with zeros and compute the correlation, and later on the convolution:

$$\widetilde{\mathbf{C}}_{1,1}^{(1)} = \begin{matrix} 0 \\ 0 \\ 0 \\ 0 \\ 0 \\ \vdots \end{matrix} \begin{bmatrix} 0 & 0 & 0 & 0 & \dots \\ 1 & 2 & 3 & 4 & \dots \\ 7 & 8 & 9 & 10 & \dots \\ 4 & 3 & 2 & 1 & \dots \\ 5 & 2 & 7 & 8 & \dots \\ \vdots & \vdots & \vdots & \vdots & \ddots \end{bmatrix} \circ \begin{bmatrix} 0.3 & 0.2 \\ 0.2 & 0.3 \end{bmatrix}$$

$$= (0 \times 0.3 + 0 \times 0.2 + 0 \times 0.2 + 1 \times 0.3) = 0.3,$$

$$\widetilde{\mathbf{C}}_{1,2}^{(1)} = \begin{matrix} 0 \\ 0 \\ 0 \\ 0 \\ 0 \\ \vdots \end{matrix} \begin{bmatrix} 0 & 0 & 0 & 0 & \dots \\ 1 & 2 & 3 & 4 & \dots \\ 7 & 8 & 9 & 10 & \dots \\ 4 & 3 & 2 & 1 & \dots \\ 5 & 2 & 7 & 8 & \dots \\ \vdots & \vdots & \vdots & \vdots & \ddots \end{bmatrix} \circ \begin{bmatrix} 0.3 & 0.2 \\ 0.2 & 0.3 \end{bmatrix}$$

$$= (0 \times 0.3 + 0 \times 0.2 + 1 \times 0.2 + 2 \times 0.3) = 0.8,$$

$$\widetilde{\mathbf{C}}_{2,1}^{(1)} = \begin{matrix} 0 \\ 0 \\ 0 \\ 0 \\ 0 \\ \vdots \end{matrix} \begin{bmatrix} 0 & 0 & 0 & 0 & \dots \\ 1 & 2 & 3 & 4 & \dots \\ 7 & 8 & 9 & 10 & \dots \\ 4 & 3 & 2 & 1 & \dots \\ 5 & 2 & 7 & 8 & \dots \\ \vdots & \vdots & \vdots & \vdots & \ddots \end{bmatrix} \circ \begin{bmatrix} 0.3 & 0.2 \\ 0.2 & 0.3 \end{bmatrix}$$

$$= (0 \times 0.3 + 1 \times 0.2 + 1 \times 0.2 + 7 \times 0.3) = 2.3.$$

Hence, our correlation matrix takes the form

$$\widetilde{\mathbf{C}}^{(1)} = \begin{bmatrix} 0.3 & 0.8 & \dots \\ 2.3 & 4.5 & \dots \\ \vdots & \vdots & \ddots \end{bmatrix}.$$

If we compute the correlation matrix with $\mathbf{K}_2$, we obtain the following result:

$$\widetilde{\mathbf{C}}_{i,j}^{(2)} = \mathbf{I} \otimes \mathbf{K}_2 = \begin{bmatrix} 0.3 & 0.8 & \dots \\ 2.3 & 4.4 & \dots \\ \vdots & \vdots & \ddots \end{bmatrix}.$$

To compute the convolution $\mathbf{C}_{i,j}^{(n)} = \mathbf{I} * \mathbf{K}_n$ we rotate the kernel matrix by 180° and perform the same operations as we did for the correlation calculation.

$$\mathbf{C}_{2,2}^{(1)} = \begin{bmatrix} 1 & 2 & 3 & 4 & \dots \\ 7 & 8 & 9 & 10 & \dots \\ 4 & 3 & 2 & 1 & \dots \\ 5 & 2 & 7 & 8 & \dots \\ \vdots & \vdots & \vdots & \vdots & \ddots \end{bmatrix} \circ \begin{bmatrix} 0.3 & 0.2 \\ 0.2 & 0.3 \end{bmatrix}$$

$$= (1 \times 0.3 + 2 \times 0.2 + 7 \times 0.2 + 8 \times 0.3) = 4.5,$$

$$\mathbf{C}_{1,2}^{(1)} = \dots.$$

We obtain the following results:

$$\mathbf{C}^{(1)}_{i,j} = \mathbf{I} * \mathbf{K}_1 = \begin{bmatrix} 0.3 & 0.8 & \cdots \\ 2.3 & 4.5 & \cdots \\ \vdots & \vdots & \ddots \end{bmatrix}.$$

And for $\mathbf{C}^{(2)}$:

$$\mathbf{C}^{(2)}_{2,2} = \begin{bmatrix} 1 & 2 & 3 & 4 & \cdots \\ 7 & 8 & 9 & 10 & \cdots \\ 4 & 3 & 2 & 1 & \cdots \\ 5 & 2 & 7 & 8 & \cdots \\ \vdots & \vdots & \vdots & \vdots & \ddots \end{bmatrix} \circ \begin{bmatrix} 0.3 & 0.2 \\ 0.2 & 0.3 \end{bmatrix}$$

$$= (1 \times 0.3 + 2 \times 0.2 + 7 \times 0.2 + 8 \times 0.2) = 3.7,$$

$$\mathbf{C}^{(1)}_{1,2} = 0.6, \ \mathbf{C}^{(1)}_{2,1} = 1.6, \ \mathbf{C}^{(1)}_{1,1} = 0.2, \ldots,$$

$$\mathbf{C}^{(2)}_{i,j} = \mathbf{I} * \mathbf{K}_2 = \begin{bmatrix} 0.2 & 0.6 & \cdots \\ 1.6 & 3.7 & \cdots \\ \vdots & \vdots & \ddots \end{bmatrix}.$$

Comparing $\mathbf{C}^{(1)}$ and $\widetilde{\mathbf{C}}^{(1)}$ we notice that they are the same. However, if we compare $\mathbf{C}^{(2)}$ and $\widetilde{\mathbf{C}}^{(2)}$ they are no longer equal. The difference is that $\mathbf{K}_1$ is symmetric and $\mathbf{K}_2$ is not. Noticing this difference and observing the indices in Equations (8.5) and (8.7), we can conclude that convolution is equivalent to cross-correlation if we flip the kernel matrix by 180°.

In general, we can make the following statement: If the kernel matrix is a symmetric matrix, the convolution and the cross-correlation operation yield the same results. Since many applications in image processing use symmetric kernels, we will find that many CNNs are utilizing the cross-correlation function, even though we have the convolution in the name of the structure. As long as the kernel is symmetric, this will not matter. In image processing, cross-correlation is usually used for image alignment and simple image matching processes.

Example 8.2

Let's take a step back and consider a one-dimensional data set, such as a time series. Suppose our time series and kernel are given by the following vectors:

$$\vec{y} = [1 \quad 1 \quad 2 \quad 1 \quad 2 \quad 2], \ \vec{k}_1 = [1 \quad 2 \quad 1] \text{ and } \vec{k}_2 = [1 \quad 2 \quad 3].$$

We can compute the correlation and the convolution as follows:

Correlation:

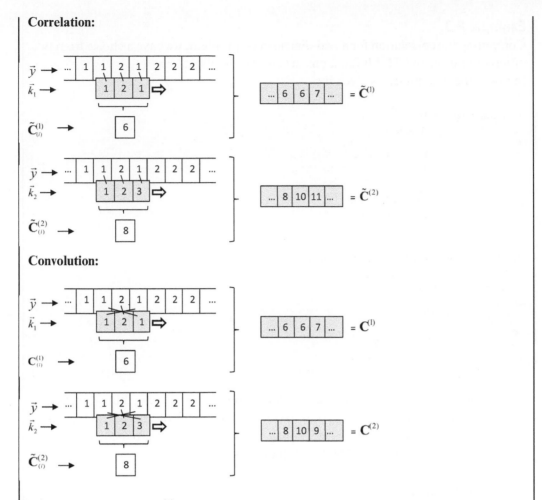

Convolution:

Comparing $\mathbf{C}^{(1)}$ and $\widetilde{\mathbf{C}}^{(1)}$, we notice again that they are equal since $\vec{k}_1$ is a symmetric vector. However, if we compare $\mathbf{C}^{(2)}$ and $\widetilde{\mathbf{C}}^{(2)}$ where $\vec{k}_2$ is no longer symmetric, the convolution computation no longer yields the same result as the correlation computation. Flipping of the kernel has a physical interpretation for the one-dimensional case. Here, we deal with time series, which represent data from a causal system – that is, a system where an input generates an output at the time of the input or after, but not before. It turns out that the convolution of such a signal will yield a causal signal. When performing the flipping of the kernel, the convolution with an impulse to a causal system will result in the response of this system. If no flip were to be performed, the computation will result in the reversal of the signal.

MATLAB supplies functions for both computations. We'll demonstrate its use with some examples.

Example 8.3

Computing the correlation for a two-dimensional problem, we have a choice from two different built-in MATLAB functions: `filter2()` and `imfilter()`.

Suppose the data matrix is given by:

```
≫ Image=magic(5)
         Image =

         17   24    1    8   15
         23    5    7   14   16
          4    6   13   20   22
         10   12   19   21    3
         11   18   25    2    9
```

In addition, we can define the kernel as:

```
≫ k= fspecial('sobel')
k =

   1   2   1
   0   0   0
  -1  -2  -1
```

The function `fspecial()` creates a two-dimensional filter with attributes for specific filter functions, such as `average`, `gaussian`, `log`, etc. One of the choices is `sobel`, which is sometimes used for edge detection in images. To find the cross-correlation, we can now use either of the two MATLAB functions:

```
≫ C=filter2(k,Image,'valid')
C =

   37  -18  -43
  -13  -38  -13
  -43  -18   37
```

You may wonder how MATLAB solved the padding issue? It turns out that MATLAB uses automatically the zero-padding method we utilized in the above discussion. The attribute `valid` will cause the function `filter2()` to return the elements of the correlation matrix that were computed without padding. If the full matrix is desired, the command should be changed as follows:

```
≫ C=filter2(k,Image,'full')
C =
```

$$\begin{vmatrix} -17 & -58 & -66 & -34 & -32 & -38 & -15 \\ -23 & -51 & -40 & -33 & -51 & -46 & -16 \\ 13 & 44 & 37 & -18 & -43 & -26 & -7 \\ 13 & 19 & -13 & -38 & -13 & 19 & 13 \\ -7 & -26 & -43 & -18 & 37 & 44 & 13 \\ 10 & 32 & 53 & 71 & 64 & 27 & 3 \\ 11 & 40 & 72 & 70 & 38 & 20 & 9 \end{vmatrix}$$

You notice that the last computation yielded a 7×7 matrix, where the core is the same as we obtained with the attribute `valid`. There is a third attribute, `same`, which returns the core of **C** with the same size as the image matrix. The function `imfilter()` is used to compute the cross-correlation for multidimensional arrays.

Example 8.4

To compute the convolution, we can use the MATLAB function `conv2()`. Suppose we utilize the same image matrix from the above example and the same kernel, MATLAB will compute the convolution as follows:

```
≫ Cv=conv2(Image,k)
Cv =

   17    58    66    34    32    38    15
   23    51    40    33    51    46    16
  -13   -44   -37    18    43    26     7
  -13   -19    13    38    13   -19   -13
    7    26    43    18   -37   -44   -13
  -10   -32   -53   -71   -64   -27    -3
  -11   -40   -72   -70   -38   -20    -9
```

So, why do we need convolution in CNNs? Recall that in the previous chapters we created our own features. Manually designed features often end up being either over-specified or incomplete. They also take time to engineer and to evaluate in terms of effectiveness and independence. The goal of DL, and hence CNNs, is to have features that are learned and are easily adaptable, are flexible and somewhat universal. This is achieved in part by using a convolution computation of the input. The convolution basically contributes a useful feature that allows accommodating inputs with varying sizes. In addition, the convolution facilitates the CNN to have sparse interaction, resulting in parameter sharing. Let's look at those attributes a bit more specifically.

In the above example for computing the convolution, we utilized a kernel that was effectively smaller in size than the image matrix. When performing the convolution operation of an image – as an example – the image may be composed of several thousand pixels. Having the kernel move across this image and perform dot multiplication is basically equivalent to a filtering process. In the above example, we used an attribute called sobel which we associated with being utilized for detecting edges in images. A CNN processes the input by first trying to detect small features such as edges. We can see that the convolution operation serves this purpose by the appropriate selection and application of the kernel function. Having kernel functions, we reduce the size of the data as well, since we chose kernel functions that are smaller than the image or input matrix. Having smaller data sets will also result in the network possessing and hence storing fewer parameters for those layers. Having fewer parameters will lead to fewer multiplications and operations. To showcase the implication of the convolution operation and the resulting reduction in parameter storage and computational effort, consider an input to a CNN of size n_i and output of size n_o. The corresponding matrix multiplication will need $n_o \times n_i$ parameters. If we can reduce n_o to k, then the matrix multiplication will reduce to $k \times n_i$ operations. This reduction in parameters is referred to as the sparse connectivity approach. Fewer multiplication operations also has an effect on the time the network needs to complete its output. Sparse connectivity refers to how many neurons are directly connected between two layers. If a kernel of size three is used, the implication is that the network effectively connects three neurons from the first layer to one neuron in the second layer of the network.

Another result of the convolution computation in a CNN is the effect of parameter sharing. Parameter sharing refers to having a weight, applied to one connection, that has always the same weight as applied to another connection. Hence, parameter sharing is often referred to as the network having tied weights. As we use kernel functions to perform the convolution computation, each element of the kernel is used at every input element. The resulting computation implies that the network will need to learn one set of parameters for all inputs. For example, suppose the first layer of the CNN is responsible for detecting certain small features, say edges in an image. The same type of edges may appear at different locations of the input image or input matrix. Hence, we can utilize the same parameters. The same is true for time series data, where a specific form of the function is detected at two or more different instances in time. To represent those features of the time series the same parameters can be utilized. As an example, Table 8.1 lists a few kernels that have different effects when applied to images in the process of convolution operation.

A CNN layer usually houses three functions. The first function is associated with the convolution operation. Often, a number of convolutions are performed to create a set of linear functions. The second function of the layer is commonly referred to as the detector stage. Here, the linear functions are evaluated against a nonlinear activation function. Finally, the third operation is the pooling process. There are a number of different pooling functions in use for CNNs. The max pooling and average pooling operations are shown in Figure 8.5. Here, the

Table 8.1 Kernel selection and effects on images

Effect	Kernel
Edge detection	$\mathbf{K} = \begin{bmatrix} 1 & 0 & -1 \\ 0 & 0 & 0 \\ -1 & 0 & 1 \end{bmatrix}$
Edge detection	$\mathbf{K} = \begin{bmatrix} 0 & -1 & 0 \\ -1 & 4 & -1 \\ 0 & -1 & 0 \end{bmatrix}$
Edge detection	$\mathbf{K} = \begin{bmatrix} -1 & -1 & -1 \\ -1 & 8 & -1 \\ -1 & -1 & -1 \end{bmatrix}$
Blurring (box blur)	$\mathbf{K} = \dfrac{1}{9} \begin{bmatrix} 1 & 1 & 1 \\ 1 & 1 & 1 \\ 1 & 1 & 1 \end{bmatrix}$
Blurring (Gaussian)	$\mathbf{K} = \dfrac{1}{16} \begin{bmatrix} 1 & 2 & 1 \\ 2 & 4 & 2 \\ 1 & 2 & 1 \end{bmatrix}$
Blurring (Gaussian)	$\mathbf{K} = \dfrac{1}{256} \begin{bmatrix} 1 & 4 & 6 & 4 & 1 \\ 4 & 16 & 24 & 16 & 4 \\ 6 & 24 & 36 & 24 & 6 \\ 4 & 16 & 24 & 16 & 4 \\ 1 & 4 & 6 & 4 & 1 \end{bmatrix}$
Unsharpening	$\mathbf{K} = \dfrac{-1}{256} \begin{bmatrix} 1 & 4 & 6 & 4 & 1 \\ 4 & 16 & 24 & 16 & 4 \\ 6 & 24 & -476 & 24 & 6 \\ 4 & 16 & 24 & 16 & 4 \\ 1 & 4 & 6 & 4 & 1 \end{bmatrix}$
Sharpening	$\mathbf{K} = \begin{bmatrix} 0 & -1 & 0 \\ -1 & 5 & -1 \\ 0 & -1 & 0 \end{bmatrix}$

feature map – as computed by the second function of the layer – is filtered using a mask of a particular size – in the figure it is a 2×2 matrix. For max pooling, for each instance the value of the element with the largest entry is retained while the other values are discarded. In the case of average pooling we compute the average value of the overlay and retain that value as the results. This process is repeated for each iteration, where we reposition the filter mask by a certain number of elements, or stride length.

A benefit of the pooling operation is the resulting invariance to small changes in the input of the CNN. We generally will obtain the same value after a pooling operation for an input that is slightly translated by a small amount from the original input. The filter size provides for this tolerance and for such translations.

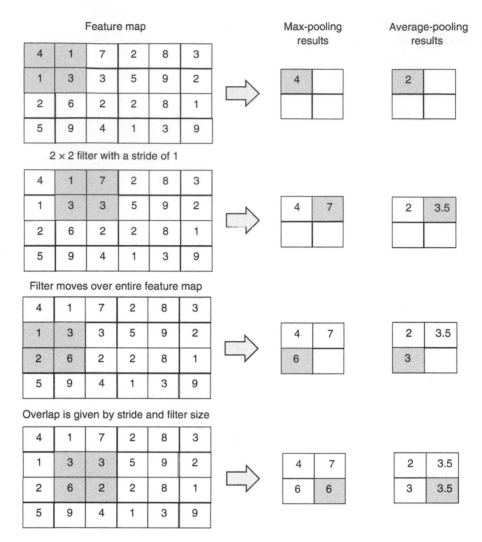

Figure 8.5 Pooling operation using two different types of implementations: the feature map is the result of the convolution and detection stages of the layer. Pooling is accomplished by applying a filter mask and determining either the maximum element or the average value of the elements covered by the mask. In the figure, a mask of 2 × 2 elements and a stride length of 1 is used. However, larger masks and stride lengths are common in CNNs. Also, there are other pooling algorithms such as L2 norm, and weighted average, among others.

Effectively, having some degree of invariance to translation in the input is equivalent to emphasizing the discovery of a feature rather than the location of the feature.

Besides the gained invariance to small changes in the input, pooling also has the effect of reducing the dimensions of the feature maps. When using max pooling we emphasize the most prominent features in the feature map.

This may result in omitting some details. However, when applying average pooling, the operation will take all features into account, effectively smoothing the features. This allows the network to recognize more complex features. Hence, the selection of what type of pooling is employed is related to the layer of the network in which this operation is to take place, and hence the degree of complexity in features to be identified. A more specific discussion on which type of pooling is effective in image recognition-based applications is detailed by Boureau et al. [3].

Having discussed the major operations that occur in CNNs, and using the knowledge of Chapter 6, we can assemble the overall structure of a common CNN. Figure 8.6 depicts all major components of a general CNN. This sample structure is suited for an image classification task.

Figure 8.6 partitions the CNN structure into two parts: the feature-learning part and the classification part. In the feature-learning portion, repeated use of the convolution and pooling operation is made. However, another operation is sometimes employed as well: the batch normalization process. This algorithm is utilized to standardize the inputs to each layer. Normalization is generally beneficial for reducing the number of epochs needed during the training stage of the deep NN and decreases the sensitivity to the initial values in the network. Using backpropagation for networks that employ activation functions that have saturated bounds, such as the sigmoid activation function, the resulting gradient may tend to go to zero. In deep NNs this tendency is compounded by the fact that these small gradients get even smaller as they are multiplied together. As we discussed for the vanishing gradient problem, the result of this tendency is a limitation on the depth of the network in terms of how many layers it can train. However, if using nonsaturated actuation functions such as the ReLU function, applying normalization still has benefits. In these cases, the normalization discourages the network weights from being moved into the zone where the same effect as observed with saturation may occur during the training process. Suppose we use one set of data N_b that is a subset of the entire data set, and refer to this as the mini-batch, which is the batch of data used for one epoch during the training of the network. Then the normalization using the mini-batch can be given by the following equation [4]:

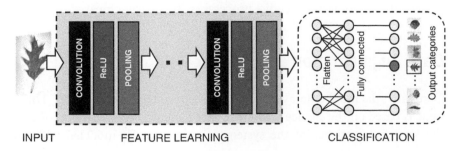

INPUT   FEATURE LEARNING   CLASSIFICATION

Figure 8.6 Structure of a convolutional neural network.

$$\widehat{x} = \frac{x - E[x]}{\sqrt{\dfrac{1}{N_b}\displaystyle\sum_{j=1}^{N_b}(x - E[x])^2}}, \qquad (8.8)$$

where x is the input to the next layer. Equation (8.8) ensures that the input to the next layer x has a mean value of zero and a standard deviation of 1.0. Applying Equation (8.8) has the added benefit of addressing the internal covariate shift problem. The internal covariate shift problem refers to the varying input distribution for each neuron in the network. During training, the weights are adjusted using different data sets – mini-batches – which cause the input to the next layer to be different each time. In deep networks these small changes get amplified by the depth of the network, as they are passed through many layers, resulting in large changes in the input distribution of the layers at the end of the network architecture. These large changes adversely affect the learning rate of the weights, and hence cause bottlenecks in the training of deep NNs. Normalization mitigates this bottleneck and allows for a more even learning rate.

The second part of Figure 8.6 includes the classification process. Here, a fully flattened NN is employed that has as many output neurons as classes. Figure 8.6 depicts a generic CNN structure. There are a number of different structures available that have been tested for specific applications, as well as that have been trained on certain data sets. We will make use of some of these trained networks when we discuss transfer learning. However, for any of these networks there are three primary hyperparameters that are responsible for the size of such structures: the number of filters we employ in the network, which has a direct effect on the depth of the network; the stride size, which is the distance the kernel moves over the input matrix or image; and padding. Large stride sizes result in smaller outputs and hence reduce the size of the network. As Figure 8.6 indicates, a CNN structure applies a ReLU function after each convolution action. The activation function allows us to include nonlinear properties into the network. Given their characteristics, CNNs have been found to be effective and are quite useful in a number of applications. For example, they are used in medical imaging applications to detect cancer cells. Other applications are found in automated driving as well as audio processing in smart homes.

8.4   Data Augmentation and Preprocessing

Generally, the more data available the better the training outcome. This simple logic is based on the chance that more data increases the likelihood of capturing the key characteristics of the system or subject to be modeled with an NN.

However, what options do we have if there is insufficient data to train a network? Suppose we rely on data that were acquired during a non-repeatable experiment, resulting in a set of data points that is insufficient for reliably training a given network. If no other data are available through simulation or existing repositories, an alternative approach can be employed to artificially increase the size of the data set. Data augmentation is a method that allows us to increase the size of the data using the existing data. The basic idea behind augmentation is to use the existing set of data and manipulate them so that they represent the key characteristics in different forms. Usually, this results in an improved generalization of the resulting network model. Besides improving the prediction accuracy, data augmentation reduces the effort of collecting and/or labeling data sets. For classification problems where some classes have many more data points than other classes, data augmentation can mitigate such class imbalances for specific less-populated classes by using the existing class data.

Data augmentation is performed in different domains. However, it is intuitive to demonstrate this method for applications found in image and vision processing tasks. In the following, we will introduce a number of position augmentation methods and their MATLAB implementations. Additionally, we will present some color augmentation schemes and their MATLAB implementations. A brief visual overview of some of the implementations of the different data augmentation techniques is presented in Figure 8.7.

For position augmentations some of the common functions employed are scaling, cropping, rotating and flipping, padding, blurring, adding noise, and translating. There is also a set of color augmentation schemes such as changing the brightness, the contrast, the saturation, and the hue of an image.

We will first make use of the MATLAB Image Processing toolbox to do manual augmentation. However, for users of MATLAB 2018 and any later versions, augmentation has been made easier by the use of datastore objects. We will cover datastore objects and their use for data augmentation in view of DL in this section.

Using the Image Processing toolbox from MATLAB, many of the above-listed augmentation methods can be realized. MATLAB treats images as matrices. Hence, many of the functions to alter an image to create a larger data set are based on matrix manipulation. A number of data augmentation effects rely on performing a translation and/or rotation of the image/data. These operations can be formulated by using a transformation matrix. To showcase how MATLAB is utilizing these transformation matrices, consider the Cartesian coordinate system as shown in Figure 8.8 and the associated rotations corresponding to the yaw, pitch, and roll angles.

MATLAB associates a matrix representing an image as being seen from the roll axis. Hence, a rotation about this axis will employ the rotation defined by

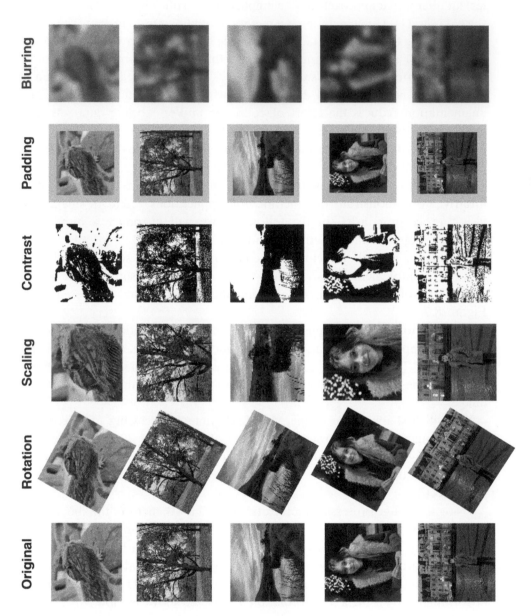

Figure 8.7 Different data augmentation implementations and their implications for a set of images.

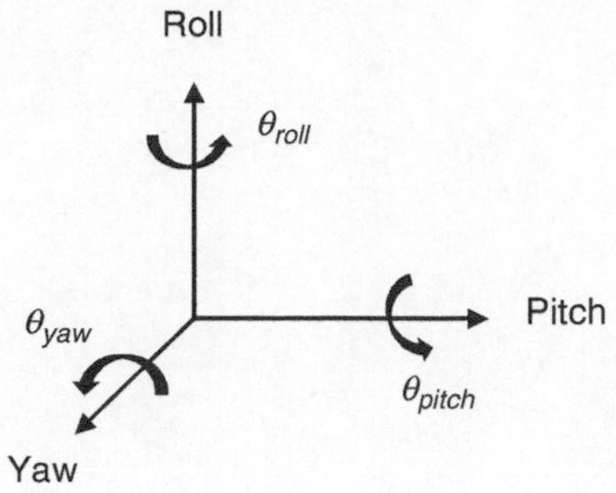

$$R_{yaw}\left(\theta_{yaw}\right) = \begin{bmatrix} 1 & 0 & 0 \\ 0 & \cos\left(\theta_{yaw}\right) & -\sin\left(\theta_{yaw}\right) \\ 0 & \sin\left(\theta_{yaw}\right) & \cos\left(\theta_{yaw}\right) \end{bmatrix}$$

$$R_{pitch}\left(\theta_{pitch}\right) = \begin{bmatrix} \cos\left(\theta_{pitch}\right) & 0 & \sin\left(\theta_{pitch}\right) \\ 0 & 1 & 0 \\ -\sin\left(\theta_{pitch}\right) & 0 & \cos\left(\theta_{pitch}\right) \end{bmatrix}$$

$$R_{roll}\left(\theta_{roll}\right) = \begin{bmatrix} \cos\left(\theta_{roll}\right) & -\sin\left(\theta_{roll}\right) & 0 \\ \sin\left(\theta_{roll}\right) & \cos\left(\theta_{roll}\right) & 0 \\ 0 & 0 & 1 \end{bmatrix}$$

Figure 8.8 Rotation transformation in three dimensions.

$R_{roll}(\theta_{roll})$, where the angle θ_{roll} is defined by the user. Let's start by reading an image and showing it to the user by employing the imread() and imshow() functions:

```
originalPicture = imread('tree.jpg');
imshow(originalPicture)
```

Example 8.5

In our case, we used a figure of a tree. MATLAB supports other file formats such as *.png. To accomplish a rotation we will perform an affine transformation with a random rotation angle θ_{roll}. The MATLAB function that defines the geometric transformation with a random attribute is called randAffine2d(). As we want to complete a rotation, we are using 'Rotation' as the first attribute, and as the second attribute we are defining the span of angles in degrees from which the random number generator is selecting the rotation angle:

```
traform = randomAffine2d('Rotation',[-65 25]);
```

The next task is to create a spatial referencing object that can be used for the MATLAB function imwarp():

```
outView = affineOutputView(size(originalPicture),traform);
```

The function imwarp() is the function that actually performs the transformation as specified by the geometric transformation object. As attributes it requires the image, the transformation object, a parameter that controls the specific aspects of the geometric transformation, and the spatial referencing object. In our case we are using the 'OutputView' to define the location and size of the output image.

```
imagAug = imwarp(originalPicture,traform,'OutputView',outView);
```

Figure 8.9 (a) original figure, (b) rotated figure using the MATLAB function `Rotation` `randomAffine2d()` with a random angle.

We can use the function `imshow()` to inspect the resulting transformation. The original figure and the transformed figure are depicted in Figure 8.9.

We can also inspect the actual transformation parameters that MATLAB used in order to perform the rotation:

```
≫ traform.T
    ans =
        0.5938  0.8046  0
        0.8046  0.5938  0
        0       0       1.0000
```

Accordingly, the angle picked by MATLAB is −53.5729 degrees, which is within the specified range [−65 25].

We use the same approach and function to perform a translation of the image; however, we specify a different type of transformation by selecting the appropriate parameters.

Example 8.6

In our example, we will perform a translation horizontally and vertically:

```
traform = randomAffine2d('XTranslation',[-2000 500],'YTranslation',[-1000 200]);
outView = affineOutputView(size(originalPicture),traform);
imagAug =imwarp(originalPicture,traform,'OutputView',outView);
imshow(imagAug)
```

Note that the function `randomAffine2d()` uses two parameters to define the shift in the x-direction and the y-direction. The value of the shift corresponds to the number of pixels, and hence is dependent on the resolution of the image. The original image for this example is 6,000 × 8,000 pixels. Our MATLAB code specifies that we allow a translation in the *x*-axis anywhere between −2,000 and +500 pixels, and a translation in the *y*-axis between −1,000 and 200 pixels. The corresponding output is shown in Figure 8.10. We can examine the actual translation value used by the random number generator of the `randomAffine2d(.)` function by using our transformation object `traform`:

```
≫ traform.T
      ans =
        1.0e+03 *
        0.0010        0         0
        0   0.0010   0
        -1.6060      0.1647   0.0010
```

Note that the main diagonal entries in the transformation matrix are populated with 1.0, implying that no rotation has occurred. The *x*-axis translation moved the image by 1,606 pixels to the left, while the *y*-axis translation moved the image down by 165 pixels.

Figure 8.10 Translation of an image in the *x*-axis and *y*-axis by a random number of pixels.

Another useful augmentation function is changing scale. Scale refers to zooming in or out of an image. MATLAB uses the same 2D utility `randomAffine2d()`. However, we need to specify that the transformation is referring to scale. The second set of parameters supplied to the transformation function is the magnitude of the zooming action. Using numbers greater than 1 will result in enlarging the image, while numbers smaller than unity will cause the image to shrink.

Example 8.7

To zoom in, we can use the following MATLAB code:

```
traform = randomAffine2d('Scale',[2, 2.2]);
outView = affineOutputView(size(originalPicture),traform);
imagAug =imwarp(originalPicture,traform,'OutputView',outView);
imshow(imagAug)
```

The numbers [2, 2.2] provide MATLAB with a range of values from which it can choose to perform the zooming action. Using our transformation object `traform`, we find that a value of 2.0092 was utilized to generate the figure shown in Figure 8.11(a).

```
≫ traform.T
     ans =
       2.0092     0    0
            0  2.0092   0
            0          0    1.0000
```

To zoom out, we change the range of values from which MATLAB can chose to numbers smaller than 1:

```
traform = randomAffine2d('Scale',[0.3, 0.5]);
```

which used a factor of 0.3681 – that is, the new image is only approximately 37% of its original size, as shown in Figure 8.11(b).

Figure 8.11 Using the attribute scale: (a) zooming into an image; (b) zooming out of an image using the default background; (c) defining the empty space with a gray background. Image (d) is the result of the flipping transformation.

In Example 8.7 MATLAB decided to pad the "empty" space with black pixels. The empty space refers to the difference between the original image size and the newly created image using scale. MATLAB provides for options to fill this new "empty" space with different pixel attributes.

Example 8.8

We can fill the empty space with a gray pixel set:

```
imagAug = imwarp(originalPicture,traform,'OutputView',outView,'FillValues',
[128 128 128]);imshow(imagAug)
```

The numbers in the array ([128 128 128]) refer to the RGB color scheme. RGB stands for *red*, *green*, and *blue*, and the protocol is basically an additive model where each color – identified by an integer – is added to create the composite color. In our case, the sequence refers to the color gray and can be seen in Figure 8.11(c).

Another manipulation of an image is the reflection operation. Reflection refers to the "flipping" of the image about any of the major axes. To accomplish this, we utilize the randomAffine2d() function. Since this function uses a random number generator, the chance that the flipping of the image occurs is 50% in each direction. The corresponding MATLAB code is:

```
traform = randomAffine2d('XReflection',true,'YReflection',true);
outView = affineOutputView(size(originalPicture),traform);
imagAug = imwarp(originalPicture,traform,'OutputView',outView);
imshow(imagAug)
```

The corresponding output image is shown in Figure 8.11(d). Looking at the transformation matrix, we find that the main diagonal elements are now negative unity, implying an inversion about the horizontal axis:

```
≫traform.T
     ans =
       1  0  0
       0 -1  0
       0  0  1
```

To shear an image allows us to move one part of an image in one direction and another part of the image in the opposite direction. MATLAB utilizes a shear angle along which this shift will occur. To perform shear, we utilize the same randomAffine2d() function to allow MATLAB to choose a random angle within a range of values:

Figure 8.12 Shearing of an image: the original image was moved along an axis defined by the random number generator.

```
traform = randomAffine2d('XShear',[-60 10]);
outView = affineOutputView(size(originalPicture),traform);
imagAug = imwarp(originalPicture,traform,'OutputView',outView);
imshow(imagAug)
```

The corresponding sheared image is shown in Figure 8.12.

Besides geometric transformations, MATLAB also allows for a number of color transformations that can be useful for augmenting a data set. The color augmentation schemes include changing the contrast, the brightness, the saturation, or the hue.

Example 8.9

To demonstrate brightness use the following MATLAB function:

```
imJittere = jitterColorHSV(originalPicture,'Brightness',[-0.3-0.1]);
```

The attribute `Brightness` can be changed to `Hue`, `Saturation`, and `Contrast`. The function works using the RGB color code on each pixel to alter the specifics based on the given attribute. For the brightness effect we specify two values to indicate the minimum and maximum value for the brightness.

To compare the altered image with the original image, we can use the MATLAB function `montage()`:

```
montage({originalPicture,imJittere})
```

The comparison is shown in Figure 8.13, where the left side is the original image and the right side is the altered image using the brightness attribute and the montage function to depict both images.

Figure 8.13 Montage of two images, the left side represents the original image, the right side is the image created by altering the brightness value.

One additional transformation that MATLAB provides – that is commonly used for data augmentation – is the addition of noise. Noise for images can be specified by several means. The MATLAB function allowing us to add noise is `imnoise()`. There are different noise characterizations possible, for example using the `salt & pepper` noise type, our MATLAB command will read as:

```
imagNoise = imnoise(originalPicture,'salt & pepper');imshow(imagNoise)
```

Other types are `gaussian`, `localvar`, `poisson`, and `speckle`. Images altered with `salt & pepper`, `speckle`, and `gaussian` are shown along with the original image in Figure 8.14.

These methods work well for data that can be represented in matrix form. However, the question remains of how to apply these augmentation techniques to time series data. It turns out that the approach is very similar to the above-listed methods, except that we have to be careful to maintain causality. As time series data represent information that can be expressed in the time domain as well as in the frequency domain, we can apply augmentation techniques in both domains, as well as in a time–frequency domain.

Time-domain augmentation techniques can be accomplished by injecting noise into the original data set. Commonly Gaussian noise is used, but also spikes or slope-like trends are utilized to alter the existing data.

The window wrapping method works by selecting a random range within the time series and either compressing or extending the data set by using downsampling or upsampling, respectively. Note, the labeling has to experience the same amount of elongation or contraction in terms of the time series. To perform downsampling we can employ MATLAB's `downsample()` function. For example, to downsample the noise-affected signal from the above example by a factor of 3, we can use the following expression:

```
yd = downsample(y,3);
```

Figure 8.14 Montage of four images with upper-left corner depicting the original image, upper-right corner the image altered by including `salt & pepper` noise, the lower-left image altered by including `speckle` noise, and the lower-right by applying `gaussian` noise.

Example 8.10

As an example, adding white Gaussian noise to a square wave form signal representing our time series is accomplished in MATLAB with the function `awgn()`:

```
k = (0:0.1:10)'; % create discrete time vector
x = square(k); % square wave form
y = awgn(x,10,'measured'); % additive white gaussian noise:awgn
figure
plot(t,x,'k.-')
hold
plot(t,y,'k-')
legend('Original Signal','Signal with AWGN')
```

The corresponding plot is shown in Figure 8.15. We made use of a signal-to-noise ratio of 10 as well as the attribute 'measured' in the MATLAB function `awgn()`. This attribute uses the `measured` signal power of the input. If this attribute is left out, MATLAB's default value for the noise power is 0 dBW.

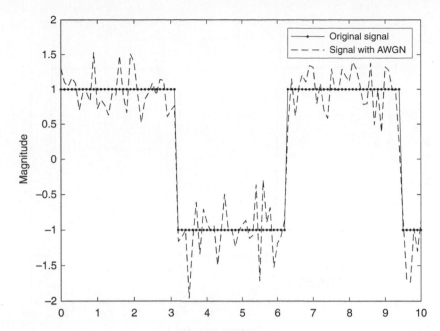

Figure 8.15 Additive white Gaussian noise with a square wave signal for time series augmentation.

For the upsampling approach, MATLAB offers a function that conveniently performs this task easily. Unfortunately, the resulting padding used by MATLAB to do the upsampling is done as a zero-padding procedure. This may introduce unrealistic signal characteristics into the time series; hence an alternate method is preferred. We suggest using the function `resample()`. This function allows for the original signal to be resampled at *p/q* times the original sampling rate. In addition, the function employs an antialiasing low-pass filter.

Example 8.11

As an example, the square wave signal without noise can be resampled at twice the original sample rate using the following MATLAB scripts:

```
fs = 10; % sampling rate
k = (0:(1/fs):10)'; % discrete time vector
x1 = square(k); % square wave signal
yu = resample(x1,2,1);
t2 = (0:(length(yu)-1))*1/(2*fs);
```

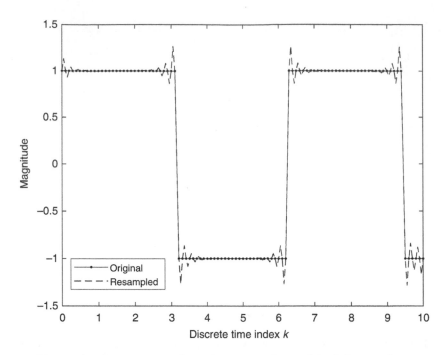

Figure 8.16 Time series data augmentation using upsampling: original discrete time square wave signal with a sampling rate of 10 is resampled using the MATLAB function `resample()` at a sampling rate of 20.

Because of the employed filter, the new signal shows the filter response superimposed onto the original signal. This can be observed in Figure 8.16 at the locations in time when the original signal instantaneously changes its magnitude. When applying up- or downsampling operations, the resulting new time series may be of different length. To restore the original data length, we can apply window cropping. Window cropping is quite similar to the cropping operation for images, where a subsample method is used to randomly select and delete continuous slices from the original data set.

Another time series data augmentation method is the flipping operation. This method is very similar to the one we discussed for images. For time series, the flipping involves the simple sign change of each data point. However, the label is kept the same.

There are also a number of frequency-domain based data augmentation algorithms in use for time series data. For example, Gao et al. [5] proposed using the Fourier transform of the time series data and randomly selecting segments of the spectrum and replacing those with Gaussian noise, where the noise is calibrated by the mean and variance of the original amplitude spectrum. The phase is treated similarly by randomly picking segments and adding zero-mean Gaussian noise to the phase spectrum.

MATLAB offers containers for mixed type time series data, such as `time-table`. These containers are composed of columns of equal length, starting with a column that includes the time stamp and followed by individual variables in each of the subsequent columns. MATLAB has some built-in tables which we can utilize to demonstrate the use and composition of these containers. One such data set MATLAB provides is a spreadsheet detailing humidity and air quality data. We can load this into a timetable container as follows:

```
indoors = readtimetable('indoors.csv');
```

The resulting timetable container includes the labels for each variable:

```
≫ indoors
  indoors =
  60×2 timetable
```

Time	Humidity	AirQuality
2015-11-15 00:00:24	36	80
2015-11-15 01:13:35	36	80
2015-11-15 02:26:47	37	79
2015-11-15 03:39:59	37	82
2015-11-15 04:53:11	36	80
2015-11-15 06:06:23	36	80
2015-11-15 07:19:35	36	80
2015-11-15 08:32:47	37	80
2015-11-15 09:45:59	37	79
2015-11-15 10:59:11	36	80
2015-11-15 12:12:23	37	80
2015-11-15 13:25:35	37	79
2015-11-15 14:38:46	36	83

```
...
```

MATLAB offers functions to synchronize different timetables even when those tables have different entries and row numbers. For example, if we wish to combine the table called `outdoors` – which is also a MATLAB-supplied sample table – with our previous table, we can utilize the MATLAB function `synchronize()`:

```
load outdoors
outdoors(1:7,:)
```

```
ans =
7×3 timetable
```

Time	Humidity	TemperatureF	PressureHg
2015-11-15 00:00:24	49	51.3	29.61
2015-11-15 01:30:24	48.9	51.5	29.61
2015-11-15 03:00:24	48.9	51.5	29.61
2015-11-15 04:30:24	48.8	51.5	29.61
2015-11-15 06:00:24	48.7	51.5	29.6
2015-11-15 07:30:24	48.8	51.5	29.6
2015-11-15 09:00:24	49	51.5	29.6

```
combtable = synchronize(indoors,outdoors);
combtable(1:7,:)

ans =
7×5 timetable
```

Time	Humidity_indoors	AirQuality	Humidity_outdoors	TemperatureF	PressureHg
2015-11-15 00:00:24	36	80	49	51.3	29.61
2015-11-15 01:13:35	36	80	NaN	NaN	NaN
2015-11-15 01:30:24	NaN	NaN	48.9	51.5	29.61
2015-11-15 02:26:47	37	79	NaN	NaN	NaN
2015-11-15 03:00:24	NaN	NaN	48.9	51.5	29.61
2015-11-15 03:39:59	37	02	NaN	NaN	NaN
2015-11-15 04:30:24	NaN	NaN	48.8	51.5	29.61

Note that the entries from the table outdoors do not align with the entries of the table indoors from a time stamp point of view. Hence, the function synchronize() utilizes a padding function where the missing information is filled with a missing data indicator (i.e., a NaN value). To fill those entries with appropriate values we can use linear interpolation, which is done by identifying this option within the synchronize() function:

```
combtable = synchronize(indoors,outdoors,'union','linear');
combtable(1:7,:)

ans =
7×5 timetable
```

Time	Humidity_indoors	AirQuality	Humidity_outdoors	TemperatureF	PressureHg
2015-11-15 00:00:24	36	80	49	51.3	29.61
2015-11-15 01:13:35	36	80	48.919	51.463	29.61
2015-11-15 01:30:24	36.23	79.77	48.9	51.5	29.61
2015-11-15 02:26:47	37	79	48.9	51.5	29.61
2015-11-15 03:00:24	37	80.378	48.9	51.5	29.61
2015-11-15 03:39:59	37	82	48.856	51.5	29.61
2015-11-15 04:30:24	36.311	80.622	48.8	51.5	29.61

Besides 'linear' interpolation, an option exists for using a spline fit. In addition, MATLAB offers a function to fill in missing values by computing replacement values using nearby data points. The neighborhood from which these values are computed can be defined by the user with attributes such as previous, next, and nearest:

```
combtable = fillmissing(combtable,'nearest');
combtable(1:7,:)

ans =
7×5 timetable
```

Time	Humidity_indoors	AirQuality	Humidity_outdoors	TemperatureF	PressureHg
2015−11−15 00:00:24	36	80	49	51.3	29.61
2015−11−15 01:13:35	36	80	48.9	51.5	29.61
2015−11−15 01:30:24	36	80	48.9	51.5	29.61
2015−11−15 02:26:47	37	79	48.9	51.5	29.61
2015−11−15 03:00:24	37	79	48.9	51.5	29.61
2015−11−15 03:39:59	37	82	48.9	51.5	29.61
2015−11−15 04:30:24	36	80	48.8	51.5	29.61

For stochastic time series data, MATLAB provides for a function that allows a times series to be filtered using options such as moving mean value and moving median value, among other options:

```
ymm=smoothdata(y,'movmean');
ymm=smoothdata(y,'gaussian');
```

For small data sets the above-discussed methods for data processing and data augmentation of images and time series data provide workable solutions. However, DL relies on large data sets, either acquired or generated using augmentation. Once large data volumes are being utilized, data storage and data handling may become a concern due to limited storage capacity and processing power. In such situations an online approach for data augmentation is preferred. An online approach refers to the processing of the data (i.e., augmentation) using mini-batches. This approach is very useful during training, as it helps to keep the size and hence storage requirements from growing beyond the limitations of the hardware.

Example 8.12

Suppose a data set is too large to be loaded into the computer's memory. However, we can create a datastore that points to the location and allows us to access, manipulate, and use the data contained at that location:

```
dast = datastore('indoors.csv');
```

The variable dast in the MATLAB script points to the location of the data entailed in the file indoors.csv. We can now create also a so-called tall table:

```
tl = tall(dast);
```

The created variable tl is an array supported by a datastore, and hence can entertain billions of rows.

Even though your computer may have some limitations on the storage capacity, using the conversion to a tall array allows you to work with this very large data set. These data sets are not limited to numerical data types; any MATLAB type can be included in these arrays.

To convert a table into a timetable container we can use the function `table2-timetable()`:

```
tita = table2timetable(t1);
```

Having a datastore, data augmentation of large data sets can now be easily accomplished. As an example, suppose we have a data set of images detailing different food items. Each item is placed in a corresponding folder that bears the category name. An example of such a data set is shown in Figure 8.17.

In order to create an `imageDatastore` object that includes the path, the subfolders, labels, and the folder names, we can use the MATLAB command `fullfile()`:

```
digDatSetPath = fullfile('C:\Users\Desktop\Food');
imDaS =
imageDatastore(digDatSetPath,'IncludeSubfolders',
true,'LabelSource','foldernames');
```

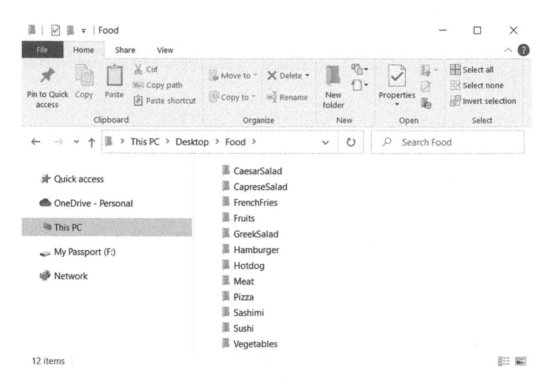

Figure 8.17 The folder *Food* with food categories, each containing images of food items belonging to that food type.

In the above code, the first instruction refers to the location of the data set on the drive. In this case we are pointing to a folder called 'Food.' The function `fullfile()` takes the location as a string and assigns it to the variable `digDatSetPath`. The second instruction entails the creation of an `imageDatastore` object called imDaS. This creation utilizes the function `imageDatastore()`, which requests information on the path where the data are located, and gives an instruction on how to label the different categories. In our case, we instruct MATLAB to use the given folder names as the label source. We can inspect the created `imageDatastore` object by calling it in MATLAB:

```
imDaS =

ImageDatastore with properties:

        Files: {
            'C:\Users\Desktop\Food\CaesarSalad\Caesar_1.jpg';
            'C:\Users\Desktop\Food\CaesarSalad\Caesar_2.jpg';
            'C:\Users\Desktop\Food\CaesarSalad\Caesar_34.jpg'
            ... and 975 more
            }
        Folders: {
            'C:\Users\Desktop\Food'
            }
        Labels: [CaesarSalad; CaesarSalad; CaesarSalad... and 975 more categorical]
    AlternateFileSystemRoots: {}
            ReadSize: 1
    SupportedOutputFormats: ["png" "jpg" "jpeg" "tif" "tiff"]
       DefaultOutputFormat: "png"
            ReadFcn: @readDatastoreImage
```

We can also inspect the number of files this new `imageDatastore` houses:

```
≫ size(imDaS.Files)
ans =

    978    1
```

To define the batch size, we set the attribute `ReadSize` of `imDaS` variable to 6:

```
imDaS.ReadSize = 6;
```

Applying a few data augmentation routines on the images contained in the datastore, we can utilize a MATLAB script that defines and performs the corresponding transformations.

Example 8.13

For example, if we want to apply noise using salt and pepper, perform a random rotation, and apply a scale, the following MATLAB script allows us to do so to the contains in the datastore:

```
function [augImag, info] = imageAugmentation(imagOrg, info)
index = size(imagOrg);
augImag = cell([size(imagOrg,1),2]);
for i = 1:index
    imInd = imagOrg{i};
    imInd = imnoise(imInd,'salt & pepper');
    transImag = randomAffine2d('Rotation',[-65
    25],'Scale',[0.7,1.25]);
    outView = affineOutputView(size(imInd),transImag);
    imInd = imwarp(imInd,transImag,'OutputView',outView);
    augImag(i,:) = {imInd,info.Label(i)};
end
```

The above **MATLAB** script accesses each image in the datastore individually, applies the `salt & pepper` property using the `imnoise()` function, applies the `randomAffine2d()` function with two attributes, one for rotation and one for scale, and defines the overall transformation object `outView`. Having defined the specifics for the set of augmentation operations, the function `imwarp()` applies the defined transformation to the images. The last line in the loop is to affix the corresponding category labels onto the newly created images.

To inspect the augmented images we use the functions `preview()` and `montage()`:

```
imPreview = preview(augData);
montage(imPreview(:,1))
```

For MATLAB 2018 and later versions there are some advanced data augmentation tools available that allow us to do wholesale type processing. The function `augmentedImageDatastore()` permits us to produce groups of augmented images by using predefined transformations. To configure the augmentation operations, the function `imageDataAugmenter()` is utilized. To demonstrate their utility we will make use of the preloaded image data set of handwritten numerical digits. You can load this data set into MATLAB using:

```
[Imags,Labels] = digitTrain4DArrayData;
```

The data set consists of a four-dimensional array, where `Imags` contains a $28 \times 28 \times 1 \times 5,000$ array. The 28×28 dimension refers to the size of the individual image, the 1indicates the array has one channel, and the 5,000 refers to the fact that there are 5,000 images of handwritten digits. The variable *Labels* is a categorical vector possessing the information on the labels for each image. To create an `imageDataAugmenter()` object for augmentation we use the following MATLAB command:

```
imageAugmenter = imageDataAugmenter('RandRotation',
[-20,20],'RandXTranslation',[-4 2],'RandYTranslation',[-4 2])
```

The object specifies that the images are randomly rotated between −20 and +20 degrees, as well as translated between −4 and 2 pixels in both the *x*-direction and the *y*-direction. We can display the properties of this transformation by calling the new object:

```
imageAugmenter =
imageDataAugmenter with properties:
FillValue: 0
RandXReflection: 0
RandYReflection: 0
RandRotation: [-20 20]
RandScale: [1 1]
RandXScale: [1 1]
RandYScale: [1 1]
RandXShear: [0 0]
RandYShear: [0 0]
RandXTranslation: [-4 2]
RandYTranslation: [-4 2]
```

Note that we did not specify any scale values or fill values, which are options with the `imageDataAugmenter()` object. Next, we can create an `augmentedImageDatastore` object by defining the size of each of the images to be a 28 × 28 pixel image:

```
imageSize = [28 28 1];
augimds =
augmentedImageDatastore(imageSize,Imags,Labels,'DataAugmentation',
imageAugmenter);
```

The resulting datastore contains the augmented images generated by the transform defined by `imageAugmenter`. To see a sample of the new images, we can use the preview and montage commands:

```
ims=augimds.preview();
montage(ims{1:6,1})
```

Example 8.14
As an alternate way to view the images, we can also use the command `read()`:

```
numBatch = ceil(augimds.NumObservations / augimds.MiniBatchSize);
for i = 1:numBatch
    ims = augimds.read();
    montage(ims{:,1});
    pause;
end
augimds.reset();
```

The loop allows you to inspect a batch of images at a time and proceed to the next by pressing the enter key on the keyboard. Figure 8.18(a) shows the resulting images generated with the `montage()` command and an image set of six, while Figure 8.18(b) shows one of the images resulting from the for-loop code.

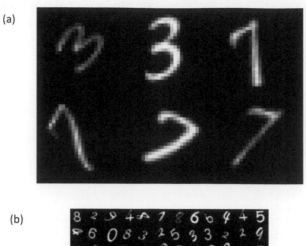

Figure 8.18 (a) using the `montage()` command, six augmented images are depicted in one composition; (b) using a for-loop and a batch size of 128 images, a subset of the augmented images is depicted.

8.5   Convolutional Neural Networks with MATLAB

We will develop CNNs in two different ways using MATLAB: (1) using MATLAB script and (2) using MATLAB's Deep Network Designer (DND) app. We start with the script approach by following the steps outlined below to construct a CNN:

Step 1: Create an `imageDatastore` object to load and label the data set. Preferably, use folders for each data set organized by individual folders for the different categories. To inspect the data, use the `imshow(.)` function:

```
digDatSetPath = fullfile('C:\Users\Desktop\Cars');
imDaS =
imageDatastore(digDatSetPath,'IncludeSubfolders',true,
'LabelSource','foldernames');
```

Step 2: Verify the number of items in each folder/category by using the function `countEachLabel(.)`:

```
labCount = countEachLabel(imDaS)
```

Step 3: For image-based CNN work, define the size of the input layer by utilizing the size of the images. If the images are gray scale the RGB value is 1:

```
imga = readimage(imDaS,1);
size(imga)
```

Step 4: Segment the data set into a set of training data and a set that is used for validation by using the MATLAB function `splitEachLabel()`:

```
numTrFiles = 1250;
[imadsTrain,imadsValid] =
splitEachLabel(imDaS,numTrFiles,'randomize');
```

Step 5: Construct the appropriate CNN architecture using the functions `imageInputLayer()`, `convolution2dLayer()`, `batchNormalizationLayer()`, `reluLayer`, `maxPooling2dLayer()`, `fullyConnectedLayer()`, `softmaxLayer`, and `classificationLayer()`:

a. `imageInputLayer()`: as we defined the size of the input by the size of the image in Step 3, the input layer has to match this size. For a 28×28 pixel image, we will define the input layer as:

```
imageInputLayer([28 28 1]);
```

Note, the last numerical value corresponds to the fact that we deal with gray scale images. If we were working with color images there would be a vector with three numbers according to the RGB values.

b. `convolution2dLayer()`: This function allows us to define the filter size (kernel), the number of filters, the padding option, and the stride size:

```
convolution2dLayer(3,40,'Padding',2);
```

The above command created a convolution layer with a 3×3 filter, 40 filters which results in 40 feature maps that are padded, and a stride size of 2. Note, MATLAB's default value for the stride size is 1.

c. `batchNormalizationLayer`: allows for normalization of the data flow through the network. This command does not entail any parameters, and is useful when applied to each layer, as well as the ReLU layer:

```
batchNormalizationLayer
```

d. `reluLayer`: creates the ReLU activation function:

```
reluLayer
```

e. `maxPooling2dLayer()`: This is the function which implements the max pooling operation. The parameters defining the operation are the pool size and stride size, using the name–value pair Stride,2:

```
maxPooling2dLayer(2,'Stride',2)
```

f. `fullyConnectedLayer()`: This layer establishes connections to all the neurons in the previous layer. The single parameter required for this layer defines the number of output neurons, which is equal to the number of categories we want to classify the input in:

```
fullyConnectedLayer(8);
```

In the above statement, we define that there are eight possible classes in which we want to categorize the given data.

g. `softmaxLayer`: The output layer is served by the data going through the softmax activation function, which is convenient as we relate the output to the probability a data point belongs to a specific category:

```
softmaxLayer;
```

h. `classificationLayer`: The classification layer uses the output of the softmax function to select the one category with the highest probability for assigning the classification of the input.

```
classificationLayer;
```

Step 5 parts (a)–(h) are packaged into a single statement where we define a variable that includes all the layer information. This can be achieved as follows:

```
cnnlayers = [
 imageInputLayer([28 28 1]);
 convolution2dLayer(3,8,'Padding',1);
 batchNormalizationLayer
 reluLayer

 maxPooling2dLayer(2,'Stride',2);

 convolution2dLayer(3,16,'Padding',1);
 batchNormalizationLayer
 reluLayer

 maxPooling2dLayer(2,'Stride',2);

 convolution2dLayer(3,32,'Padding',1);
 batchNormalizationLayer
 reluLayer

 fullyConnectedLayer(8);
 softmaxLayer
 classificationLayer];
```

When executing this script in MATLAB we create an array called cnnlayers. The content of this variable is:

```
cnnlayers =

15×1 Layer array with layers:
1 '' Image Input   28×28×1 images with 'zerocenter' normalization
2 '' Convolution   8 3×3 convolutions with stride [1 1] and padding [1 1 1 1]
3 '' Batch Normalization   Batch normalization
4 '' ReLU   ReLU
5 '' Max Pooling   2×2 max pooling with stride [2 2] and padding [0 0 0 0]
6 '' Convolution   16 3×3 convolutions with stride [1 1] and padding [1 1 1 1]
7 '' Batch Normalization   Batch normalization
8 '' ReLU   ReLU
9 '' Max Pooling   2×2 max pooling with stride [2 2] and padding [0 0 0 0]
10 '' Convolution   32 3×3 convolutions with stride [1 1] and padding [1 1 1 1]
11 '' Batch Normalization Batch normalization
12 '' ReLU   ReLU
13 '' Fully Connected   8 fully connected layer
14 '' Softmax   softmax
15 '' Classification Output   crossentropyex
```

Step 6: Define the training options for the constructed network. This includes what type of optimization algorithm, the learning rate, how many epochs are used for a training cycle, the specification of the validation data, how often these data are accessed, and whether we shuffle the training data during the training. The MATLAB script that defines all of these training options employs the function trainingOptions():

```
options =
trainingOptions('sgdm','InitialLearnRate',0.01,'MaxEpochs',
4,'Shuffle','every-epoch','ValidationData',imadsValid,
'ValidationFrequency',30,'Verbose',false,'Plots','training-
progress');
```

Step 7: Train the network using the training options defined in Step 6 using the function trainNetwork():

```
cnn_net = trainNetwork(imadsTrain,cnnlayers,options);
```

The network cnn_net will be trained using the training data contained in imadsTrain, by means of the architecture defined in cnnlayers, with the training options specified in options. When executing the above command a new window will open, depicting the training progress. Figure 8.19 shows the MATLAB window with the details of the training progress.

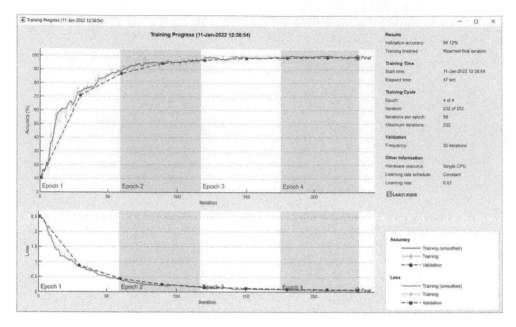

Figure 8.19 Progress indication during training of the created CNN.

The monitoring window MATLAB provides during the training process allows for inspection of a number of success indicators. These results are graphically shown on the left of the figure and by category on the right. For example, the 'Results' category will list the validation accuracy, which is the accuracy measured solely by the data we set aside for validation – that is, in the above example the data contained in `imadsValid`. It also shows whether the final iteration was reached. The other categories listed on the right of the figure are 'Training Time', 'Training Cycle', 'Validation', and 'Other Information.' The plots themselves represent the training accuracy over each individual mini-batch. In the example we have four epochs and the accuracy is given as a plot for each iteration, as a smoothed curve, and as a line approximation at a certain number of epochs. The second graph shows the loss at each mini-batch iteration, as well as a smoothed function and the loss based on the validation data.

Step 8: Use the trained network to compute the final validation accuracy. The validation accuracy is given by the ratio of the number of labels that the trained network predicts:

```
predClass = classify(cnn_net,imadsValid);
outVal = imadsValid.Labels;
accur = sum(predClass == outVal)/numel(outVal)
```

The variable `accur` provides for how many predicted labels are correctly classified using the validation data set.

An alternative way of constructing deep NNs is by using the DND app. This tool is an interactive visual guide to build, import, and edit networks. So far we only looked at CNNs; however, the DND app allows us to design, modify, and work with all types of network architectures. In addition to the construction, the DND app offers also analysis tools to inspect the created networks and to work with the corresponding data to do augmentation and visualization. Another useful feature of the DND app is the functionality of exporting networks not only to the workspace of MATLAB, but also to Simulink and to generate MATLAB code for the construction and training of those networks.

Example 8.15

There are two ways to start the DND: (1) use the command `deepNetworkDesigner` in the MATLAB command window, or use the Apps tab to find the DND app. When starting the DND app we are confronted with the initial window with three tabs: Getting Started, Compare Pretrained Networks, and Transfer Learning. In addition, the initial window lists an array of structure categories. The initial window of the DND is shown in Figure 8.20.

Selecting the "Blank Network" icon will open a new window that serves as a canvas for drawing a network by selecting and dropping components onto the canvas. We will recreate a very similar network to what we designed in the first part of this section,

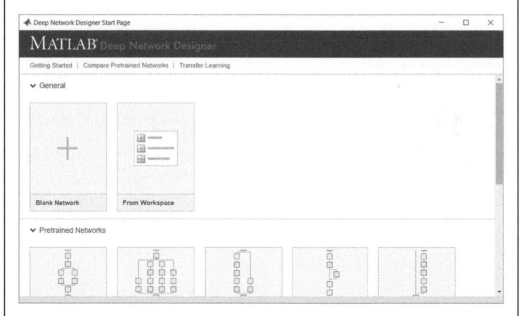

Figure 8.20 MATLAB's DND: the initial window with three tabs for selecting the type of work to be done with the DND as well as a selection of different network structures.

using many of the same parameters; however, we will do this using the DND and designing for 10 classes.

Step 1: Select the *imageInputLayer* icon from the input section of the Layer Library and drag it to the canvas (Figure 8.21). In the right column of the window we can specify the image input size. As we are designing for an image data set composed of gray scale images with a pixel size of 28 × 28, select the input size to be 28, 28, 1.

Step 2: Select the convolution layer `convolution2dLayer` (Figure 8.22) and set the size of the kernel to 3 × 3 (i.e., 3,3) and the number of filters to 8. Note that the "Padding" option lists as value "same," which corresponds to the default value of 1.

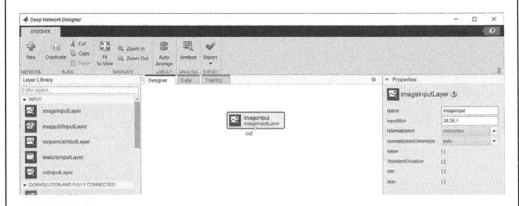

Figure 8.21 Input layer selection and parameter settings for input data.

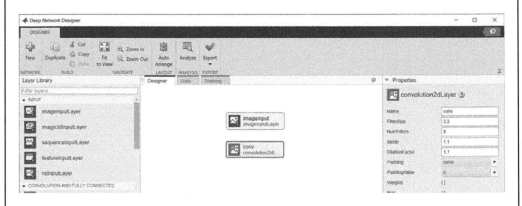

Figure 8.22 Convolution layer with selected properties on the DND canvas.

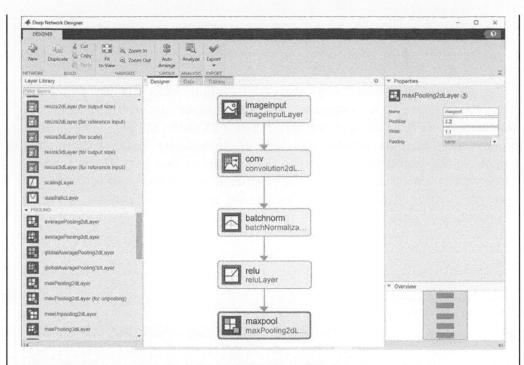

Figure 8.23 Initial network with one input layer, a convolution layer, a `batchnorm` layer, and an activation function layer using the ReLU activation function followed by the `maxpool` layer.

Step 3: Add a `batchnorm` layer, a `reluLayer`, and a `maxpool` layer to the network and connect the inputs and outputs of each element. Select the `maxpool` parameters from the previous example (Figure 8.23).

Step 4: Repeat the sequence of layers matching the architecture of the prior example. Add a `fullyconnectedLayer`, a `softmaxLayer`, and a `classificationLayer` to the end of the network flow chart. The final architecture is shown in Figure 8.24, with the initial parameters used.

Step 5: Obtain data for training the constructed network. Using the Data tab on the DND app window, training and validation data can be imported. The DND app supports both image data and datastore objects from the workspace. When selecting *Import Image Data* a new window appears with options to designate the source of the data, the folders with classes, and augmentation options. For this example we will reuse the files employed in Section 8.4 – that is, the preloaded data set provided by MATLAB detailing handwritten numbers:

```
dataFolder =
fullfile(toolboxdir('nnet'),'nndemos','nndatasets','DigitDataset');
imDas =
imageDatastore(dataFolder,'IncludeSubfolders',
true,'LabelSource','foldernames');
```

To import the data into the DND app, select Data source as `imageDatastore` in workspace and select the `imDas` object. The import window also allows for data augmentation. Figure 8.25 shows the import image data window with the data augmentation options. In addition to the augmentation functions, the validation data set is defined in the second part of the window. Here we can amend more data and categorize it as validation data, or we can split the current loaded data stored in the datastore object by providing a percentage of data to be used for the validation process.

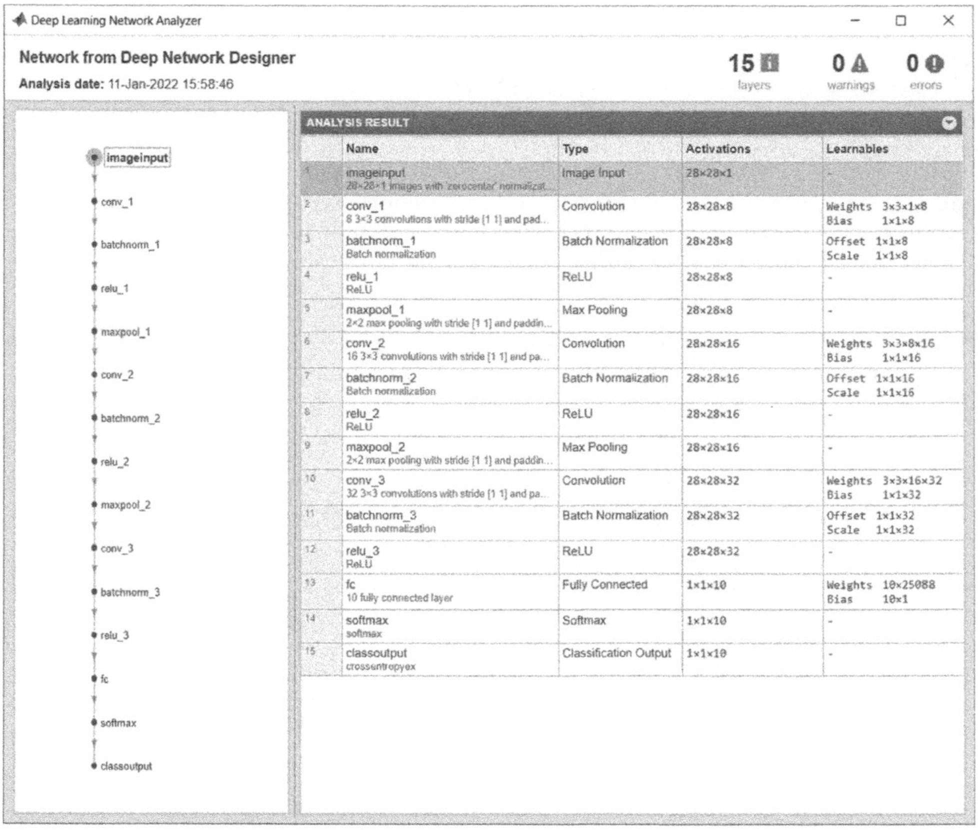

	Name	Type	Activations	Learnables
1	imageinput 28×28×1 images with 'zerocenter' normalizat...	Image Input	28×28×1	-
2	conv_1 8 3×3 convolutions with stride [1 1] and pad...	Convolution	28×28×8	Weights 3×3×1×8 Bias 1×1×8
3	batchnorm_1 Batch normalization	Batch Normalization	28×28×8	Offset 1×1×8 Scale 1×1×8
4	relu_1 ReLU	ReLU	28×28×8	-
5	maxpool_1 2×2 max pooling with stride [1 1] and paddin...	Max Pooling	28×28×8	-
6	conv_2 16 3×3 convolutions with stride [1 1] and pa...	Convolution	28×28×16	Weights 3×3×8×16 Bias 1×1×16
7	batchnorm_2 Batch normalization	Batch Normalization	28×28×16	Offset 1×1×16 Scale 1×1×16
8	relu_2 ReLU	ReLU	28×28×16	-
9	maxpool_2 2×2 max pooling with stride [1 1] and paddin...	Max Pooling	28×28×16	-
10	conv_3 32 3×3 convolutions with stride [1 1] and pa...	Convolution	28×28×32	Weights 3×3×16×32 Bias 1×1×32
11	batchnorm_3 Batch normalization	Batch Normalization	28×28×32	Offset 1×1×32 Scale 1×1×32
12	relu_3 ReLU	ReLU	28×28×32	-
13	fc 10 fully connected layer	Fully Connected	1×1×10	Weights 10×25088 Bias 10×1
14	softmax softmax	Softmax	1×1×10	-
15	classoutput crossentropyex	Classification Output	1×1×10	-

Figure 8.24 Constructed CNN architecture using the DND app and the corresponding layers with the selected parameters.

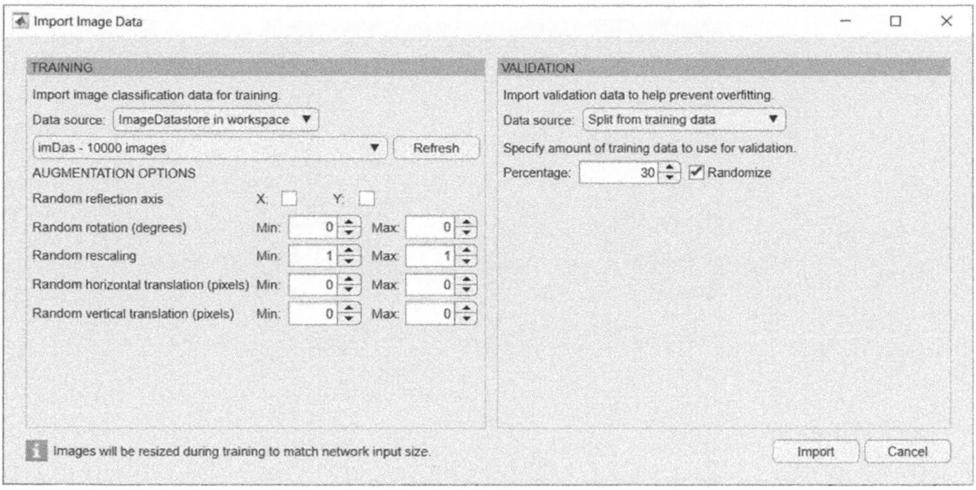

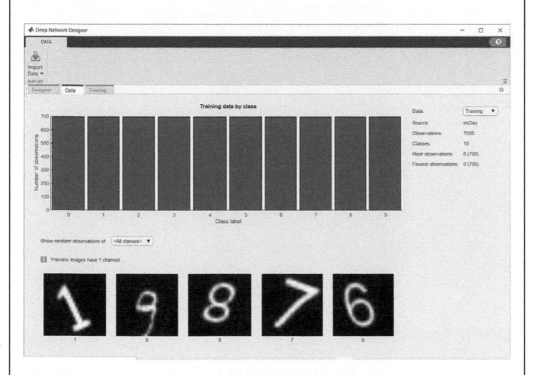

Figure 8.25 Import Image Data window for selecting training and validation data. In addition to loading data, the tool allows for data augmentation.

Figure 8.26 Import data with statistics using the DND app.

Once imported, the DND tool provides for a histogram of the training data by class as well as some basic information on the loaded data. This information is shown in Figure 8.26.

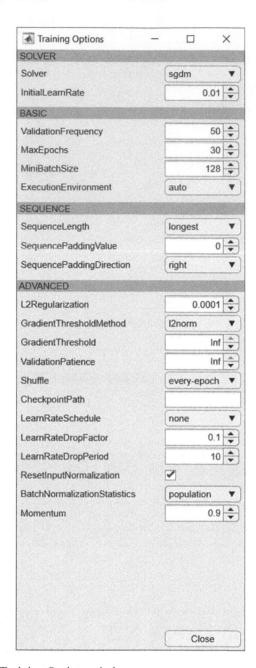

Figure 8.27 DND app Training Options window.

Step 6: Train the network: Using the third tab on the DND app, Training, we select the training options. These options are shown in Figure 8.27.

Once the training options are selected we can use the DND app and train the created network using the "Train" button (Figure 8.28).

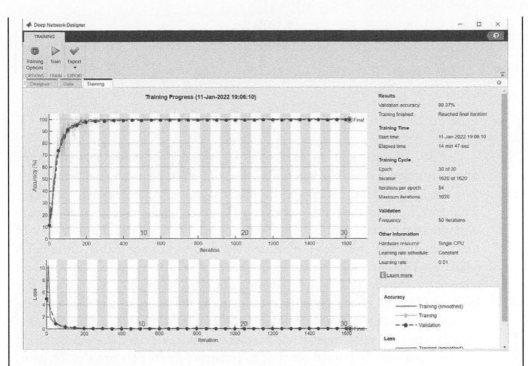

Figure 8.28 DND tool window after the training is completed.

Step 7: Exporting the trained network: In addition to training the created network, the DND tool also allows for creating the MATLAB script to import the data and train the network, as well as for exporting the trained network to the MATLAB workspace using the "Export" button in the DND tool.

Step 8: Test the trained network on sample data: for this step we will load a new picture, ensure that it is a gray scale image, and resize it to match the input layer of the trained CNN. If you are unsure about the input layer size, you can inspect this by going back to the DND tool and selecting the input layer to find the size in the Properties column. The following MATLAB script performs these tasks, including the conversion to a gray scale picture if it is an RGB image. The script also resizes the test image to be 28×28 pixels:

```
testImag = imread("testImage.png");
size(testImag)
testIg = rgb2gray(testImag);
tImg = imresize(testIg, [28 28]);
[pred,prob] = classify(trainedNetwork_1,tImg);
imshow(tImg)
label = pred;
title(string(label) + ", " + num2str(100*max(prob),3) + "%");
```

The corresponding output is depicted in Figure 8.29.

6,100%

6

Figure 8.29 Predicted number and prediction probability based on trained CNN using the DND app.

The development and, in particular, the training of deep NNs is heavily dependent on the available data set. As large data sets have a higher probability to embed the general underlying characteristics, network designers use large databases to train their DL networks. However, such undertakings require large computational capacities for processing and storage, and consume a lot of time. An alternate approach is often employed when neither of such resources is available. This alternative approach utilizes an already trained DL network and alters it for the new task or new data domain. This approach is termed transfer learning (TL). TL effectively attempts to expand the utility of a trained existing DL network so that it can be deployed to other data sets or solve different tasks than it was originally trained on. In simple terms, this is achieved by using common parts of the trained network to construct a new network that contains a few new parts or layers that need to be trained. As the number of parameters that need to be trained is greatly reduced, TL is rather effective in creating a new DL system with few resources.

Let's define the domain **D** as a two-element vector containing the feature space and the marginal distribution of the sample data:

$$\mathbf{D} = \{\mathbf{F}, P(X)\}. \tag{8.9}$$

where $\mathbf{F}$ is the feature space and X is the data set with $X = \{x_1, x_2, \ldots, x_N\} \in \mathbf{F}$. Graphically, we can distinguish between two different domains $\mathbf{D}_1$ and $\mathbf{D}_2$, as shown in Figure 8.30. In Figure 8.30 the domain $\mathbf{D}_1$ is much larger than the domain $\mathbf{D}_2$. The two domains have some commonalities, but they are not exactly the same. Hence, if a DL system was constructed and trained based on $\mathbf{D}_1$ it seems plausible to utilize this network and apply it to the domain given by $\mathbf{D}_2$. However, without any alteration the performance of this network operating on $\mathbf{D}_2$ is expected to suffer. TL can be applied when $\mathbf{D}_1$ and $\mathbf{D}_2$ are the same (or very similar) but the task the network solves is different. This type of TL is referred to as inductive transfer learning (ITL). When the domains are different but the tasks are the same or very similar, we speak of transductive transfer learning (TTL). For DL we use the ITL approach. Hence, ITL will use the same feature space $\mathbf{F}$ but update the marginal distribution $P(X)$ of the data set to predict the new label set. This is symbolized in Figure 8.30, where we have a similar feature space and two different tasks; task 1, which identifies what sound an animal makes, versus task 2, where we label where the animal is most likely found in the world. The

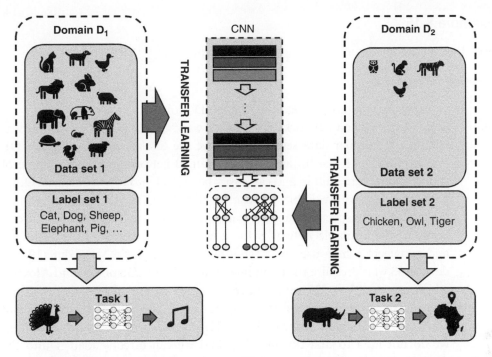

Figure 8.30 Two different domains entailing two sets of similar data. Each data set is aligned with a label set. One domain contains a much larger set than the other. Each domain is associated with a task, which can be the same or different.

simplest form of TL is where we copy the original DL system and substitute the last layer with a new one that is trained on the new task.

Transfer learning as described in the above paragraph is simple to implement in MATLAB, which we will demonstrate on an example below. However, for interested readers, the literature details more complex TL approaches that are based on fine-tuning of a trained network to allow for solving a new task. The steps for performing TL using MATLAB's DND app are as follows:

Step 1: Select a pre-trained network from the DND library.
Step 2: Use the DND app to select the new data set for which the modified network is adapted.
Step 3: Delete the last few layers and substitute new layers corresponding to the dimension of the new data set.
Step 4: Using the Training Option tab, change the learning rates for the new layers to be considerably faster than the ones used for the trained network.
Step 5: Use the DND app to train and export the altered network.

In the following we will demonstrate TL using MATLAB's DND tool with a MATLAB example of classifying food images that we store in different folders based on their category.

Example 8.16

Step 1: Start the DND tool by typing in the command line:

```
deepNetworkDesigner
```

or using the app tab and selecting the DND tool.

Step 2: Select a pre-trained network. For this we want to select a network that was trained on a similar data set as the one we are to classify with the new DL system. MATLAB has an array of pre-trained networks available. Some of these networks are built in and can easily be loaded into the workspace. Other networks that are available for use to classify or for TL require you to download the respective support packages using the "Add-On-Explorer." Some of the available networks for download are ResNet-18, ResNet-50, and googlenet, which use an image input size of 224 × 224, darknet19 and darknet53 which require the input images to be 256 × 256 pixels, and AlexNet which uses a 227 × 227 pixel image as input. In our example we will use SqueezeNet, which consists of 18 layers, has 1.24 million parameters and entertains images with a size of 227 × 227 pixels. As this network is built in we do not need to first download it, and we can directly load it into the workspace, as shown in Figure 8.31.

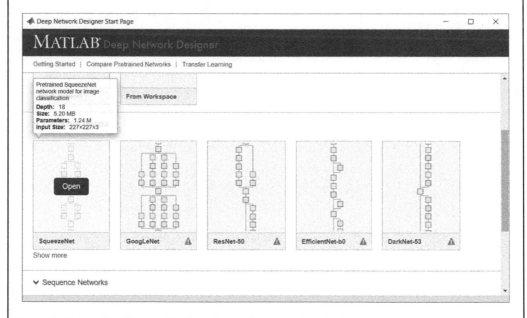

Figure 8.31 Loading SqueezeNet into the workspace using the DND app.

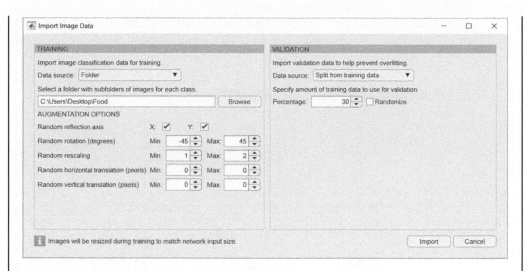

Figure 8.32 Loading of the data set using folders and data augmentation.

Step 3: Load the new data set. Use the DND Data tab to import the data. The *Import Image Data* window allows us to modify the current data set by using augmentation. As we have a limited data set for TL, we allow DND to augment the data set using random rotation, random rescaling, and random reflection on the *x*- and *y*-axes. The settings are shown in Figure 8.32. Note that we have a choice to select the percentage for splitting the data for training and for validation. Once you have finalized the selection, click the "Import" button. MATLAB will produce a summary page with a histogram detailing how many items are contained in each category and how many categories there are.

Step 4: Edit the network to conform to your data set. As the pre-trained network most likely has a different number of categories, we will need to adjust this to the new network we are creating using TL. This is accomplished by removing the final layer used for classification as well as the convolutional layer prior to this. This convolutional layer is referred to as the "last learnable layer" by MATLAB documents and needs to be identified when using other pre-trained networks. SqueezeNet's last learnable layer is labeled as `conv10`. In place of these layers we add a new `convolution2dLayer` from the Layer Library and change `FilterSize` to 1,1 and the number of filters detailed by `NumFilters` to the number of classes our new data set contains. Delete the output layer and add a new `classificationLayer`. The layers are identified in Figure 8.33 along with the new parameters. Note that our example uses nine classes, hence the number of filters in the new convolution layer is set to 9. In addition to the number of filters, it is common to increase the `WeightLearnRateFactor` as well as the `BiasLearnRateFactor` as we have most of the network already trained and need to train the new convolution layer and the output layer only. However, for the classificationLayer we do not need to specify the number of classes, as this is automatically detected by MATLAB and adjusted.

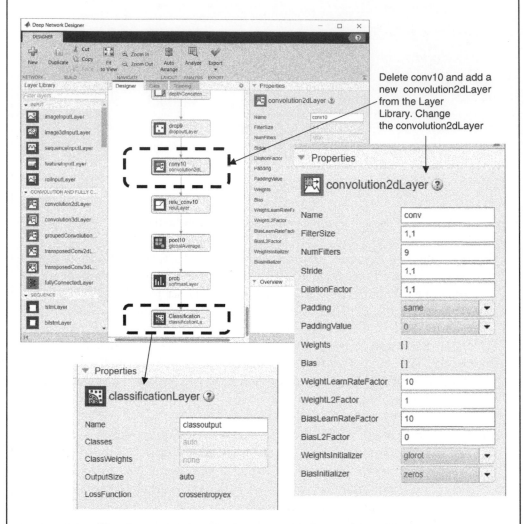

Figure 8.33 Modification to the pre-trained network "SqueezeNet" by deleting the last trainable layer as well as the classification layer and replacing them with a new convolution2dLayer as well as a new classificationLayer, while adjusting the given parameters.

Step 5: Test the new network for errors prior to training. Using the Analyze tool of the DND we can check for any errors prior to training of the new network. The test results are presented in a new MATLAB window with indication of any errors or warnings. Note that when inspecting the new elements, the new size – in our case nine different output classes – is used in the new network, as shown in Figure 8.34.

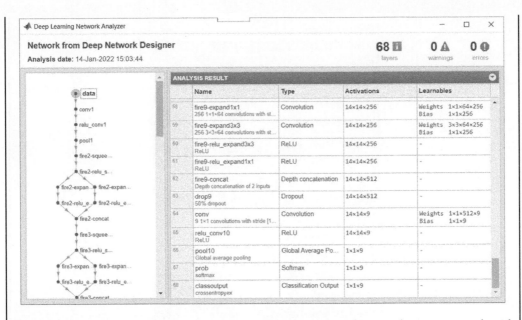

Figure 8.34 Using the Analyze tool of the DND app, we can inspect the new network and identify any errors or warnings. In addition, the analysis results also present us with the parameters for each layer, including the newly substituted layers.

Step 6: Train the new network with the given data set. As we utilize a pre-trained network we want to retain the knowledge from the prior training to some extent. This knowledge is represented by the trained features, weights, biases, etc. Hence, we will set the training options of the transferred layers so that the initial learning rate is very small. In addition, we will select a small number of epochs as less training is required due to the pre-training. For this we also need to set the mini-batch size to preferably be proportional to each category – that is, since we have 9 categories and about 700 images for the new data set, we divide 700 by 9 and set this to be our new mini-batch size (i.e., 78). The settings for the training of this example are shown in Figure 8.35. To initiate the training, press the "Train" button in the DND window.

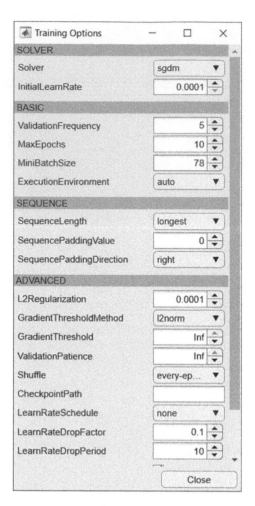

Figure 8.35 Training parameters for the new network. Using a small initial learning rate allows us to retain the pre-learned information of the SqueezeNet.

In certain cases the training results may not meet the expected validation accuracy. To improve this there are a number of options and different possible causes. The following is a small list of items to account for when performing TL:

- A common mistake when using pre-trained networks for TL is to mismatch the size of the new data set (i.e., for images the number of pixels of the new data set has to match the number of pixels of the images used to generate the pre-trained network originally). An easy and convenient approach to avoid input size mistakes is to use the `augmentedImageDatastore` function:

```
augimDas = augmentedImageDatastore(inputSize,imDaS)
```

- Having different numbers of items in each class will result in the trained network having a bias toward the class with the most items used in the training process. Besides data augmentation to equalize the number of items in a class, one can also use the `ClassWeights` option given in the `classificationLayer`.
- Training of the network can be extended by increasing the number of epochs. This is done in the `trainingOptions` by changing the `MaxEpochs` settings.
- Selecting the appropriate validation data size is important: If the validation data size is too large, the training may slow considerably. If the validation data size is too small, the validation may not represent the actual accuracy of the network. The validation data size `ValidationData` can be changed in `trainingOptions`.

With regard to the utilized pre-trained network, it is of note that the original SqueezeNet [6] was introduced in 2016 for computer vision applications, and primarily compared with AlexNet in terms of parameters and accuracy.

8.6 Long Short-Term Memory Network

One of the more popular RNNs is the LSTM network. It was first introduced by Hochreiter and Schmidhuber in 1997 [7] and features a memory and several gate components. Generally, for RNNs we connect the output of a neuron in the hidden layer with the input of the same neuron, hence the name *recurrent*. This loop allows us to access the data from the previous time step when dealing with time series data. However, a typical RNN is challenged when dealing with time series and systems that require storage of data that span more than 10 time steps. The popularity of LSTMs is based on their capacity to entertain many more time steps than a regular RNN. The concept of recurrence is also present in LSTM networks; however, a key modification is found in the memory unit of an LSTM network. For RNNs a simple activation function is employed to process the recurring data points. For LSTMs a rather complex architecture of processes is used. LSTMs employ a cell state that can be modified by deleting or adding new information. This modification is controlled by gates. Gates in the LSTM network are constructed by using a sigmoid NN layer that generates an output that is bound by [0 1]. Here, a 0 represents a gate closure, while a value of 1 implies that the gate is open. For example, to remove – or forget – a state the LSTM makes use of a dedicated forget gate. We can express the functioning of an LSTM cell by considering a data sequence we want to model – for example, the time series sequence $\vec{x} = \{x_1, x_2, \ldots, x_m\}$. When considering an LSTM cell we also need to include the hidden state vector of a neuron, which is given by $\vec{h} = \{h_1, h_2, \ldots, h_m\}$. Then an LSTM with a forgetting gate computes its activation vector according to:

$$f_k = \sigma_g\left(\mathbf{W}_f^{(1)}x_k + \mathbf{W}_f^{(2)}h_{k-1} + b_f\right), \tag{8.10}$$

where $\mathbf{W}_f^{(i)}$ and b_f are weight matrices and bias vectors, respectively, that are learned during the training phase of the LSTM, and k is the discrete time index. Similarly, the input i_k, output o_k, and cell activation vectors $\widetilde{c}_k$ at time k are governed by gates as well and can be computed as:

$$i_k = \sigma_g\left(\mathbf{W}_i^{(1)}x_k + \mathbf{W}_i^{(2)}h_{k-1} + b_i\right), \tag{8.11}$$

$$o_k = \sigma_g\left(\mathbf{W}_o^{(1)}x_k + \mathbf{W}_o^{(2)}h_{k-1} + b_o\right), \tag{8.12}$$

$$\widetilde{c}_k = \sigma_c\left(\mathbf{W}_c^{(1)}x_k + \mathbf{W}_c^{(2)}h_{k-1} + b_c\right), \tag{8.13}$$

where σ_g is the sigmoid activation function and σ_c is the hyperbolic tangent activation function. With these quantities we can update the cell state vector

$$c_k = f_k \circ c_{k-1} + i_k \circ \widetilde{c}_k. \tag{8.14}$$

Using the newly computed cell state the associated hidden state vector is also updated according to:

$$h_k = o_k \circ \sigma_c(c_k). \tag{8.15}$$

The corresponding weights and biases are learned during the training process and the initial values for the state and neuron outputs are set to zero. Figure 8.36 depicts the simplified LSTM cell given by the above version of the LSTM network.

There are many nuances and modification found in the literature addressing different aspects and improvements of the LSTM. However, our objective is to use MATLAB's implementation for time series data that are commonly found in control applications. Figure 8.37 shows two simple architectures used for implementing an LSTM in MATLAB for two different tasks: classification or regression. The classification task employs the softmax activation function for computing the probabilities of each class, which is not needed for the regression task using LSTM networks.

Many of the MATLAB-generated DL algorithms can be deployed onto hardware in different ways. Since MATLAB R2021a it is possible to generate C/C++ code that is independent of third-party libraries. To do so you will need to set the configuration as follows:

```
cfg.DeepLearningConfig = coder.DeepLearningConfig('TargetLibrary', 'none');
```

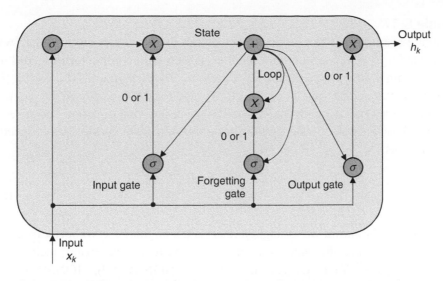

Figure 8.36 LSTM cell with three gates: one for the input, one to regulate when to forget the current data point, and one for controlling the output of the cell.

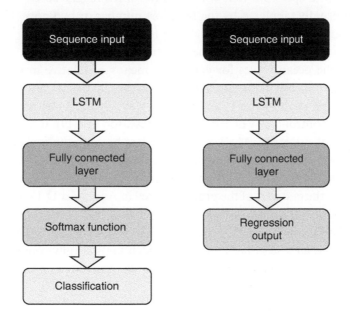

Figure 8.37 LSTM implementation using MATLAB for classification or regression tasks.

Once generated, you can use different blocks to implement the code in Simulink, and then transfer it to your hardware. For example, the C-Function block allows you to access external C code from the Simulink environment. See the MATLAB documentation for the latest options for hardware implementation.

Example 8.17

We will use a simple LSTM network to model and predict ocean waves. Ocean waves are irregular waves and depend on a number of factors, including location and wind speed. A prominent wave spectrum was developed by Pierson and Moskowitz [8] to model ocean waves. The foundation for this wave model is the assumption that the sea is fully developed (i.e., a steady-state situation when the wind direction and speed has not changed for waves to find an equilibrium state). The Pierson–Moskowitz spectrum can be given as:

$$S_{PM} = \frac{5}{16} H_S^2 \omega_p^4 \omega^{-5} e^{\left[-\frac{5}{4}\left(\frac{\omega}{\omega_p}\right)^{-4}\right]}, \tag{8.16}$$

where H_S is the significant wave height and ω_p is the angular spectral peak frequency. There are various modifications and simplifications to this spectrum, including the Joint North Sea Wave Observation Project (JONSWAP). JONSWAP uses a modification factor to accommodate a not fully developed sea state. For the purpose of demonstrating an LSTM network we employ a simple implementation of the Pierson–Moskowitz spectrum that computes the wave height based on wind speed. Wind speed is usually tabulated by the Beaufort number or by knots. The characterization and associated wind speeds are shown in Table 8.2.

Wave height information is useful when controlling wave energy converters (WEC). WECs harvest the ocean waves for energy by means of relative motion. There are a number of different designs currently in use. However, many of these benefit from the knowledge of wave heights prior to an incoming wave, hence short-term prediction of wave heights for controlling WECs is an active research field [9]. In the following, the implementation of a Pierson–Moskowitz wave height computation in MATLAB is presented, followed by the construction and training of a simple LSTM network. Note that we choose to use the MATLAB script option for realizing this example. With the data generated in the first part of the program listed in Appendix D.13, one can also use the DND app to construct the LSTM.

In the MATLAB script detailed in Appendix D.13 we employ an LSTM network with one input layer and an LSTM layer with 250 hidden units, a fully connected layer, and a regression output layer. The training window is also generated and depicts the RMS

Table 8.2 Wind speed and associated characterization labels

Characteristics	Beaufort no.	Knots	km/h	m/s
Calm	0	1	1.855	0.515
Light breeze	2	5	9.275	2.576
Moderate breeze	4	15	17.28	4.800
Strong breeze	6	25	46.375	12.882
Moderate gale	7	30	55.65	15.458
Strong gale	9	45	83.475	23.188
Storm	11	60	111.3	30.917

error value as well as the loss value during each epoch of the training, as shown in Figure 8.38. The network is used to predict approximately 100 time steps of wave heights, which corresponds to a prediction horizon of 10 seconds. This prediction is compared with the actual wave forms, as shown in Figure 8.39. Note that the prediction could be greatly improved by optimizing the hyperparameters of the network as well as the training options. However, the goal of this example is to show the simplicity of generating an LSTM network and how to train such a network using time series data.

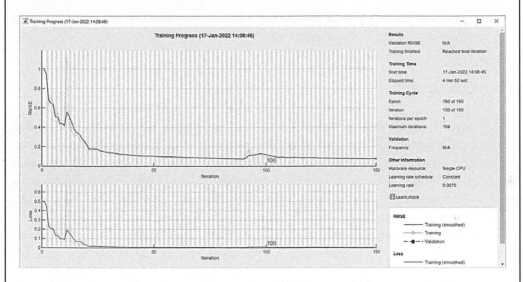

Figure 8.38 Training of a simple LSTM network depicting the RMSE and loss value over the number of iterations.

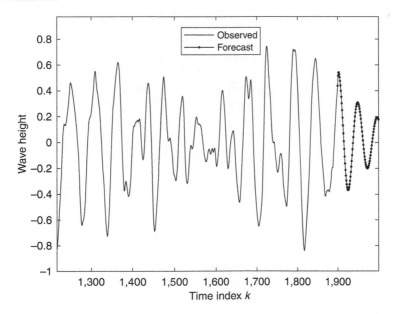

Figure 8.39 Wave height data generated using the Pierson–Moskowitz spectrum and predicted wave height data using the LSTM network.

When experimenting with deep CNNs a useful resource is a set of large data repositories. There are a number of tools available online. A small subset of these resources is summarized in Table 8.3. Note that some of these resources allow you to find data libraries, while others are data repositories.

In addition to the online data search tools and repositories, MATLAB provides for a rich set of data, which is preloaded with your MATLAB installation. A selected few are listed in Table 8.4. To load any of these data sets, use:

```
load datasetname.mat
```

Most of the listed *.mat files in Table 8.4 contain descriptions of the different variables and the data themselves.

8.7   Conclusions

Based on our knowledge from Chapter 6, in this chapter we expanded the size and capabilities of NNs by introducing DL concepts. There are four basic steps involved when creating DL systems: (1) data collection; (2) data analysis; (3) DL structure selection and training; and (4) DL deployment. This chapter included discussions and introductory material covering many of these concepts needed to create a DL system using MATLAB. As backpropagation is utilized in the training of DL networks, the existence of multiple hidden layers caused issues with the updates of

Table 8.3 Internet-based data set resources

Type	URL (as of 2022)
Search	https://datasetsearch.research.google.com/
Search	https://visualdata.io/discovery
Search	www.kaggle.com/datasets
Autonomous driving dataset	https://bdd-data.berkeley.edu/
COVID-19 dataset	www.semanticscholar.org/cord19
Education statistics	https://datacatalog.worldbank.org/search/dataset/0038480
Finance data	https://finance.yahoo.com/
Global data	https://ourworldindata.org/
Human Genome data set	www.hagsc.org/hgdp/files.html
Images	https://image-net.org/
Images of dogs	http://vision.stanford.edu/aditya86/ImageNetDogs/
Images of faces	https://data.vision.ee.ethz.ch/cvl/rrothe/imdb-wiki/
Images of pets	www.robots.ox.ac.uk/~vgg/data/pets/
Materials data set	https://materialsdatafacility.org/
Medical information data set	https://mimic.mit.edu/
US government data	https://datausa.io/
Videos of human actions	https://deepmind.com/research/open-source/kinetics
World Bank data set	https://datacatalog.worldbank.org/search/dataset/0037800

Table 8.4 MATLAB-provided data set resources

File name	MATLAB's description
arrhythmia.mat	Cardiac arrhythmia data from the UCI machine learning repository
carbig.mat	Measurements of cars, 1970–1982
cereal.mat	Breakfast cereal ingredients
cities.mat	Quality of life ratings for US metropolitan areas
examgrades.mat	Exam grades on a scale of 0–100
gas.mat	Gasoline prices around the state of Massachusetts in 1993
hospital.mat	Simulated hospital data
humanactivity. mat	Human activity recognition data of five activities: sitting, standing, walking, running, and dancing
lawdata.mat	Grade point average and LSAT scores from 15 law schools
mileage.mat	Mileage data for three car models from two factories
morse.mat	Recognition of Morse code distinctions by non-coders
polydata.mat	Sample data for polynomial fitting
popcorn.mat	Popcorn yield by popper type and brand
spectra.mat	NIR spectra and octane numbers of 60 gasoline samples
stockreturns.mat	Simulated stock returns

weights and bias terms during the training process. In particular, the gradient information became either too large, which we termed exploding gradient, or too small, which is referred to as the vanishing gradient problem. Solving such issues contributed greatly to the development of DL and hence its popularity. The solutions to these problems entail the use of specific activation functions, primarily the ReLU activation function, and a concept called dropout. The latter term refers to the process where we randomly leave out nodes in the training process, and effectively train at a given instance a smaller network. One of the more popular DL structures is the CNN, which is often used for image processing. A typical CNN for classification is structured in two parts: a feature-learning part, which is composed of a deep NN, and a classification part, which is constructed by a flattened and fully connected network. As DL concepts require large amount of data, we introduced a number of data augmentation methods in the form of a MATLAB script as well as embedded in MATLAB's datastore objects. Augmentation is easy to visualize for image-based data; however, we also found ways to expand the data set for training of time series data using augmentation with simple MATLAB commands.

Transfer learning is particularly useful when adapting an existing DL system to a new data set or a new task. We labeled these two types of TL as transductive transfer learning and inductive transfer learning, respectively. There are an ever-increasing number of pre-trained deep NNs to choose from, and we found that MATLAB provides easy access to many of these networks with its DLN tool. A specific DL system for time series data is the RNN. The RNN employs a feedback structure to accommodate a history of time samples. However, its time horizon into the past is limited. Hence, LSTMs networks are rather useful for data representing long time histories. Such data may be found in control applications, but also in applications of time series modeling for processes related to economics, environmental engineering, biology, and other domain-specific fields.

In the next chapter we will look at ML, where the feature engineering process is left for the ML designer. This is in contrast to the concept we encountered in this chapter, where CNNs were able to create their own features automatically.

SUMMARY

In the following, a number of key concepts of deep learning as treated in this chapter are summarized.

Machine Learning and Deep Learning

Machine learning and deep learning are part of artificial intelligence. They use data to construct models that are capable of making predictions or rendering decisions. The data for the training of machine learning models is first processed using manually derived features. For developing deep learning models those features are automatically constructed within the DL algorithm. Deep learning is considered part of machine learning, with the distinction of having automatic feature generation and being constructed as deep neural networks – that is, neural networks with more than one hidden layer.

Vanishing and Exploding Gradient

During the training of deep neural networks, the error of the output of the network is backpropagated through the network using the gradient of the loss function. Backpropagation is used to update the weights and biases of the network. When the gradient becomes too small to affect the updates of the weights and biases we refer to this as the vanishing gradient problem. When the gradient grows without bounds, we are dealing with the exploding gradient problem. To remedy either situation one can utilize different activation functions, such as the ReLU function.

Regularization

Regularization is used to address the generalization problem in deep neural networks. Generalization is the ability of a network to perform well when used on new data. Poor generalization is often caused by overfitting. Regularization is essentially the modification of the loss function by adding a suitable term. In deep neural networks the additional term often addresses the complexity of the network. There are two popular regularization methods in use for deep neural networks: L1 regularization (sometimes also referred to as Lasso regularization) and L2 regularization (also known as Ridge regularization). When using L1 regularization the outputs of the network's features are scaled to be within 0 and 1, and essentially address the number of features in the network. L2 regularization attempts to include the weights of the features of the network in the loss function, causing some of the weights to approach zero and hence reduce the complexity of the network.

Dropout

Dropout is another method to address the overfitting problem in deep neural networks. The method randomly selects a set number of neurons in the hidden layers and sets these to zero during the training phase. The effect of a neuron being set to zero is equivalent to a simpler overall network; however, at each iteration different neurons are selected to be dropped out of the training process.

Convolution

Convolution in deep neural networks allows for the automatic generation of features. The process, when applied to matrices or image data, performs a convolution operation on the matrix or image along with a kernel to produce another matrix, which represents the extracted feature. Different kernels have different effects and produce different features.

Pooling

Deep neural networks tend to utilize a large number of parameters. The pooling operation in a network attempts to reduce the number of parameters of the network by minimizing the spatial size in the network. Pooling layers are often used after a convolution layer.

Structure of a CNN

The structure of a convolutional neural network is composed of several segments that entail each a convolution layer, a batch normalization layer, an activation function layer, and a pooling layer. This is followed by one or more fully connected layers.

Data Augmentation

When there is insufficient data to train a network, data augmentation allows for the addition of data that is derived from the existing data set. The new data is generated by modifying the existing data in small ways, such as flipping, adding noise, or geometric changes to matrix- or vector-based data. Data augmentation is also a form of regularization and hence has an effect to reduce overfitting.

Epochs, Batches, and Mini-Batches

During training of a deep neural network the training data are used to update the different parameters of the network. When the training algorithm utilizes the entire training data once, we refer to it as one epoch. When we cannot utilize all of the training data in one iteration we use a batch of the training data. Hence, a batch or a mini-batch is a subset of the training data which is used to compute the gradient and hence the weights and biases of the network.

Batch Normalization

To create some independence between the layers of a deep neural network during the training phase, batch normalization is used. Batch normalization adjusts the mean values and variances of each of the inputs to each layer in a deep neural network, resulting in a centering and scaling of these signals. It has the effect of generating a faster training phase by using higher learning rates.

Transfer Learning

Transfer learning allows us to utilize prior-trained networks to compose and train new networks that can solve similar tasks as the original network. Usually, the original network is trained on a large data set, while the new network is generated by utilizing a smaller data set.

LSTM Network

Long short-term memory networks are a special type of recurrent network that use a number of gates in a cell structure to accommodate long time sequences. LSTM networks are used in time series modeling and sequence prediction problems.

REVIEW QUESTIONS

1. What is the difference between ML and DL?
2. What do we refer to when we talk about domain knowledge?
3. What are the four steps in the workflow of a DL system?
4. What is achieved by the reduction of dimension when applied to an image data set?
5. What is the curse of dimensionality?
6. What is the exploding gradient problem?
7. What type of activation functions are prone to resulting in the vanishing gradient problem during training of DL systems?
8. What is an L1 regularization term?
9. When is a convolution operation equal to cross-correlation computation?
10. What is the sparse connectivity approach?
11. What does it mean if parameter sharing is used in a CNN?
12. What is the difference between max pooling and average pooling?
13. What is the batch normalization process and what effect does it have on the learning process of a CNN?
14. What is data augmentation?
15. Changing the scale of an image will result in what type of effects?

16. What does shearing of an image result in?
17. What is downsampling?
18. What does it mean when we "flip" a time series?
19. What is a timetable in MATLAB?
20. In MATLAB, what is a datastore object?
21. What kind of data types can a tall array store?
22. What is an `augmentedImagestore()` function?
23. What are the components of a CNN?
24. Given an image data set of color images each having 128×128 pixels, what is the size of the input layer of the CNN trying to model the data set?
25. What is the difference between an ReLU and a Softmax activation function? When are they used and why?
26. What do we define with the function `trainingOptions()`?
27. What kind of networks can the DND app construct?
28. Can a trained classification network be used for another classification task?
29. When doing transfer learning, what components of the original network are kept and what components are newly constructed?
30. When training a network that was established using transfer learning, why are we using a slower learning rate for pre-existing layers compared to the new layers?

PROBLEMS

8.1 Consider the following data matrix and kernel:

$$
\mathbf{I} = \begin{bmatrix}
-1 & -1 & -1 & -1 & -1 & -1 & -1 & -1 \\
-1 & 0 & 1 & 4 & -1 & 2 & 3 & 0 \\
-1 & 1 & 2 & 0 & 4 & 8 & 2 & 0 \\
-1 & 4 & 1 & 3 & 6 & 3 & 1 & 0 \\
-1 & 3 & 2 & 5 & 9 & 7 & 1 & 0 \\
-1 & 5 & 7 & 8 & 5 & 3 & 0 & 0 \\
-1 & 3 & 3 & 2 & 3 & 1 & 1 & 0 \\
0 & 0 & 0 & 0 & 0 & 0 & 0 & 0
\end{bmatrix}, \mathbf{K} = \begin{bmatrix}
1 & 1 & 2 \\
1 & 2 & 3 \\
2 & 3 & 4
\end{bmatrix}.
$$

    a. Compute the convolution of **I** and **K** using adjacent value padding.
    b. Compute the correlation of **I** and **K** using adjacent value padding.

8.2 Compute the convolution and correlation of the matrices given in Problem 8.1 using zero padding and compare the results with MATLAB.

8.3 Consider the following time series:

$$\vec{x} = \begin{bmatrix} -1 & 1 & 2 & 0 & -1 & -2 & 4 & 2 & 5 & 7 & 3 & 1 & -2 & -4 & -1 & 2 \end{bmatrix}.$$

Using $\vec{k} = \begin{bmatrix} 1 & 4 & 3 \end{bmatrix}$, compute and plot

a. the correlation;

b. the convolution.

8.4 Using the kernel for edge detection (items 1–3 in Table 8.1) and MATLAB, compute the resulting convolution with the data matrix as $|\mathbf{I}|$ (absolute value of matrix given in Problem 8.1). Compare the results with regard to the objective to detect edges.

8.5 Using the absolute value of the data matrix given in Problem 8.1, compute first the convolution using the 3×3 Gaussian blurring kernel given in Table 8.1, followed by a second computation of the convolution using the sharpening kernel. Compare the initial matrix with the resulting matrix after first blurring than sharpening.

8.6 Repeat Problem 8.5 using MATLAB with the 5×5 kernel for Gaussian blurring and the 3×3 kernel for sharpening.

8.7 Using your phone, webcam, or any digital camera, take a "selfie" and load it as an image file into MATLAB.

a. Determine the size in terms of pixels.

b. Apply a random rotation to the image and depict it along with the original image using the MATLAB command montage.

c. Apply noise using Gaussian, Poisson, speckle, and salt & pepper noise. Depict all images in one image using the MATLAB command montage.

d. Using the original image apply random x and y translation as well as shear and depict the image along the original using the MATLAB command montage.

8.8 Using the MATLAB script below, generate an input and output sequence using a noise level of 0.01. Note, the data generated have a sampling time of 0.02 seconds. Use the generated output and

a. downsample the output to have a sampling time of 0.04 seconds;

b. upsample the output so the output sequence has a sample time of 0.01 seconds.

```
%problem_P8_8
% Data generation using dyn. system, plant & and measurement noise
%_____
%
% System description
Ac = [0 11;-1.8861-11.826145];Bc = [0;760];C = [1 0];D = 0;
TS = 0.02; % sampling time
nd = 1000; % number of samples
[A,B] = c2d(Ac,Bc,TS);
% Additive measurement and process noise computations
[~,ni] = size(B);
[no,n2] = size(C);
X = dlyap(A,B*B');
M = sqrt(diag(C*X*C'+D*D'));
X = sqrt(diag(X));
% Initialize vectors and matrices
```

```
x = zeros(n2,1);
u = zeros(ni,nd);
y = zeros(no,nd);
pn1 = input('standard deviation ratio of process noise [.1] = ');
pn=pn1*X;
mn1 = input('standard deviation ratio of measurement noise [.1] = ');
mn=mn1*M;
% Simulate model with noise and random input
for i=1:nd
    u(:,i) = randn(ni,1);
    y(:,i) = C*x+D*u(:,i)+randn(no,1).*mn;
    x = A*x+B*u(:,i)+randn(n2,1).*pn;
end
figure(1)
subplot(2,1,1);plot(y,'-');grid;xlabel('k');ylabel('y');title('Output')
subplot(2,1,2);plot(u,'-');grid;xlabel('k');ylabel('u');title('Input')
```

8.9 Using the MATLAB script from Problem 8.8, alter the script so that the input is a sine wave, and the noise computation is done using the `awgn()` function from MATLAB. Compare the output with the one generated from Problem 8.8 (using a sine wave as input as well) and adjust the noise level to match both outputs.

8.10 Using the MATLAB command window, load the following data:

```
T_1 = readtable('airlinesmall_subset.xlsx','Sheet','1996','Range','A1:E11')
T_2 = readtable('airlinesmall_subset.xlsx','Sheet','2008');
```

  a. Create a new variable that stores only the data from October 1996.
  b. Create a new variable that stores only the data from October 2008.
  c. Combine the generated table from parts (a) and (b).
  d. Create a `spreadsheetDatastore` for the years 1996 and 2008 using `UniqueCarrier` as the category.

8.11 Consider a deep neural network with one input layer with one neuron, two hidden layers each with two neurons, and one output layer with one neuron. Outline the steps for computing the weights using backpropagation. Use a sigmoid activation function.

8.12 Load the data contained in cities.mat using MATLAB.
  a. Create a datastore object that holds the economic and education data for all the cities.
  b. Create a datastore object that holds the crime information as well as recreation data for all the cities.

8.13 Load the data contained in `morse.mat` using MATLAB. Create a convolutional neural network and train the network to recognize the sequence "SOS".

8.14 Go to: www.robots.ox.ac.uk/~vgg/data/pets and download the pets image library. Create a folder named "dogs" and 25 subfolders with each of the listed dog breed images stored in these folders. Develop a CNN that is capable of classifying a dog picture from the downloaded repository. Adjust your hyperparameters as well as CNN structure until you achieve 90% accuracy.

8.15 Using the data and folders generated in Problem 8.14, use SqueezeNet to perform transfer learning. Compare the accuracy of the modified SqueezeNet with the results obtained in Problem 8.14

8.16 Using the data and folders generated in Problem 8.14, perform data augmentation so that each class of dogs (breeds) has 250 images. Retrain your constructed CNN from Problem 8.15 and compare the results.

8.17 Generate a new folder named "cats" and store the 12 categories of cat images from www.robots.ox.ac.uk/~vgg/data/pets. Construct a CNN based on the ResNet-50 network (use the DND tool and download the necessary files). Adjust the image size to match the input layer of the ResNet-50 network. Perform transfer learning and determine the accuracy of predicting cat breeds.

8.18 Using the MATLAB script from Problem 8.8, generate a sequence of 2,000 output data points using 0.001 for the standard deviation of the process and measurement noise. Construct an LSTM and predict the output for the last 10 data points.

REFERENCES

[1] Simonyan K, Zisserman A (2015). Very deep convolutional networks for large-scale image recognition. *Proceedings of ICLR Conference*, pp. 1–14.

[2] He K, Zhang X, Ren S, Sun J (2016). Deep residual learning for image recognition. *Proceedings of the IEEE Conference on Computer Vision and Pattern Recognition*, pp. 770–778.

[3] Boureau Y-L, Ponce J, Yann L (2010) A theoretical analysis of feature pooling in visual recognition. Proceedings of the 27th International Conference on Machine Learning, pp. 111–118.

[4] Ioffe S, Szegedy C (2015). Batch normalization: accelerating deep network training by reducing internal covariate shift. *International Conference on Machine Learning*, pp. 448–456.

[5] Gao J, Song X, Qingsong W, et al. (2020). Robusttad: Robust time series anomaly detection via decomposition and convolutional neural networks. *MileTS'20: 6th KDD Workshop on Mining and Learning from Time Series*, pp. 1–6.

[6] Iandola FN, Han S, Moskewicz MW, et al. (2016). SqueezeNet: AlexNet-level accuracy with 50x fewer parameters and <0.5mb model size. arXiv:1602.07360.

[7] Hochreiter S, Schmidhuber J (1997) Long short-term memory. *Neural Computation*, 9 (8): 1735–1780.

[8] Pierson WJ, Jr., Moskowitz LA (1964). Proposed spectral form for fully developed wind seas based on the similarity theory of S. A. Kitaigorodskii. *Journal of Geophysical Research*, 69: 5181–5190.

[9] Schoen MP, Hals J, Moan T (2011) Wave prediction and robust control of heaving wave energy devices for irregular waves. *IEEE Transaction of Energy Conversion*, 26 (2): 627–638.

9   Machine Learning

9.1   Feature Engineering

As we explore some of the more popular ML algorithms in this chapter, we find ourselves dealing with large amounts of data. When not using deep learning, we generally are responsible for extracting features from the data before being able to train and apply ML algorithms. In this section we will review some of the more common feature engineering issues and solutions that can provide guidance when constructing an ML algorithm. As there is a vast number of ML algorithms, our objective is to highlight, explain, and use some of these algorithms and showcase how to employ MATLAB, as well as what to watch for when dealing with data and extracting suitable features.

When we talk about feature engineering, we are referencing the process that is responsible for extracting suitable characteristics from acquired process data and presenting it in a form that enhances the chances of obtaining and training good ML models. Feature composition and extraction is usually an iterative process. To introduce some of the issues and possible approaches to prepare the data accordingly, we will make use of an example.

Example 9.1

MATLAB provides for a number of sample data sets, which we will utilize for this example. The data set for large cars is called `carbig` and is in this example used to demonstrate data cleaning and data preparation for ML applications. Accessing the data in MATLAB is accomplished by the following command:

```
load carbig;
```

This data set contains a number of car brands with properties such as weight, number of cylinders, horsepower, model name, origin, etc. We will use only two features for this example: the weight and the horsepower of a car. Our objective is to prepare the data so it can be used for a classification algorithm to predict the miles per gallon (MPG) of a car. First, we construct a table variable from the data we loaded onto the workspace:

```
classTable = table(Weight,Horsepower,MPG);
```

It is a good habit to inspect the data set we plan on using for creating our model as the data may contain outliers or may have missing values, or we may see that some properties have very large magnitudes compared to others. In this case, if we print the table, we find that there are a number of rows that contain the value NaN (i.e., missing values). We can remove those rows using the following script:

```
compTable = rmmissing(classTable);
```

An alternative way to remedy the missing data problem is to fill in data points using derived data, such as the mean value. In our example the horsepower data have some NaN entries. We substitute the mean value computed by all data points for those instances using the following script:

```
hpm = mean(Horsepower,'omitnan');
Horsepower = fillmissing(Horsepower,'constant',hpm);
```

The first line computes the mean value from the horsepower data, and the second line fills in this value in the spots where we are missing values, using the `fillmissing()` function. For non-numerical data sets with missing entries we first have to convert the data into a categorical data type using `categorical()` and then apply the `fillmissing()` function.

Often, an ML algorithm employs geometric features that are extracted from the data to construct our classification model – for example, support vector machines. It turns out that it is good practice to scale our features in order to preserve geometric sensitivity. Consider the two features of our data set we are planning to use for creating our ML model: weight and horsepower. Suppose the weight is in grams and the horsepower data is given in hp units. Computing the Euclidean distance of any of our data points, where the x-axis is representing the horsepower and the y-axis the weight, we quickly find that the weight dominates the distance as the weight scale is so much larger than the scale for the horsepower. This unbalance usually results in a poor ML model for predicting the MPG. A common practice in ML is to scale the feature variables to possess the same range or to normalize it. For example, we can standardize the variable by computing the following transformed variable having a mean value of zero and a standard deviation of unity:

$$w_{\text{standardized}} = \frac{w - \text{mean}(w)}{\text{standard deviation}(w)}. \tag{9.1}$$

Additionally, normalization of a variable can be achieved by the following computation:

$$w_{\text{normalized}} = \frac{w - \min(w)}{\max(w) - \min(w)}, \tag{9.2}$$

where normalization scales the variable to be in the range of 0 to 1. Another data-handling consideration that should be taken into account is the issue of data outliers. An outlier may cause the ML model to generate poor predictions, hence we hope to remove any such data point from the set prior to the training process of the ML model. The difficulty in removing outliers from a data set is the recognition and identification of such outliers. MATLAB provides a simple algorithm to aid in the discovery of such data points. The method is embedded into the function isoutlier(), which has a Boolean return value when applied to individual data points. For example, we can investigate our data set for the weight to find outliers by using the following MATLAB script:

```
outliers = isoutlier(Weight);
```

This will result into a vector with 0s and 1s – though for our example all entries are 0s as there are no outliers. The outliers are identified by computing the median data point in the data set and using a threshold value of three times this value to label a data point as an outlier. However, this margin can be changed by the user of isoutlier() by including additional arguments in the MATLAB function. For example, instead of using the median value, we could specify the mean value:

```
outliers = isoutlier(Weight, 'mean');
```

There are a range of other options to choose from with this particular outlier detection function from MATLAB, which can be explored using the help feature in the command window or by consulting the MATLAB documentation.

Creating features from data can be accomplished in a number of different ways, sometimes using statistical measures, but spatial, temporal, and other attributes can be utilized as well. The number of features generated and used in an ML algorithm refers to the dimension of the ML process. Having many features may help in detecting the underlying characteristics of the data set, but also may degrade the performance of the constructed ML algorithm due to excessive computational resource demand. Hence, reducing the dimension of an ML algorithm and using only the essential features that contribute to the classification or regression problem is preferable. We may ask ourselves how we know which feature is contributing more than another feature in order to rank and ultimately select features. This question has been dwelled upon in the research community for some time and yielded an abundant number of approaches. Generally, there are three feature selection approaches: using filters that employ statistical tests to evaluate the correlation of features with the output variable; using wrapper methods which test every combination of the feature set using a predefined evaluation criterion for a specific ML algorithm; and embedded type feature selection, which is a combination of the filter and the wrapper type feature selection procedures. MATLAB's Statistics and Machine Learning toolbox hosts a number of feature selection algorithms for both classification and regression tasks.

We briefly highlight here one such feature selection function offered by this toolbox. The sequential forward selection (SFS) algorithm starts by using an empty set and updates the feature set iteratively by considering the best-performing feature defined by the maximum score achieved:

$$SFS_{k+1} = SFS_k \cup \underset{\text{arg max}}{G} (SFS_k \cup f, D, M), \tag{9.3}$$

where SFS_k is the set of features at iteration k; $D = [X, y]$ is the data set; M is the ML classification model; and G is the evaluation criterion. This algorithm is available as a MATLAB function:

```
SFS_keep = sequentialfs(@Gfun,X,y)
```

When using MATLAB's implementation of the sequential feature selection algorithm, G is a handle to an error function. A possible implementation of an error function is

```
function error = @Gfun(Xtrain,ytrain,Xval,yval)
svmModel = fitcsvm(Xtrain,ytrain);
ypred = predict(svmModel,Xval);
error = nnz(ypred~=yval);
```

Note, for the error function we included a prediction matrix and its response vector as well as a validation data set. The error function yields a scalar output (i.e., the prediction error). When dealing with categorical data sets, one common approach to accommodate the `sequentialfs()` function is by assigning a number to each category. Unfortunately, such approaches may lead to a distorted view of the data set. Consider numbers 0 through 4 being assigned as five predictor categories. When computing the distance, categories 0 and 4 are the furthest apart, even though categorically this is not necessarily correct. A better approach would be to impose some equidistance between each of the categories. This can be accomplished by introducing so-called dummy predictors for each category. MATLAB facilitates this approach with specific functions. For example, we can introduce a matrix of dummy variables as follows:

```
duva = dummyvar(cat);
```

Here, `duva` entails as many columns as `cat` stores categories. Rows in this matrix correspond to the number of observations, where one entry in each row has a value of 1 and all other entries hold 0s. The 1s in each row identify which observation is associated with which category. Hence, this matrix becomes our categorical information vector in ML algorithms, where each column corresponds to a specific predictor variable that uses the unit entry to indicate the presence of this category and 0 to indicate no presence of this category.

As an active research field we find a great multitude of other algorithms that allow for selecting the most pertinent features for an ML project. Some of these methods employ ML algorithms. For example, the random subset feature (RSF) selection algorithm uses a k-nearest neighbor classifier to find the most probable feature set in an iterative fashion [1]. Others, such as the mutual information (MI) feature selection method, are based on probability theory. The MI gauges the dependence between two random variables, in our case two features [2]. For this assessment, the MI is computed by utilizing the probability density functions of each feature. Suppose we have feature α and feature β, then the MI can be computed as

$$\text{MI} = \sum_{\alpha \in A} \sum_{\beta \in B} P(\alpha, \beta) \log \left(\frac{P(\alpha, \beta)}{P(\alpha)P(\beta)} \right). \tag{9.4}$$

Fortunately, MATLAB provides a convenient app that allows you to explore and optimize your feature selection for ML applications. The Diagnostic Feature Designer app is found in the Control System Design and Analysis section of the Apps tab. In the following we demonstrate the use of this app for selecting features using our simple gas mileage example.

Example 9.2

After opening the Diagnostic Feature Designer app, import the data entailed in `carbig`. For this, we will use a few more entries included in this data set:

```
cTable = table(Weight,Horsepower,cyl4,Cylinders,Displacement,MPG);
```

Also, apply the `rmmissing(cTable)` command to remove any missing values. We can now open a new session in the Diagnostic Feature Designer app and select our corrected table under the Source category. After importing this data set we have an opportunity to assign data properties to this set. For example, the data detailing the MPG is our independent variable as we want to predict this quantity. Hence, highlight the MPG entry in the Source variable section and select in the Source variable properties section the variable type as independent variable (IV). You can also add unit information to each of the variables. For *weight and horsepower*, *cycl4*, *cylinders*, and *displacement* select the variable type as feature. The selection is shown in Figure 9.1. Once the data are imported and the variables have received proper role assignment, we are ready to import the information into the Diagnostic Feature Designer app.

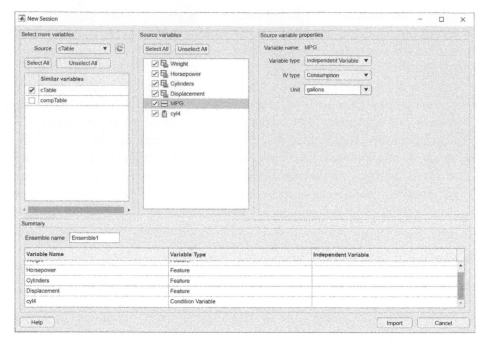

Figure 9.1   Diagnostic Feature Designer app: importing data using a new session.

With the Select Feature option and using the Feature Trace plot, we find that there is good sensitivity between *horsepower* and *MPG* as well as *displacement* and *MPG*. Note that we did not scale or normalize our data and hence the weight data are dominant when compared to the other quantities. Often we use derived features using statistics or time and frequency domain quantities. They can be used in the same fashion as shown in the example above and compared to each other. To determine which feature has the most relevance you can rank the features using a number of different methods such as the T-test, entropy, and others. Finally, once you have decided which features to use, you can export them to the workspace in MATLAB.

In the following we will introduce a few ML methods and demonstrate their implementation in MATLAB, as well as provide examples using both MATLAB code and the Classification Learner app.

9.2   Support Vector Machines

Although support vector machines (SVMs) can be used for regression tasks, they have been used extensively in a number of classification problems, including facial recognition, text categorization, bioinformatics, and image classification [3,4]. As a supervised learning algorithm its basic functioning is by sorting the presented data into one of two categories. The sorting is accomplished by introducing a decision boundary that partitions the data geometrically into two *n*-dimensional spaces. The boundary, also called hyperplane, is found by maximizing the distance between this hyperplane and any of the closest data points. For a two-dimensional problem (i.e., having two features to classify a set of data) we can depict the geometrical configuration of the classification tasks as shown in Figure 9.2. The data in Figure 9.2 are farm animals being recognized by two quantifiable features – perhaps weight and height. The task is to distinguish the two categories of animals based on those two features alone. For this example, SVM constructs a line between the two groups of animals and maximizes the margin defined by this line and the closest animals, which are the extreme data points. Those closest animals or data points, identified by circles in Figure 9.2, are called the support vectors. Finding this hyperplane represents the learning phase of an SVM algorithm. To apply SVM for finding the classification outcome of a new data point, we simply map the new data point into the learned graph using its feature values and determine on which side of the hyperplane this new data point is located. Note, the larger the margin the lower the chance of misclassification. Hence SVM is an optimization problem. Our approach of fitting a straight line into our data set in order to create classification boundaries experiences some problems when the data set presents itself in different forms – for example, as shown in Figures 9.3(a) and 9.3(c). Suddenly, a straight line is insufficient and hence the data set is no longer linearly separable.

We have the option of creating boundaries that possess more complexity; in our case we could use circles, as indicated by the dashed lined circles in Figures 9.3(a) and 9.3(c). However, since we are no longer able to separate the data set with linear functions, what if we were to create an additional dimension

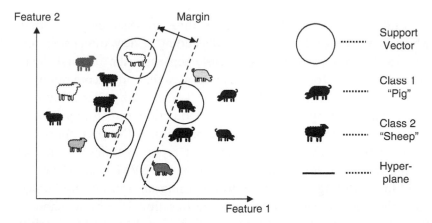

Figure 9.2 SVM of a two-dimensional classification problem, indicating the support vectors, the margin, and the hyperplane.

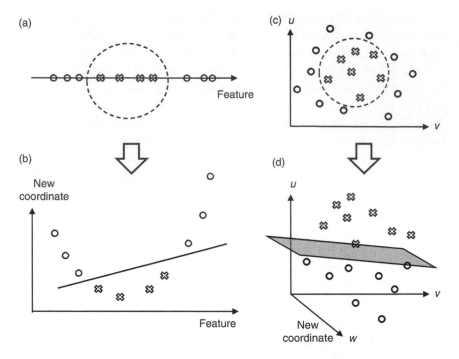

Figure 9.3 (a) One feature data set not linearly separable; (b) transformation using a kernel function, resulting in a linearly separable data set; (c) two-feature data set which is not linearly separable; (d) converted data set using a kernel function.

for plotting the data set? Consider the data set in Figure 9.3(c). If we introduce a new coordinate, for example

$$w = u^2 + v^2, \tag{9.5}$$

we can plot the given data points in three dimensions rather than in two, which is shown in Figure 9.3(d). By adding a new dimension we are now able to find a hyperplane that linearly separates the original data set. This approach is true for higher dimensions as well, and is commonly referred to as the kernel trick [5]. Specifically, the kernel trick takes a low-dimensional input space and by applying a kernel function transforms it into a better dimension that results in a separable problem. SVM algorithms make use of this kernel trick by employing different types of functions. In the case of having more features than data points in an SVM task we are confronted with an overfitting problem. However, using an appropriate kernel function may also resolve such situations. The question is how to choose a kernel function. A very popular kernel function in use for SVM algorithms is the radial basis function (RBF):

$$K(u, v) = \exp\left(-\gamma \|u - v\|^2\right). \qquad (9.6)$$

There are other kernel functions available as well, including

$$\text{Linear: } K(u, v) = u^T v, \qquad (9.7)$$

$$\text{Polynomial: } K(u, v) = \left(u^T v + 1\right)^d, \qquad (9.8)$$

$$\text{Sigmoid: } K(u, v) = \tanh\left(a u^T v + b\right). \qquad (9.9)$$

Unfortunately, it is rather difficult to find out which will work best for your data set in advance. A guiding principle would be to recognize that the kernel trick is essentially a similarity measure, and hence we would choose a kernel according to such prior information on invariance characteristics.

Considering translational invariance, we see that the RBF kernel is the only kernel from the above-listed functions that possesses a translational invariance:

$$K(u, v) = K(u + a, v + a). \qquad (9.10)$$

Here, a is some arbitrary vector. Note, the RBF kernel is a function utilizing the Euclidean distance between the data points. This is not true for the other kernels listed – that is, they are functions of the inner product of the data points. The similarity originates here from the fact that points that are closer – measured by the Euclidean distance – to each other may be more similar. Points computed by the inner product may be close to each other but have a lower value and hence similarity is not recognized. As an example, consider two points close to the origin but on opposite sides: Such a configuration of data points will yield a low value from a kernel function employing the inner product. These properties are the reasons that RBF kernels are used so widely for SVM applications.

MATLAB implementation of SVM can be achieved in multiple ways. One approach is to code it as a script file *.m using the functions fitcsvm() and fitrsvm(). The first of these functions is for classification and the second is for regression problems. Another approach is the use of the Classification Learner app. We will demonstrate both approaches in the following.

Example 9.3

Suppose you have a set of input data X and the corresponding measurement y which you believe is contaminated with noise. Your objective is to use the data and to extract an SVM regression model that relates the input variable X with the output variable y. The function fitrsvm() utilizes a linear default kernel (i.e., $X_i X_j$). Since there seems to be no linear relationship between the input and the output we will employ a Gaussian

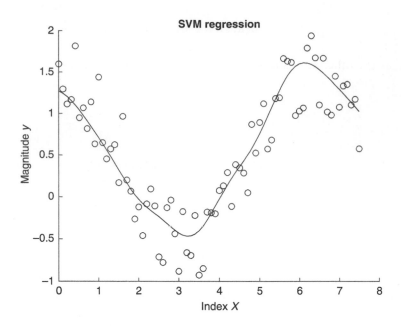

Figure 9.4   Regression SVM model prediction (line) and data (circles).

kernel function. We simulate the data using a simple cosine wave and a random additive component. The corresponding MATLAB code is

```
X = 0:0.1:7.5;
y = cos(X)+rand(1,length(X));
X = X';y = y';
data = table(X,y);
MSVD = fitrsvm(data,'y','KernelFunction','gaussian');
YFit = predict(MSVD,data);
scatter(X,y);
hold
plot(X,YFit,'r')
title('SVM regression')
ylabel('Magnitude y')
xlabel('Index X')
hold
```

Note that we again use the table function to submit the data, as the MATLAB functions fitrsvm() and fircsvm() only accept data in table format. The corresponding plot of the prediction and the regression model is shown in Figure 9.4.

MATLAB made it quite convenient to explore different options, including kernels, using an app that includes most major ML algorithms, including the SVM method. We will highlight the major steps in the use of this app with an example.

Example 9.4

We shall utilize one of MATLAB's built-in data sets to demonstrate the Classification Learner app:

```
fishertable = readtable('fisheriris.csv');
```

As with the prior examples, we note that data sets in ML applications are handled in the form of MATLAB table objects. Since the data set for this example is in the form of a spreadsheet (i.e., a CSV file), the function `readtable()` will automatically convert it to a table object.

From the Apps tab select the Classification Learner app, which will yield the main window of MATLAB's classification computation. To begin a new ML investigation, select New Session and chose to load the data from the workspace. You will be presented with a summary of the data that are considered for your ML modeling project. Figure 9.5 depicts the data set as it has been collected by the Classification Learner app.

Once all the information is inspected and deemed correct, continue by selecting the "Start Session" button to navigate into the main Classification Learner app. Under the

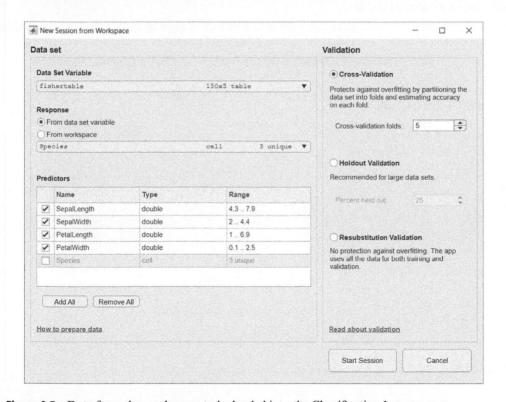

Figure 9.5 Data from the workspace to be loaded into the Classification Learner app.

Model Type tab, select the preferred ML method. In our case, as we are working with SVM, we find an entire selection of SVM algorithms. To find some of the specifics of each SVM method, hover with your mouse over the corresponding SVM icon to be presented with a brief description of the type of SVM under consideration. For example, the first entry is linear SVM, which uses linear kernel functions. The All SVMs option allows for testing all SVM methods entailed in MATLAB to be employed and tested on the data set currently loaded. Once the appropriate ML method is selected, train the model by pressing the green "Train" button in the Training tab of the app. MATLAB will compute the selected model parameters. In our example, if we choose the All SVMs model option, we are presented with the results of each SVM model that the app offers. The performance of each of the models is listed in the Models section on the left side of the app. Each model is given with a validation accuracy (Figure 9.6). In our case there are two models that achieve the highest validation accuracy, the quadratic SVM and the medium Gaussian SVM algorithm, with each achieving 96.7% validation accuracy. Model details are depicted in the Current Model Summary section.

The plot controls are located on the right-hand side of the app, allowing for plotting prediction or the data, identifying mis-classifications by the use of different symbols and the color-coded classes.

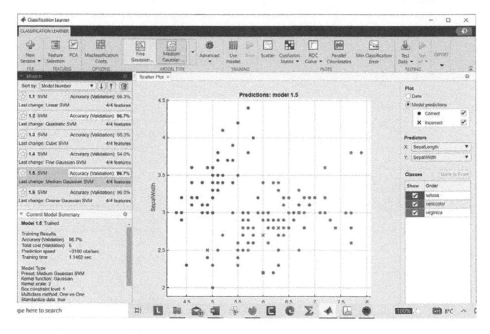

Figure 9.6   SVM training results based on model selection.

You also have access to the confusion matrix plot and the ROC curve. For details of these assessment plots, see our discussion in Chapter 6. Finally, you have the option to export the model to the workspace as a compact form or as a function. If the model lacks accuracy or performance you also have the option to further optimize it, utilizing a manual selection of kernel functions (use the "Advanced" button), or to test the model on new test data using the "Test Data" button.

For training ML and DL algorithms we employ the validation data to detect overfitting and to assess the accuracy of the developed and trained model. So far we selected a percentage of the data to be kept for validation purposes, while the majority of the data was utilized for the training process. One lingering question we have yet to answer is: How do we know that the validation data are representative of the overall data set? If the selected data are representative, we do not have to worry about this question. However, we have the choice to randomly select data points or we can select a block of data for this purpose. In either case, we do not control the selection by the information contained in the data that is dedicated to validation. Suppose we partition the data into k equal sets, and use $k-1$ data sets for training and the kth data set for validation. Once completed we repeat this for each of the other $k-1$ data sets, until each of the k sets has been used once for validation. This process is called k-fold cross-validation [6]. Symbolically, we can depict the selection process as shown in Figure 9.7.

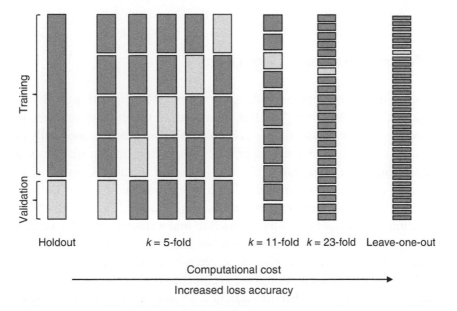

Figure 9.7 Cross-validation, using one segment of the data set for validation and the other for training, or using k-fold cross-validation. The higher k, the larger the computational cost but also the higher the accuracy of the loss estimate. If k = number of data points, cross-validation is called leave-one-out validation.

In Figure 9.7 we introduce also the leave-one-out cross-validation, where k is equal to the number of data points in the data set. Logically, the leave-one-out cross-validation setup results in the most accurate estimation of the loss, but also in the most computational costly validation computation.

The Classification Learner app allows us to choose which type of cross-validation we want to use, including the number k. This can be seen in Figure 9.5 on the right-hand side of the app window. However, if we use a MATLAB script to achieve cross-validation we can utilize the functions `crossval()` and `kfoldLoss()` to compute the generalization error.

Example 9.5

Using the model generated in Example 9.3:

```
MSVD = fitrsvm(data,'y','KernelFunction','gaussian');
```

we can compute a 10-fold cross-validation error:

```
crossMSVD = crossval(MSVD) % 10-fold cross-validation
kfoldLoss(crossMSVD) % resulting error
```

Using 25% of the data as holdout, we can compute the cross-validation error as

```
crossMSVD025 = crossval(MSVD,'Holdout',0.25) % 25% holdout cross-validation
kfoldLoss(crossMSVD025) % resulting error
```

9.3   *K*-Nearest Neighbor Algorithm

One of the more popular and rather simple ML algorithm is the k-nearest neighbor (KNN) method [7,8]. It is often used for solving classification problems; however, it is just as useful for regression problems. KNN is essentially nonparametric (i.e., it does not assess or use information on the distribution of the underlying data). Instead, it utilizes learned instances without imposing a model structure. It accomplishes its classification or regression task by recalling training instances from its supervised learning process. The K in KNN stands for the number of neighbors, which is used to make classifications or predictions. If a new data point is presented to the KNN algorithm with the objective to assign a new class label, distances to the k nearest neighbors are computed. KNN's philosophy is that similar things are in close proximity. Hence, using this distance we can count the number of instances or data points for each class in this neighborhood. The majority count (i.e., the category with the most instances in the neighborhood) is determinant in the class assignment of the new data point. Besides selecting the appropriate distance measure, the number of neighbors

k has a great influence on the performance of the KNN algorithm. Suppose we choose to utilize a small number of neighbors, then noise may have a larger influence on the performance of the algorithm. On the other hand, if we choose a large number of neighbors we impose a greater computational burden on the classification problem algorithm.

An example of $k = 12$ is shown in Figure 9.8, where we have two classes: the stars and the circles. The new data point – shown as a square – is placed into the data set and the 12 nearest neighbors' classification statuses are counted in order to assign the new class label to the new data point. In this example we used the simple Euclidean distance also known as the L_2 norm. However, there are other distance measures available and used in KNN algorithms, such as the Hamming distance, Manhattan distance, and the Minkowski distance [9].

The Euclidean distance or L_2 distance can be computed in a Cartesian coordinate system for an n-dimensional problem between two points $\vec{x}$ and $\vec{y}$ as follows:

$$d(\vec{x},\vec{y}) = \sqrt{(x_1 - y_1)^2 + (x_2 - y_2)^2 + \cdots + (x_j - y_j)^2 + \cdots + (x_n - y_n)^2}. \quad (9.11)$$

The Manhattan distance gets its name from finding the driving distance through a town, such as Manhattan, whose layout is dominated by square city blocks. Navigating through such towns involves straight-line movements and turns that are changes in directions of about 90°. Mathematically, we can compute this distance as

$$d(\vec{x},\vec{y}) = \sum_{i=1}^{n}(x_i - y_i). \quad (9.12)$$

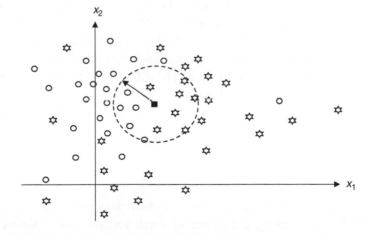

Figure 9.8 KNN example, where the square symbol represents the new data point that needs to be classified into the class circle or the class star. Using $K = 12$, we find that in this neighborhood there are five circles and seven stars; hence the new data point is classified as a star.

Another form of computing the neighborhood is using the Minkowski distance. For our two points, the p-order Minkowski distance is evaluated as

$$d(\vec{x}, \vec{y}) = \left[\sum_{i=1}^{n} |x_i - y_i|^p \right]^{\frac{1}{p}}. \tag{9.13}$$

In ML we may deal with a nonnumeric problem; computing the distance for such an occasion needs to be accomplished with different formulas, such as the one defined by the Hamming distance. For example, if we want to compute the Hamming distance between two strings of characters, each having the same number of characters, the distance is found by counting the number of positions at which the corresponding characters differ. We can illuminate this by an example.

Example 9.6

Given the two binary sequences, we can compute the Hamming difference as "00110101" and "10100101". The Hamming distance is 2, since the first entry and the fourth entry are different between the two sequences.

Suppose you have the two names "Mario" and "Marco" – the corresponding Hamming distance between these two words is 1, as only one character differs in the two sequences.

To implement KNN into MATLAB we have two choices: using script or MATLAB's Classification Learner app. As we demonstrated the Classification Learner app in the SVM discussion, we will focus on the MATLAB script implementation using an example. However, if using the Classification Learner app, instead of selecting the SVM option we can choose among a number of KNN algorithms. To start the KNN modeling using the app, first define the data using a new session and then select the KNN model you want to use for the classification task. The different options for KNN using the app are highlighted in Figure 9.9.

We demonstrate the KNN implementation using MATLAB script based on a standard MATLAB example below.

Example 9.7

MATLAB's Fisher's iris data set is a pre-installed data set used in MATLAB's documentation for showcasing a number of classification tasks. We will recycle this approach and highlight a few commands for implementing the KNN algorithm in MATLAB script. The primary command to fashion a KNN model is the `fitcknn()` function. The argument delivered to this function is the data set in the form of the predictor and the corresponding response. The function `fitcknn()`

tolerates a wide range of data set types, including arrays and cell arrays, logical and numerical vectors and matrices. In this example we load the Fisher's iris data where the observations are in a matrix composed of four petal measurements for 150 irises, while the response is housed in a cell array detailing the iris species. To load and assign the data to the corresponding variable names to be used with the KNN algorithm, use

```
load fisheriris
X = meas;
Y = species;
```

To train a KNN model with the number of nearest neighbors $k = 3$, we can use the following MATLAB script employing the fitcknn() function:

```
KNNMmodel =
fitcknn(X,Y,'NumNeighbors',3,'NSMethod','exhaustive','Distance', ...
'minkowski','Standardize',1);
```

The above evaluated model uses an exhaustive search algorithm – as specified by NSMethod, 'exhaustive' – to compute all the distances from all the data points. We also specify the distance measure to be used for finding the nearest neighbors as the Minkowski distance. The final argument in the above command centers and divides the columns of the data by their standard deviation. This command is activated by the logical TRUE integer 1, as the default value for this argument is false. Having an initial KNN model, we can use cross-validation and test for classification errors. This is accomplished as follows:

```
CVKNNmodel = crossval(KNNMmodel);
classError = kfoldLoss(CVKNNmodel)
```

The crossval() function default setting incorporates a 10-fold cross-validation algorithm. Those settings can be edited to utilize different error functions. The kfoldLoss() function evaluates the loss for data points that were not utilized in the training of the KNN model. We can also optimize the hyperparameters of the inferred KNN model to reduce the cross-validation loss:

```
OptKNNModel = fitcknn(X,Y,'OptimizeHyperparameters','auto', ...
    'HyperparameterOptimizationOptions', ...
    struct('AcquisitionFunctionName','expected-improvement- plus'))
```

Executing the above command, we will obtain the following response:

```
Optimization completed.
MaxObjectiveEvaluations of 30 reached.
Total function evaluations: 30
Total elapsed time: 36.8418 seconds
Total objective function evaluation time: 3.321
```

```
Best observed feasible point:

NumNeighbors    Distance
      3         correlation

Observed objective function value = 0.026667
Estimated objective function value = 0.028825
Function evaluation time = 0.12581

Best estimated feasible point (according to models):

NumNeighbors    Distance
      3         cosine

Estimated objective function value = 0.026954
Estimated function evaluation time = 0.07115
```

In addition, MATLAB offers graphing tools depicting the evaluation results for each of the different distance measure and the function evaluation plot. For brevity, we only show the main results in the above section, where we find that the cosine distance with one neighbor yields the best estimated feasible point.

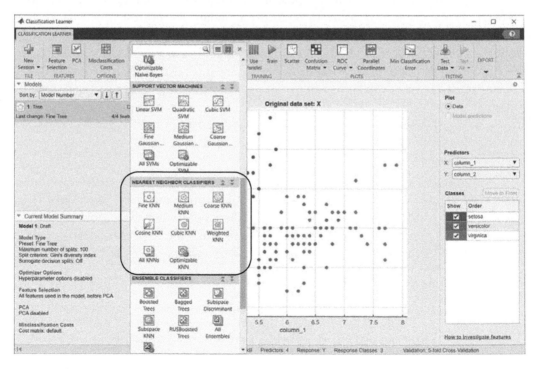

Figure 9.9 Classification Learner app for KNN modeling. The app allows for choosing among a number of different KNN model forms, including the option of selecting all.

9.4   Linear Regression Algorithms

A widely used statistical technique to predict outcomes from data is linear regression [10]. Assuming we have a set of outputs/targets, $\vec{y} = [y_1 \quad y_2 \quad \dots \quad y_n]^T$ and we want to fit a set of predictors

$$\mathbf{\Phi} = \begin{bmatrix} 1 & x_{11} & x_{12} & \dots & x_{1p} \\ 1 & x_{21} & x_{22} & \dots & x_{2p} \\ \vdots & \vdots & \vdots & \ddots & \vdots \\ 1 & x_{n1} & x_{n2} & \dots & x_{np} \end{bmatrix} = \begin{bmatrix} \vec{x}_1 \\ \vec{x}_2 \\ \vdots \\ \vec{x}_n \end{bmatrix}$$

in order to build a model of the form

$$\vec{y} = \mathbf{\Phi}\vec{\theta} + \vec{\varepsilon}, \tag{9.14}$$

where $\vec{\theta} = [\theta_1 \quad \theta_2 \quad \dots \quad \theta_p]^T$ are the coefficients of the predictor values and $\vec{\varepsilon} \in \mathbb{R}^{n \times 1}$ is a residual vector with the property of being white and having a zero mean. Equation (9.14) represents the output of a process as a weighted sum of predictors where $p + 1$ is the number of regression parameters. The predicted output is therefore

$$\hat{\vec{y}} = \mathbf{\Phi}\hat{\vec{\theta}}. \tag{9.15}$$

To find the corresponding weights, or coefficients $\vec{\theta}$, we can compute the least-squares solution:

$$\vec{\theta} = \arg\min_{\theta} \sum_{i=1}^{n} (y_i - \theta x_i)^2 = \arg\min_{\theta} \|\vec{y} - \theta x\|_2^2 = \sum_{i=1}^{n} (y_i - \hat{y}_i)^2 = \sum_{i=1}^{n} \varepsilon_i^2. \tag{9.16}$$

The model parameters $\vec{\theta}$ in Equation (9.16) are found by solving this optimum problem, which leads to the well-known ordinary least-squares solution:

$$\hat{\vec{\theta}} = \left(\mathbf{\Phi}^T \mathbf{\Phi}\right)^{-1} \mathbf{\Phi}^T \vec{y}. \tag{9.17}$$

Note in Equation (9.17) we make the assumption that all columns in $\mathbf{\Phi}$ are linearly independent. Suppose we define the following quantities:

$$v_{SST} = \sum_{i=1}^{n} (y_i - \bar{y})^2, \tag{9.18}$$

$$v_{SSE} = \sum_{i=1}^{n} (\hat{y}_i - \bar{y})^2, \tag{9.19}$$

where v_{SST} is the total sum of squares computed utilizing the mean value $\bar{y}$ and v_{SSE} is the residual sum of squares. With the quantities as defined by Equations (9.18) and (9.19), we can express the goodness-of-fit value of a regression model:

$$R^2 = 1 - \frac{v_{SSE}}{v_{SST}}. \tag{9.20}$$

The goodness-of-fit value expresses the fraction of the data variation in y that can be explained by the regression model. In Equation (9.20), the R^2 value is limited to the range of 0 and 1. As the size of the regressor matrix influences the R^2 value in Equation (9.20), an adjusted R^2 value computation is often employed:

$$R^2_{adj} 1 - \frac{n-1}{n-p-1} \frac{v_{SSE}}{v_{SST}}. \tag{9.21}$$

Our estimation model may suffer from bias error. Bias errors are a result of erroneous assumption about the given data set we compute – or train – our model on. If we are confronted with a high bias, we miss relevant relations between the data Φ and the output y. This is essentially an underfitting problem. The overfitting problem results from having a high variance due to the inclusion of random noise. The problem of finding a balance between the influence of bias and variance can be addressed by regularization, which allows us to ignore the finer details in the data and focus more on the "big picture."

The ridge regression modulates the bias to reduce the variance in our regression problem. The corresponding ridge regression coefficients can be computed as follows:

$$\vec{\theta}_{Ridge} = \arg \min_{\theta} \sum_{i=1}^{n} (y_i - \vec{x}_i \theta)^2 + \lambda \sum_{j=1}^{p} \theta_j^2$$

$$= \arg \min_{\theta} \|\vec{y} - \Phi\vec{\theta}\|_2^2 + \lambda\|\vec{\theta}\|_2^2. \tag{9.22}$$

Equation (9.22) minimizes the squared error in addition to the magnitude of the model coefficients by the addition of a penalty term. The weighting factor λ can be used to control how much of a penalty we want to impose on the magnitude of the model parameters. When $\lambda = 0$ we favor overfitting, when $\lambda \to \infty$ we approach underfitting.

The least absolute shrinkage and selection operator (LASSO) is another regularization method [11]. LASSO helps with cases when the data set exhibits high multicollinearity or if we desire to infer simpler and more interpretable regression models. The LASSO coefficients are computed as

$$\vec{\theta}_{LASSO} = \arg \min_{\theta} \|\vec{y} - \Phi\vec{\theta}\|_2^2 + \lambda \sum_{j=1}^{p} |\theta_j|$$

$$= \arg \min_{\theta} \|\vec{y} - \Phi\vec{\theta}\|_2^2 + \lambda\|\theta\|_1. \tag{9.23}$$

We notice that Equations (9.22) and (9.23) are very similar, with the only difference given by which norm is applied to the penalty term. Having the L_1 norm employed in the LASSO regression method allows for some of the coefficients to approach zero. Having a zero value for the tuning parameter is equivalent to setting the model coefficient to zero and hence reduces the complexity of the model.

Example 9.8

Suppose we have a data set containing 250 measurements, a target vector and want to infer a regression model using both the ridge regression method and the LASSO regression method. The following MATLAB code simulates the data set using a model with five coefficients, three of which are zero. We also construct the corresponding matrices for the regression problem. For the ridge regression we will use MATLAB's function `fitrlinear()`, while for the LASSO regression model we can employ the function `lasso()`:

```
Phi = randn(250,5);
Theta = [0;1.95;0;2.9;0];
Y = Phi*Theta + randn(250,1)*0.05;
[Thetahat,Info] = lasso(Phi,Y);
Info.MSE(1)
```

The last line in this program extracts the mean square error for the first fitted model. `Thetahat` stores 42 models that the LASSO regression algorithm infers from the simulated data set. Comparing the first model with the original model given by `Theta`, we find $\theta = \begin{bmatrix} 0 & 1.95 & 0 & 2.0 & 0 \end{bmatrix}^T$ and $\vec{\theta}_{LASSO} = \hat{\theta} = \begin{bmatrix} 0 & 1.8802 & 0 & 2.8318 & 0 \end{bmatrix}^T$, with a mean square error of 0.0114. Using the ridge regression, the following script is employed:

```
RegMod = fitrlinear(Phi,Y,'Regularization','ridge','Lambda',0.4);
predictOut = predict(RegMod,Phi);
RegErr = mse(predictOut,Phi);
RegMod.Beta
```

The resulting model is given by $\vec{\theta}_{Ridge} = \hat{\theta} = [-0.0517 \quad 1.0484 \quad -0.0031 \quad 1.6978 - 0.0039]^T$, which yields a mean square error of 4.4238.

An alternative way of solving regression problems using MATLAB is by using the Regression Learner app. The app is structured in the same way as the Classification Learner app, but houses regression tools such as the LASSO and ridge regression.

Example 9.9

After starting a new session from the Regression Learner app, load in the data from Example 9.8 from the workspace. The selection is shown in Figure 9.10.

After loading the data you will have the option of selecting different model structures, including linear regression models and SVM models. If unsure which model to choose, you have the option of selecting all. If only linear regression models are under consideration, an option of training all linear regression models is available as well. Figure 9.11 depicts the outcome of this last selection option. In the same fashion as the Classification Learner app, the Regression Learner app shows the resulting models and their performance characteristics on the left-hand side of the app window.

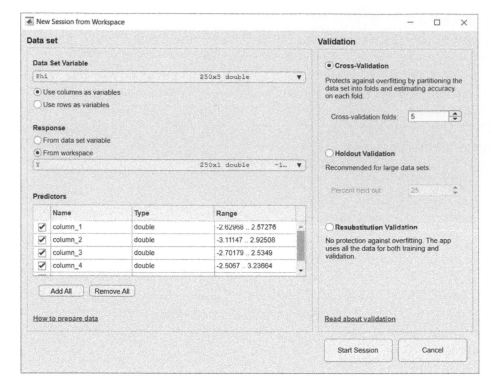

Figure 9.10 Regression Learner app interface for the selection of a data set.

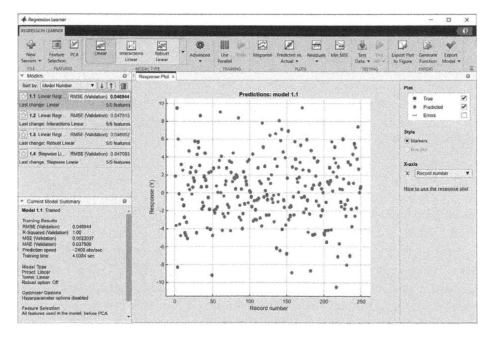

Figure 9.11 Regression Learner app main window with linear regression model results using the data from Example 9.8.

9.5 Naïve Bayes Algorithm

The naïve Bayes (NB) classification algorithm is a statistical method based on Bayes' theorem and the assumption that the features in any class are unrelated or independent from each other [12]. This independence gives rise to the term naïve in the title of this algorithm. For example, a classification of messages containing multiple words will not depend on the order of the words appearing in the text, but just on the existence of each of the words. NB models are especially suitable for large data sets to predict class membership probabilities. The underlying theorem utilizes the following probability measures:

- posterior probability $P(C|x)$: probability of instance x being in class C;
- likelihood $P(x|C)$: probability of generating an instance x given class C;
- class prior probability $P(C)$: probability of the occurrence of class C;
- predictor prior probability $P(x)$: probability of the occurrence of the instance x.

Bayes' theorem, utilizing the above introduced probabilities, can be given as

$$P(C|x) = \frac{P(x|C)P(C)}{P(x)}. \tag{9.24}$$

Bayes' theorem stipulates that the probability of an occurrence alters if we have prior information about the event. Let's use an example that showcases Bayes' theorem.

Example 9.10

Assume we have two classes: C_1 = apples and C_2 = pears. As both of these fruits can be green, among other colors, classifying a fruit based on its color can be difficult. Let's find out what the probability is of having a green apple if we pick up a green fruit from some delivery service that sells local farm products in baskets containing green and brown pears but may include also some red and green apples. We have the posterior probability defined as

$$P(C|x) = P(apple|green),$$

which reads as the probability of the fruit we pick being an apple when the color is green. The likelihood is given by

$$P(x|C) = P(green|apple).$$

Statistically we have observed from prior deliveries that approximately 210 fruits are contained in a basket, generally including about 10 apples of which 4 are green. Of the approximately 200 pears, 20 are green and the rest are brown. Hence, the likelihood expressed utilizing the given statistics is the probability of having a green fruit given that the fruit is an apple: 4/10 = 0.4.

The class prior probability is hence

$$P(C) = P(apple) = \frac{10}{210} = 0.0476,$$

and the predictor prior probability is computed as

$$P(x) = P(green) = 24/210 = 0.1143.$$

Hence, applying Equation (9.24) we find that the probability of a green apple in our hands is

$$P(apple|green) = \frac{P(green|apple)P(apple)}{P(green)} = \frac{0.4 \times 0.0476}{0.1143} = 0.1666.$$

We can also solve this problem using intuition by laying out all the pears and apples on a flat surface and defining the ratios of green fruits to brown or red fruits. This is shown in Figure 9.12.

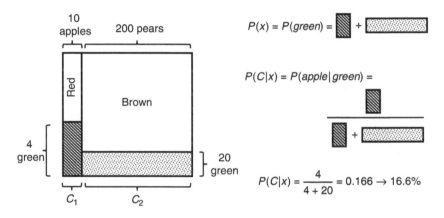

Figure 9.12 Graphical depiction of Bayes' theorem applied to the apples and pears problem.

Example 9.10 only considers the case where we have one attribute (i.e., fruit is green). However, many classification problems have many features, and hence we need to expand our NB method for cases with many attributes/features and many classes. Remember, the term naïve originates from the assumption that all features or attributes are independent distributions. Hence, we could state

$$P(x|C_j) = P(x_1|C_j) \times P(x_2|C_j) \times \cdots \times P(x_n|C_j). \tag{9.25}$$

Equation (9.25) computes the probability of class C_j generating an instance x. This can be computed by the product of each of the conditional probabilities of class C_j for each feature. To compute the conditional probability of a feature x_i

given a class, we can use the relative frequency (i.e., N_{x_i}), the number of instances belonging to class C_j having attribute x_i, and the total number of members in this class N_{C_j}:

$$P(x_i|C_j) = \frac{N_{x_i}}{N_{C_j}}. \tag{9.26}$$

However, what happens when there are no instances of attribute/feature x_i for one class? In such a case, Equation (9.25) would equate to 0. To avoid such cases, the class conditional probability can be re-estimated by adding two constants to Equation (9.26):

$$P(x_i|C_j) = \frac{N_{x_i} + 1}{N_{C_j} + n_C}. \tag{9.27}$$

Equation (9.27) is referred to the Laplace estimate using n_c as the number of classes [13]. Alternatively, the m-estimate is given as

$$P(x_i|C_j) = \frac{N_{x_i} + mp}{N_{C_j} + m}, \tag{9.28}$$

where m is the equivalent sample size and p is the prior estimate of the probability. In the absence of other information, one can assume $p = 1/n_x$, where n_x is the number of values that attribute/feature x_i can take.

The conditional probability given by Equation (9.26) does not entail any assumption of the underlying distribution – that is, we do not specify the probability distribution of each feature. Equation (9.26) is convenient when we count instances. However, if we assume a distribution such as a Gaussian probability density function, we deal with the Gaussian NB classification algorithm.

We can implement the NB algorithm in MATLAB using script or the Classification Learner app. For the script approach we use the function `fitcnb(X,y)` which allows us to use a table or matrix **X** representing the training data and the predictor variables y. Cross-validation instruction can be included in the NB function – for example, `fitcnb(X,y,'CrossVal','on')`. This will produce a 10-fold cross-validation using the data in **X** and y. When using the Classification Learner app, you will have choices of Gaussian NB, kernel NB, optimizable NB, and an option to test all available NB algorithms. The kernel NB algorithm uses a nonparametric estimation of the distribution.

9.6 Decision Tree Algorithms

A decision tree algorithm is a supervised learning method that is often employed in classification problems [14]. The name originates from the appearance of its structure when mapped out as a flow chart. For classification tasks this structure embeds a hierarchical strategy where each attribute undergoes a simple test and, based on the results of this test, is either presented with a new test on the next attribute or a classification assignment. This repetition of tests allows for a recursive partitioning of the input data set.

Consider the case of classifying the data set as depicted in Figure 9.13. The algorithm starts with the first test: $y > 0.5$, which yields two outcomes. The test is graphically depicted in the data set configuration (b), showing a dashed line at $y = 0.5$. This test effectively partitions the set into two parts, and can be represented by a binary outcome value. Figure 9.13(c) shows the next test: If the first test resulted in a YES, we gauge whether the remaining data points can be partitioned into items below $x = 0.5$ or above. As there are no more partitions needed, we have a classification of the upper part of the data set. The lower part proceeds in the same way until we have tests that result in pure classes. In the graphical version of the decision tree algorithm, as shown on the right-hand side of Figure 9.13, a test is referred to as a node. The outcome of such a test of an attribute is referred to as a branch. Once we have a pure class, we denote it as a leaf. Although our example was simple and employed only Boolean functions, decision trees are known to be able to represent any function of an input attribute.

You may ask yourself, how can we use this algorithm for regression tasks? We can generalize our observation from the above example by noting that we will obtain discrete output quantities. A leaf value is commonly set to the most dominant value of the corresponding output data set. However, for regression tasks we will have a continuous output. Hence, regression problems that are solved with a decision tree algorithm will yield leaf values that are commonly set to the mean value of the training data set. In the example we were able to plot the data set and derive simple tests in order to construct our decision tree algorithm. However, for more complex data sets, how would we arrive at those tests? The guiding principle that decision tree algorithms incorporate to develop the set of tests is based on the resulting information gain. Commonly, for this to be quantified, we use entropy, a measure of disorder in random variables. Suppose we have a binary classification task, as illustrated by Figure 9.13. Each test resulted in two sets, a positive P and a negative N set, where $S = \{P, N\}$ is the set presented to the test. The entropy E of S can be computed as

$$E(S) = -\frac{P}{S}\log\left(\frac{P}{S}\right) - \frac{N}{S}\log\left(\frac{N}{S}\right). \tag{9.29}$$

Suppose the resulting set N is empty and hence all samples belong to the set P. The corresponding entropy, according to Equation (9.29), is zero. However, if we have an equal split between the two sets, Equation (9.29) will yield an entropy of 1.0. We can now derive the formulation necessary to

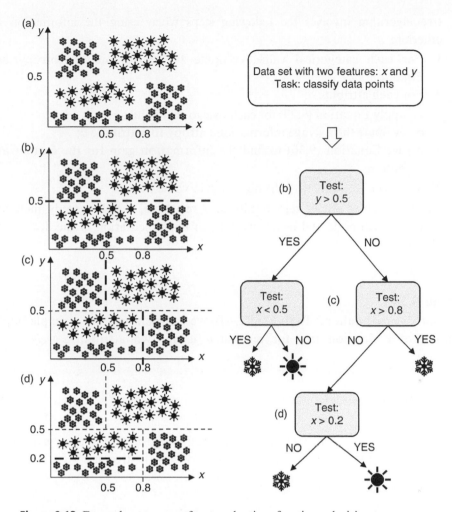

Figure 9.13 Example sequence of test evaluations forming a decision tree.

express our objective of maximizing the information gain for a test within a decision tree.

The information gain from a test is given by the expected reduction in entropy. The reduction is a result of the proposed partitioning that ensues from the proposed test or attribute. We can formally express this gain for a specific attribute α as

$$\text{Gain}(S, \alpha) = E(S) - \sum_{v \in values(\alpha)} \frac{|S_v|}{|S|} E(S_v). \tag{9.30}$$

In Equation (9.30) the set for which α has a value v is denoted as S_v, whereas the number of elements in S_v is $|S_v|$ and the number of elements in S is $|S|$. The decision tree algorithm involves the following steps when using the information gain criterion:

1. For each categorical value, compute the corresponding entropy using Equation 9.29.
2. For each attribute α:
   a. apply Equation (9.29) for each categorical value;
   b. evaluate the average information entropy for attribute α;
   c. use Equation (9.30) to find the information gain for the current attribute α.
3. Select the attribute with the highest gain value.

Repeating these three steps results in a recursive, simple, greedy method to establish each node and hence one type of implementation of the decision tree algorithm.

Example 9.11

Using our `carbig` data set from MATLAB, we can demonstrate a simple implementation of a decision tree for regression using two attributes: *weight* and *horsepower*:

```
load carbig;
X = [Weight,Horsepower];
treeMdl = fitrtree(X,MPG);
```

The function `fitrtree()` will train a decision tree model using the data set given in the matrix X and fit it to the output variable MPG. To view the tree, use

```
view(treeMdl,'Mode','graph')
```

which will produce a new window with options to depict the entire tree with each decision attribute and trained values. In order to use the trained model, use the function `predict()`:

```
pred = predict(treeMdl,mean(X))
```

where we used the mean value of the input data to generate a predicted value from the trained model `treeMdl`.

To solve classification problems using MATLAB script, use the function `fitctree()`. In addition to using the command line to train decision trees, you also have the option of using the Classification Learner app as well as the Regression Learner app, as indicated in Figure 9.14.

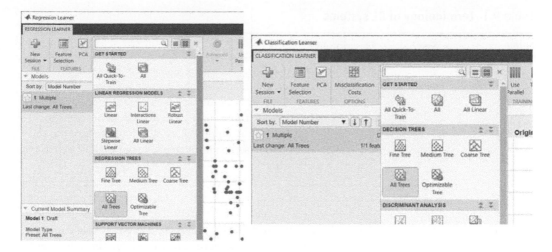

Figure 9.14 Decision trees for regression and classification problems available with the Regression Learner and the Classification Learner apps.

9.7 Reinforcement Learning Control

Next to supervised and unsupervised learning, the third category in ML is reinforcement learning (RL). Unlike unsupervised or supervised learning, RL does not provide for a solution instantly; rather, it iterates over time using trial and error to find a solution [15]. RL offers to solve a range of problems where methods from supervised learning or unsupervised learning may struggle to perform or perform well. In this section we will briefly discuss RL with a goal to introduce deep reinforcement learning (DRL), as applied to controls. In itself, RL is a rather large topic, and a section highlighting its basic functioning applied to controls does not do justice to the many facets and nuances this ML category offers. However, our aim for this section is to provide a rather brief introduction to RL as applied to controls and provide some simple examples using MATLAB's Reinforcement Learning Designer toolbox.

As a number of intelligent systems-based algorithms were inspired by observation in other fields such as biology, social behavioral science, etc., RL has partly oriented itself from psychology, where the concept of discounting is extensively used. The theory of Markov decision processes is part of RL's development. However, since we only provide an introductory treatment of RL, we will leave those underlying foundational concepts to specific references, such as [15].

When describing RL, we articulate its functioning by utilizing a specific terminology, such as *agent*, *reward*, and *policy*. Hence, we start by defining those basic terms, before employing them to describe and generally define RL control systems (Table 9.1).

Table 9.1 Terminology of RL systems

Name	Symbol	Description
Action	a_k	An action in RL defines a specific set of moves or interventions the system is allowed to undertake. Examples of actions in RL control are often associated with the actuator actively interacting with the system to be controlled, such as a valve that closes or opens to a certain degree, a power supply changing its voltage input to the circuit it controls, or a servo motor changing the angle of a revolute joint of some robot.
Agent		Just like in spy movies, we use an agent to explore and interact. In the case of RL, the agent explores the system dynamics by interacting with it using a specific action from the list of possible actions.
Reward	r_k	A well-chosen term to express its function: the reward is one of the results of an agent taking an action. It is a function that allows us to provide feedback based on how successful the action was.
State	s_k	In control systems, a state is often used to describe the current situation of a system. It has a similar definition in RL, where we use the state to describe the situation an agent currently finds itself in.
Environment		Unlike the state, the environment makes use of not only the state an agent finds itself in, but also by the action the agent undertakes. Having an environment and these two descriptors as input, the environment generates a response, which is the reward and the next state the agent will find itself in after taking the chosen action. Examples of the environment in controls are the system itself, which we try to control, and the governing equations of the dynamics – that is, the model or plant of the control system.
Policy	π	Having a set of actions to choose from, the agent employs a policy to make its selection on the specific action, hence defining the agent's behavior. To engage the policy, we need the current state as input. Generally, the policy is based on generating the largest reward. In controls, we can think of the policy as the control law we employ to guide the system in performing to our specifications.
Value function	$Q(s,a)$ $V_\pi(s)$	Value functions are used to generate estimates of future rewards. A common value function is the Q-function, which provides for an expected total reward considering the state and the action.

Hence, using these terms, we can summarize the basic idea behind RL: For an RL to function, we have an agent who choses from a list of actions to engage with the environment. The action performed by the agent will alter the future state as defined by the environment and generate a reward as feedback. The ultimate goal of RL is to maximize this reward for future steps. By utilizing this framework, the RL system is capable of learning about the system itself. RL does not need much domain knowledge, as it

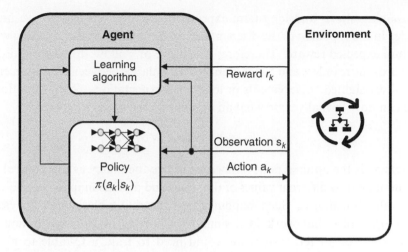

Figure 9.15 Basic interaction between agent and environment for an RL system.

learns about the domain while it is operating. The basic schematic of an RL system is symbolized in Figure 9.15.

From a controls point of view, the policy determining the agent's action is the controller while the environment represents the system we are seeking to control. The system itself responds by adjusting its states, which are observed over time and used to generate a reward. The scalar reward is then the basis for the next action taken by the agent. If we define a horizon h, the return is computed as the sum of all the separate rewards awarded from the steps within the time horizon. We can quantify this return as

$$G_k = \sum_{i=k+1}^{k+h} r_i. \qquad (9.31)$$

An example of a reward is $r_k = 1$ if the control objective is met, and $r_k = -10$ if the control action does not achieve its desired output. An RL control system's objective is to maximize the expected return $E[G_k]$. To accomplish this, we employ a policy π that determines the action (controller output) $a = \pi(s)$ based on the states of the environment s. We can expand on the expected return $E[G_k]$ by extending the horizon to infinity using a discount factor $0 < \gamma < 1$:

$$V_\pi(s_k) = E\left[\sum_{i=0}^{\infty} \gamma^i r_{k+i}\right]. \qquad (9.32)$$

The discount factor γ is responsible for Equation (9.32) staying finite, provided the individual rewards are bounded. In addition, the discount factor allows for managing short- and long-term expectations of the reward. As we utilize the index in the exponent of the discount factor, its magnitude is decreasing with each future expected reward. Therefore, the choice of γ allows for discounting future rewards more or less and hence manages the balance between having an emphasis on short-sighted achievements or long-term expectations. With Equation (9.32) we can now formally state what an optimal policy needs to achieve:

$$V_{\pi^*}(s_k) > V_\pi(s_k), \ \forall \ s, \ \pi, \tag{9.33}$$

where π^* is the optimal policy. Since we utilize the policy as our control law, we could map how different values of the states and actions map to specific values of the value function V_π. Such mappings are termed Q-tables, which are essentially look-up tables that work in a similar fashion as in gain scheduling control schemes. For an RL system we would need to train a Q-table to map every combination of state and action to a value. We can extend this concept by defining the policy as a mapping of state values to actions based on the expected value – as defined by the Q-table. However, for continuous control operation we find that the number of different combinations of states, actions, and values become excessively large. Hence, we seek to represent this mapping using a function. If we consider an instant in time k and the action–value function as defined in Equation (9.32), we can restate the Q-function in the following form:

$$Q_\pi(s_k, a_k) = E\left[\sum_{i=0}^{\infty} \gamma^i r_{k+i}\right]. \tag{9.34}$$

Equations (9.34) and (9.33) are related by

$$V_{\pi^*}(s_k) = \max_a Q_{\pi^*}(s_k, a_k). \tag{9.35}$$

Figure 9.15 uses a neural network symbol within the agent block to represent the policy, indicating that we could use such a structure to allow for learning the mapping function. However, there are different implementations of the policy itself. If we model a function that maps the states of the environment to the actions to be taken by the agent, we deal with a policy function-based learning structure. Often, this function is given by a deep neural network, acting as the policy itself. For such a structure, the neural network is termed as actor. For a value function-based learning structure we utilize the states of the environment and a choice of an action to determine the value of that action. In this case, the neural network evaluates how good the action is based on the state. The corresponding neural network embedding this mapping is termed a *critic*, since its primary mode of operation is to evaluate or criticize the choice.

In order to find the optimal policy π^*, Equation (9.35) implies that we need a model of the environment. Despite this apparent need, one of the key advantages of RL and RL control systems is that we do not need any prior information of the system we are attempting to control. An RL control system iteratively learns this model while in operation and interacting with it. This iterative learning process is built on the assumption that the system we control is based on a Markov process. The Markov decision process has a property that for an instant in time k, the system's history is defined by the set of state variables. As such, the system's distribution is defined by the current state and the action taken at the current time step k:

$$s_{k+1} \sim P(s_{k+1}|s_k, a_k). \tag{9.36}$$

A generalization of our control policy function-based learning interaction with the environment can be made by treating it as a probability distribution of actions that are a result of the current state of the environment:

$$a \sim \pi(a_k|s_k). \tag{9.37}$$

Provided the system is based on a Markov decision process, the reward r_k and the next state s_{k+1} are a function of the current state s_k and current action a_k only. To find the optimal policy π^* assuming a Markov decision process, a number of different approaches have been proposed, such as dynamic programming or Monte Carlo methods [16]. The dynamic programming approach involves the well-known Bellman equation, while the Monte Carlo method engages the RL system for a large number of times to determine an estimate of the return $V_\pi(s_k)$. However, as we only provide a brief introduction to RL, we will limit our discussion on Q-learning to finding the optimum policy π^* [17].

The Q-learning algorithm is initiated by using a random initial value function Q_0, which is iteratively updated based on the achieved reward and the corresponding action taken. The greedy policy π_Q^* is given by

$$\pi_Q^* = \arg \max_a Q(s, a). \tag{9.38}$$

The algorithm uses future states to find $Q_\pi(s, a)$ iteratively. The iteration itself is based on the Bellman equation, which can be stated as

$$Q_\pi(s, a)_{new} = Q_\pi(s, a) + \alpha[r(s, a) + \gamma \arg \max_a Q_\pi(s', a') - Q_\pi(s, a)], \tag{9.39}$$

where α is the step size parameter responsible for updating the value function. Equation (9.39) utilizes the applied action based on the current state to find the reward $r(s, a)$ and adds the maximum expected value resulting from the state s'. This value undergoes the discounting to allow for balancing instant and future rewards using the discount factor γ. The reward $r(s, a)$ and $\gamma \arg \max Q_\pi(s', a')$ are our new estimate for the value we expect from applying the action. Using this value and the previous estimate $Q_\pi(s, a)$, we have the updating error, which is modulated

Table 9.2 Q-learning algorithm

Step	Description
Step 1	Initialize value function $Q_0(s, a)$.
Step 2	Assess initial state of environment s_0 and update $Q^*(s, a)$.
Step 3	Determine action $a \leftarrow \pi(Q)$.
Step 4	Apply action and make observation of s' and reward r'.
Step 5	Update $Q_0(s, a)$ using Equation (9.39).
Step 6	Repeat Steps 2–5 until convergence.

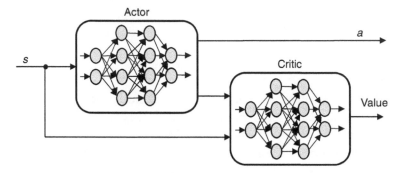

Figure 9.16 Schematic of the actor–critic RL method.

with the step size parameter α. When added to the prior estimate we find the updated Q-function $Q_\pi(s, a)_{new}$. The Q-learning algorithm is given in Table 9.2.

Recall that the policy function-based learning structure utilizes a neural network that acts as the policy, called the actor, while for a value function-based learning structure the corresponding neural network maps the states and the selected action to a value. The responsible network was termed critic. There also exists a method that combines both learning structures, the actor–critic (AC) method. Figure 9.16 shows the basic concept of the AC method.

In Figure 9.16 both networks receive the current state or observation from the environment. The actor network utilizes this information to project the best action using the policy function-based method. The critic network receives the projected action and along with the state computes an estimate of the value using the value function-based method. To train these networks the actual reward is used to compute the error between the predicted value and the reward. The error can then be used for training the critic network as discussed in Chapter 6 using backpropagation, with the objective to improve its accuracy in predicting the value it generates. The actor network also benefits from this error computation by improving the probabilities that determine the action. With this update, we effectively modify and hopefully improve the policy of the RL system.

As indicated, our discussion on RL and RL control is focused on some of the basic concepts rather than the details and underlying mathematical foundations. Therefore, in the following we will emphasis the implementation issues of such an algorithm with a focus on control problems. As the Q-learning and the AC schematic start with untrained networks, a common practice is to create a simulation of the environment and use this for training of the networks offline. Once sufficient performance has been achieved, the RL network can be deployed by exchanging the simulation model with the physical environment. In many cases the training is continued as the simulation model may not have accounted for all the nuances in terms of dynamics, as well as for any changes over time that the real environment experiences. In the real-world environment the learning can proceed in two ways: updating the networks as you go at each iteration, or keeping the networks static while improving them on the side, only to incorporate updates at certain time intervals.

Another consideration in designing and implementing RL control systems is the reward computation. The reward is our feedback on how well our policy works, hence such assessment can happen instantly or after many iterations of taking actions on the environment, depending on the goal of the controller. As an example, if the goal of the controller is to achieve a certain temperature in a room, we could structure the reward as $r = 1$ if temperature was achieved and $r = -1$ if it was not achieved. However, since it takes time to heat or cool a room, the reward is not used in intermediate steps and hence no feedback is provided to the RL system. Without feedback, the likelihood of achieving our ultimate goal of some temperature is rather small. A better approach for such instances is to provide a reward in smaller pieces that allows for guiding the system to the desired outcome (i.e., the desired temperature in the room). This approach is commonly referred to as reward shaping. Reward shaping should be based on domain knowledge as there are pitfalls when utilizing this approach that could result in less than optimal solutions. Focusing purely on the reward may also cause the RL system to not explore the environment sufficiently and ultimately potentially arrive at suboptimal solutions. By pursuing actions that may return lesser rewards, we may acquire information about the environment that ultimately leads to a better optimum solution. Obviously, there is a balance between pursuing maximum rewards (exploitation) and the exploration of the environment. Similar to our discussion on enhanced tabu search algorithms in Chapter 5, the balance is in favor of exploration at the beginning of the RL operation, while focusing more on exploitation by maximizing the reward toward the end of the operation.

MATLAB provides for a number of toolboxes to implement the concepts introduced here, as well as many other realizations of RL, and allows you to construct specific RL solutions to complex control problems. MATLAB also provides for convenient apps that simplify the construction, training, and simulation of RL systems. As our treatment of RL-based control has been of an introductory type, we will utilize those apps to provide examples for the material covered in this section. Generally, the workflow for developing an RL system is to create an environment, define a reward function, develop the policy and agent networks, define the training hyperparameters, train the RL system and finally deploy the RL system. We will provide examples for these steps in the following.

Example 9.12

In this example we will look at how to load a predefined environment for use with RL projects. A convenient way to obtain a predefined environment is by using MATLAB's Reinforcement Learning Designer app. Once started, you can select an environment for MATLAB or Simulink, using the New > Environment tab, as shown in Figure 9.17.

Once selected, the loaded environment appears in the left column of the app under the Environment section. You can also inspect the dimensions of the loaded environment. For example, choosing the Continuous Simple Pendulum Model for Simulink, the Preview window describes the environment as having a [3 1] observation vector and actions (torque) to be a [1 1] vector.

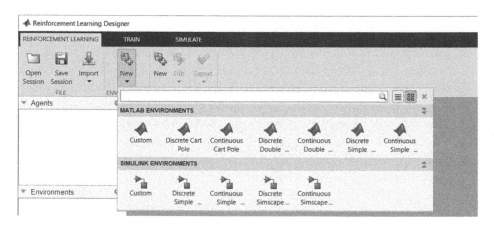

Figure 9.17 Reinforcement Learning Designer app with selection of predefined environments for MATLAB and Simulink.

Example 9.13

Another resource for utilizing predefined environments is the open-source library openAI Gym. This repository hosts a number of standard environments used in the research community to test and compare RL systems. However, the available environments are written in Python. To access and utilize these predefined environments in MATLAB we need to install the openAI Gym on our local Python installation and then import the selected environment into MATLAB. For this, Python 3.X needs to be installed. The Python version supported by MATLAB depends the version of MATLAB. Table 9.3 lists some of the more recent MATLAB installations and the corresponding Python versions.

To test whether MATLAB recognizes the Python installation, use `pyenv` in MATLAB's command window. If MATLAB does not find Python, you can direct MATLAB to it using the path to where Python is installed:

```
pyversion C:\Users\python.exe
```

To install the openAI Gym in Python, use the following command within the Windows Command window after starting Python:

```
pip install gym
```

OpenAI Gym has a number of environments for different areas, such as robotics and games. We are interested in their Classical Controls library. This library contains five environments: the Acrobot-v1 environment, which is a two-link inverted pendulum with the objective to bring up and maintain the linkage; the CartPole-v1 environment, which is a beam on a cart balancing control problem environment; the MountainCar-v0 environment, which has a car attempting to climb a mountain through oscillation in the valley; the MountainCarContiuous-v0 environment, which is the same MountainCar-v0 except that we deal with a continuous time environment; and the Pendulum-v0 environment, which is a simple pendulum. To load any of these environments into MATLAB, use the following command:

```
nv01 = py.gym.make('Pendulum-v1')
```

Table 9.3 Corresponding Python version for MATLAB installation

MATLAB	Python
R2022a	3.8, 3.9
R2021b	3.7, 3.8, 3.9
R2021a	3.7, 3.8
R2020b	3.6, 3.7, 3.8
R2020a	3.6, 3.7
R2019b	3.6, 3.7
R2019a	3.5, 3.6, 3.7

which loads the Pendulum example. To load any of the other environments, substitute their names in this command.

To find the specifics of the environment in terms of the action space and the observation space, follow the link on OpenAI's webpage to the GitHub repository. For example, the Pendulum-v0 environment defines the action space to be a torque on the pendulum with minimum and maximum values of −2.0 and +2.0. If the system is discrete it will be indicated by the type of actions it can apply to the system. The Python script also defines the observation space. For our case of the pendulum system we have three quantities: the x and y coordinates and the angular velocity of the beam. Each observation is also listed with the minimum and maximum values they can attain during the simulation. Another important detail to obtain from the Python script is the rewards function and the initial state.

Example 9.14

In our third example dealing with environments we utilize a discrete model from the openAI Gym library and make use of the visualization the library includes. The model we are using is the cart-pole model, which has a cart moving either right (action = 1) or left (action = 0). The associated MATLAB code to simulate and render the simulation is shown below.

```
env01 = py.gym.make('CartPole-v1')
env01.reset(); % initialize/reset the environment
for k = 1: 1000  %how many iterations
  action = int16(round(rand(1))); % define left or right
  env01.step(action); % apply action to environment
  env01.render() % show simulation
end
env01.close() % close simulation window
```

When executing the action using the env01.step(action) function, a return value is generated that entails the next observed states (cart position, cart velocity, pole angle, and pole angular velocity), the reward, and a Boolean value if the goal has been achieved. In the MATLAB code above we have applied a random binary action using the int16 function to cast it to the format the environment expects. However, we have not engaged the environment in a way to do any type of learning, which we will discuss in the next example.

Example 9.15

The Reinforcement Learning Designer app is a convenient way to implement reinforcement learning for control projects. Once the app is started, load the Discrete Simple Pendulum Simulink environment and select New in the Agent tab. This will lead us to define the agent. The resulting window is shown in Figure 9.18, where we have choices for the learning method of DQN and PPO. DQN stands for *deep Q-learning network*, where a deep neural network is used. PPO stands for *proximal policy optimization* algorithm. As we have discussed Q-learning, select the DQN option and leave the number of hidden units of the network as 256. After creating the agent, the main window of the Reinforcement Learning Designer app will update with options on the remaining hyperparameters of the RL learning implementation. The corresponding window is depicted in Figure 9.19. Here we can select our discount factor γ as well as the learning rate α. If a GPU is available, select GPU for faster processing. The exploration section features a parameter entitled Epsilon. This parameter balances the exploration and exploitation during the optimization. Initially, we want to focus on exploration, hence Epsilon is large (i.e., equal or close to 1). As the training proceeds, Epsilon decays according to the Epsilon decay factor, until a minimum value is achieved. The progression of the magnitude of Epsilon is shown in the corresponding graph. Using the Train tab of the app you have access to the training hyperparameters, such as episodes, settings for the stopping criteria, etc. Once all parameters have been selected, the Train button initiates the training of the created agent. During training the Episode Reward, the Average Reward, and the Episode Q0 will be tracked over the duration of the training. A sample training session is shown in Figure 9.20.

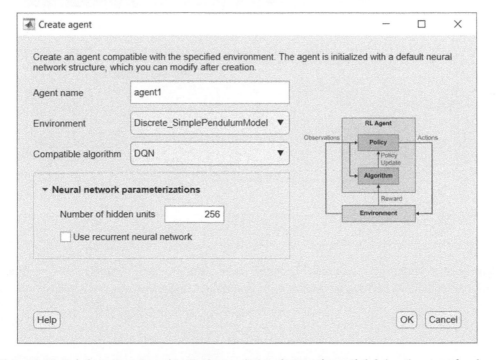

Figure 9.18 Reinforcement Learning Designer window for creating and defining the agent for deep Q-network learning and the discrete time simple pendulum environment.

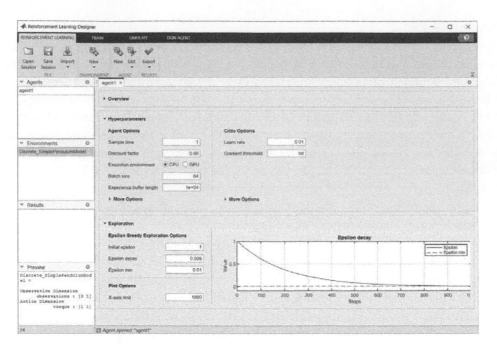

Figure 9.19 Reinforcement Learning Designer main window with hyperparameter selection for the agent as well as for balancing exploration and exploitation.

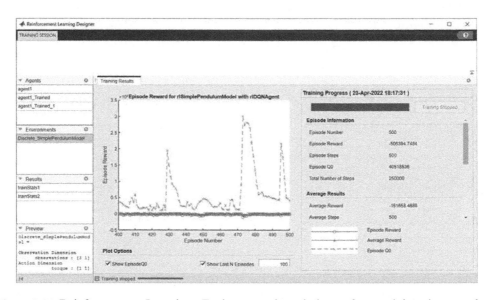

Figure 9.20 Reinforcement Learning Designer main window after training is complete. Information on the selected agent, environment, and training hyperparameters is depicted.

Example 9.16

In order to use an environment from openAI Gym with MATLAB, we need to create a wrapper function. For this we follow MATLAB's documentation [18]. The wrapper function consists of a class definition which we can inherit from an existing class in MATLAB, the `rl.env.MATLABEnvironment`. Our environment will consist of three parts: a constructor, a step function, and a reset function. The constructor allows us to define the states and dimensions as well as the actions. The set of actions we define is accomplished by the use of the following function:

```
rlFiniteSetSpec()
```

The argument for this function is a row vector with the set of actions. For the mountain car example from openAI Gym, the actions are 0 for accelerating to the left, 1 for not accelerating, and 2 for accelerating to the right. The observations are obtained using the function `rlNumericSpec()`. As there are two observations for this example, the x-axis position of the car and its velocity, our observation vector is [2 1]. The corresponding MATLAB code, which is based on MATLAB's documentation page, is given in the Appendix D.14.

We can create an environment object in MATLAB by calling the wrapper function:

```
myEnv = mountainPythonCar
validateEnvironment(myEnv)
```

Deep Learning Network Analyzer

Model — Analysis date: 24-Apr-2022 18:33:35 — **7** layers — **0** warnings — **0** errors

ANALYSIS RESULT

	Name	Type	Activations	Learnables
1	input_1 2 features	Feature Input	2	-
2	fc_1 24 fully connected layer	Fully Connected	24	Weights 24×2 Bias 24×1
3	relu_1 ReLU	ReLU	24	-
4	fc_2 48 fully connected layer	Fully Connected	48	Weights 48×24 Bias 48×1
5	relu_2 ReLU	ReLU	48	-
6	output 3 fully connected layer	Fully Connected	3	Weights 3×48 Bias 3×1
7	RepresentationLoss mean-squared-error	Regression Output	3	-

Figure 9.21 Reinforcement Learning Designer will make use of the created neural network, fitting the environment's number of inputs and outputs.

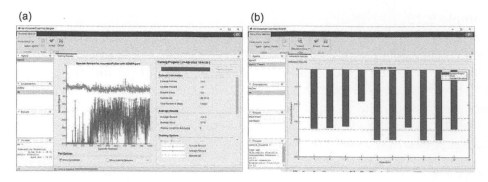

(a)   (b)

Figure 9.22 Reinforcement Learning Designer: (a) results window after training; (b) inspection window for simulation data.

To validate the environment (i.e., test if there are any issues with our created program), we use the function `validateEnvironment()`. For applying Q-learning we also need a neural network. For this, you can use the Deep Network Designer app. For our mountain car example the network will need to have two inputs for the input layer and three outputs for the last `fullyConnectedLayer`, as we generate three values for actions. We also will need a `regressionLayer` serving as the last layer in the network. To function with MATLAB's Reinforcement Learning Designer app you will have to label this regression layer as `RepresentationLoss`. We will utilize this network as our critic network in the Reinforcement Learning Designer app (Figure 9.21).

After training we have the option to accept the training results and inspect the trained network simulation data. The training outcome is shown in Figure 9.22(a) and the inspection of the simulation data is shown in Figure 9.22(b). Note, our trained network has 9 out of 10 simulations achieving the goal of moving the mountain car to the top of the hill.

Example 9.17

Another approach to create a simulation environment for use with RL, and in particular with the Reinforcement Learning Designer app, is by constructing a suitable Simulink program. The Simulink program will need to create the observations, the reward, and the action taken. We will demonstrate this approach with a simple control problem, where we want to control the response of a mass–spring–damper system due to a random step input. The corresponding Simulink diagram is shown in Figure 9.23(a), where we use a mass of 1.0 kg, a damping coefficient of 0.2 Ns/m, and a spring constant of 1.0 N/m.

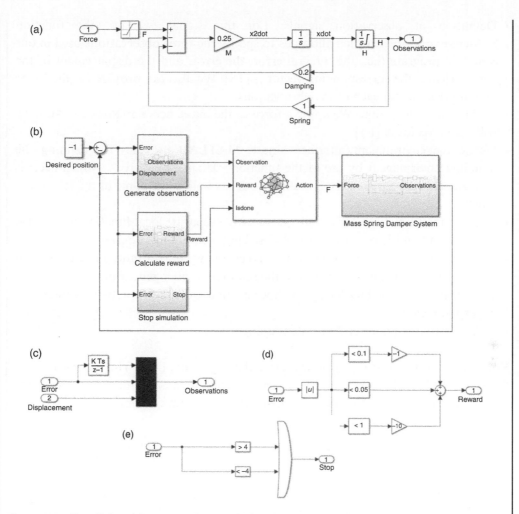

Figure 9.23 Simulink environment for RL simulation: (a) Simulink model implementation of a mass–spring–damper system; (b) overall Simulink model with the block representing the mass–spring–damper from part (a), as well as the observation block, the reward block, and the stop simulation block. The details about the observation block are shown in (c), the reward computation block is given in (d), and the stopping criterion block is shown in (e).

Figure 9.23(b) depicts the overall Simulink model with the RL Agent block. The observation, the reward, and the stopping function are also constructed using subsystems and are depicted in Figure 9.23(c)–(e). The RL Agent block will only accept single signals. As you may have multiple observations, use Mux blocks to combine them into one signal. In addition to the simulation parameters defined for the Simulink program, you also need to define the details of the RL Agent. The MATLAB documentation provides an excellent guide for this. For this example, we compose a MATLAB script to define the RL Agent. This program consists of the following sections:

- Definition of observation signals: For this signal, we use the function `rlNumericSpec()`, which allows us to specify the three observations used in our Simulink program (i.e., the integral error, the error, and the displacement of the mass). Hence, the corresponding vector is [3 1]. We also can provide for the lower and upper limits for each of the observations.
- Definition of the action: We apply a force to the mass, hence `rlNumericSpec()` will take a vector of [1 1].
- Define the environment interface object: MATLAB provides for defining the Simulink environment by use of the function `rlSimulinkEnv()`, which accepts as arguments the name of the Simulink model as well as the name of the RL Agent block.
- Define a reset function: This can be done with a separate local function which also can be used to initialize various blocks used in the Simulink program.
- Seeding the random number generator: To ensure we have a random seed for each episode of our simulation, we re-seed the random number generator.
- The creation of the DDPG agent: Specification of the AC networks including hyperparameters.
- Define training options and execute training.

For simplicity, a version of this script is found in Appendix D.15. The program is also found on MATLAB's documentation page for creating a Simulink environment for RL applications. To test the trained system, use the following commands:

```
simOpts =
rlSimulationOptions('MaxSteps',maxsteps,'StopOnError','on');
experiences = sim(env,agent,simOpts);
```

9.8   Conclusions

In this chapter we introduced some of the more popular ML algorithms with the knowledge that there are many nuances, different versions, and much underlying fundamental theory that still can be studied and explored. Our objective was to provide you with the basic concepts and main ideas, explain how to utilize these algorithms using MATLAB, and show some examples. MATLAB includes many additional algorithms in the respective apps dealing with ML and DL. It also maintains a documentation library with examples and example script. However, the question that remains to be addressed is: When to use which algorithm? ML has been found to be useful in an ever-expanding range of applications, well beyond engineering and science, and hence to answer this question is not trivial.

Table 9.4 Advantages and disadvantages of selected ML algorithms

Method	Advantages	Disadvantage
SVM	Effective method for data with dimensions greater than the sample set.	Simple SVM algorithms are not capable of generating a probabilistic confidence value of prediction.
SVM	Can be used for labeled as well as for unlabeled data.	Difficulty in choosing the correct kernel for a specific data set.
KNN	Computes prediction without any required training.	High computational cost when dealing with large data sets due to distance calculations.
KNN	Easy to use as there are only two hyperparameters: number of neighbors and choice of distance metric.	High computational cost when working with a large number of dimensions.
Regression	Effective for linearly separable data.	Susceptible to overfitting and influence of noise in the data set.
Regression	Can easily accommodate more than two independent variables.	Does not perform well for data that are linearly correlated.
Naïve Bayes	Allows for quick class prediction, including multi-class prediction.	Unobserved incidence of a category in the training data will result in zero probability (zero frequency).
Naïve Bayes	Requires less training data compared to some other ML methods.	Assumes independent predictors, which is often not the case.
Decision tree	Interpretable results with little data preparation requirements.	Susceptible to overfitting for problems with small data sets.
Decision tree	Applicable to linear and nonlinear problems.	Unsuitable for very large data sets as tree complexity becomes excessive.
RL	Applicable to complex, nonlinear systems.	Complex design requirements, with no performance assurances.
RL	Allows for accommodating complex systems as a black-box.	Computational excessively expensive and time consuming.

Instead of attempting to provide such a guide, we shall document a few advantages and disadvantages of the introduced methods. Table 9.4 lists the treated algorithms from this chapter and some of their advantages and disadvantages. This is by no means a complete list, but rather a general assessment of the base functionality of the selected ML methods. Nuances and different versions of these algorithms may change some of the listed properties, and hence alter the characteristics listed in Table 9.4.

Having only an introductory background presented on those methods, it seems adventurous to utilize them for real problems. However, you are not alone in this adventure, and MATLAB provides an automated functionality that explores which method and what hyperparameters are best used for a given data set. The functionality is AutoML, or automated machine learning. With AutoML you are able to generate good models with less time investment. MATLAB implements AutoML with the function fitcauto(). When supplied with the data and the categories, it will generate an exhaustive list of all methods, an associated performance evaluation identifier, information on the computational cost, and the optimized hyperparameters.

As an example, following MATLAB's documentation for the AutoML function, we can classify cars based on whether they were made in England or not using a table composed of horsepower, model year, MPG, weight, and origin data. MATLAB will produce its results in a table formatted and populated with the different classification methods and their performance on the given data set, as shown in Figure 9.24.

Similar to the AutoML classification function, MATLAB also provides such a tool for regression problems. The corresponding function is fitcauto(). As arguments it requests a table object and the corresponding name(s) of variables in the table.

```
Total iterations (MaxObjectiveEvaluations): 150
Total time (MaxTime): Inf
```

Iter	Eval result	Validation loss	Time for training & validation (sec)	Observed min validation loss	Estimated min validation loss	Learner	Hyperparameter:	Value
1	Best	0.0030769	20.198	0.0030769	0.0030769	ensemble	Method:	Bag
							NumLearningCycles:	201
							MinLeafSize:	7
2	Accept	0.0031847	1.0003	0.0030769	0.0030769	knn	NumNeighbors:	3
3	Accept	0.015385	0.76267	0.0030769	0.0030769	nb	DistributionNames:	normal
							Width:	NaN
4	Accept	0.0030769	10.798	0.0030769	0.0030769	ensemble	Method:	LogitBoost
							NumLearningCycles:	274
							MinLeafSize:	15
5	Accept	0.0031847	0.24982	0.0030769	0.0030769	knn	NumNeighbors:	4
6	Accept	0.015385	0.1276	0.0030769	0.0030769	nb	DistributionNames:	normal
							Width:	NaN
7	Best	0.0030769	0.30222	0.0030769	0.0030769	tree	MinLeafSize:	105
8	Best	0	0.88344	0	0	svm	BoxConstraint:	0.022186
							KernelScale:	0.085527

Figure 9.24 Partial table generated using MATLAB's AutoML function for a classification problem.

Another good resource to tackle ML problems is MATLAB's training portal Onramp. It allows users to learn and practice any of the methods contained in their toolboxes, including ML and DL algorithms.

SUMMARY

In the following, a number of key concepts of machine learning as treated in this chapter are summarized.

Feature Engineering
Before applying ML algorithms, the raw data need to be conditioned with the objective to highlight and emphasize key characteristics. Often using domain knowledge and statistical measures, these features allow for improvement of the quality of the resulting ML training.

Support Vector Machines
SVM is a supervised ML algorithm that can be used for classification problems as well as regression analysis. A support vector is a data point closest to the hyperplane that divides a class of data points from another class. The hyperplane is found by maximizing the distance to the support vectors.

K-Nearest Neighbor
KNN is a nonparametric supervised learning algorithm. Using distance measures, classification, or prediction about an individual data point is achieved by association to a group of data points that share similar distance characteristics.

Linear Regression
Belonging to the class of supervised ML algorithms, linear regression utilizes a linear relationship between the variables as base model and estimates its parameter statistically from the given data set.

Naïve Bayes
Classification using the NB algorithm is a probabilistic ML method developed on the principles of Bayes' theorem. The key assumption of this classification method is that the features used for classification are statistically independent.

Decision Tree
Decision trees are nonparametric supervised learning methods used for classification and regression problems. For classification, decision trees apply

sequentially a series of tests to the presented data to assign the data to a group. The end points (the leaves) are the categories, while the branches are the pathways for different outcomes of the applied tests.

Reinforcement Learning

Reinforcement learning is an iterative algorithm that learns from interaction with the environment. It uses an agent to explore and engage the environment, and receives rewards based on how well the stated goal is achieved. The agent's interaction is based on a policy that is part of the RL learning process and is guided by the goal of maximizing the total reward it achieves through the interactions.

REVIEW QUESTIONS

1. What is a goal of feature engineering?
2. What is data cleaning? Provide an example.
3. Why do we standardize data in ML? What kind of problems may occur if we do not standardize data?
4. What do we mean when referring to the dimension of an ML process?
5. What is the sequential forward selection (SFS) algorithm used for?
6. What is a support vector as used in SVM algorithms?
7. What principle is used to define the hyperplane in an SVM process?
8. What is the kernel trick and why do we use it?
9. For what purpose do we use cross-validation and for what kind of data would cross-validation make a difference?
10. What is leave-one-out validation?
11. How is a new data point classified using KNN, where $k = 9$?
12. What is the difference between the Manhattan distance and the Euclidean distance?
13. What is the goodness-of-fit measuring in a linear regression model?
14. When having a high variance in the data, does the resulting regression model tend to be overfitted or underfitted?
15. Why do we use regularization for estimating linear regression model parameters?
16. Which of the two regression methods yields generally a simpler regression model: LASSO or ridge regression?
17. Why do we call naïve Bayes algorithm naïve?
18. Why do we add a constant to the class conditional probability?
19. What is a node and a branch in a decision tree?
20. What role does an agent play in an RL system?
21. What function does the policy have in an RL system?
22. How are the value function and the Q-function related?
23. What is a Markov decision process?

24. What role does the critic play in an actor–critic method?
25. What is reward shaping as used in an RL system? What kind of problem does reward shaping aim to address?

PROBLEMS

9.1 The data set cities.mat is a pre-installed resource detailing the ratings of 329 US cities. The data set is based on nine criteria: climate, housing, healthcare, crime, transportation, education, arts, recreation, and economics [19]. The ratings are tabulated in a 329×9 matrix. Load this data set and perform the following ML classification tasks:

  a. Survey the data and perform data cleaning operations such as removing or substituting missing data points and applying standardization.
  b. Use the Diagnostic Feature Designer app and investigate which feature(s) have better sensitivities than others. Rank them from best to worst.
  c. Use the following MATLAB script

```
boxplot(ratings,0,'+',0)
set(gca,'YTicklabel',categories)
```

  Comment on the variability of the different classes. Which classes should be used and which should be left out to build an ML model that predicts accurately the rating of a city?

  d. Pick the two best-performing classes and develop an ML prediction model. Explain your choice of ML model structure and compare with other model structures you believe are less predictive.

9.2 Using the MATLAB data set called `carsmall`, perform the following tasks:
  a. Inspect the data and apply data cleaning operations.
  b. Generate a plot that maps the MPG data versus the weight, and comment on the relationship.
  c. Use LASSO and ridge regression to find a regression model. Compare each outcome and comment on any differences.

9.3 Using the data set from Problem 9.2, perform a classification using "All" in the Classification Learner app and compare the results based on accuracy with the choices you made in Problem 9.2.

9.4 Write a MATLAB script that asks for two words of equal length and in response computes the Hamming distance.

9.5 Use the following script to generate a set of data:

```
k = 0:0.1:3.5;
y = square(pi*k)+0.25*rand(1,length(k));
plot(k,y)
k = k';y = y';
data = table(k,y);
```

Using the generated table object, perform the following investigations:

a. Use SVM to find a regression model that fits the data contained in the table.
b. Perform a 10-fold cross-validation error test.
c. Perform a cross-validation error test with 40% of the data being holdout.
d. Using Equation (9.17) and a MATLAB script, compose the associated matrices and vectors and compute the least-squares estimate of the data k and y.
e. Compute the associated v_{SST}, v_{SSE}, and R^2 values.

9.6 Use the data generated in Problem 9.5 and the Regression Learner app to find the best regression model the app offers.

9.7 Suppose the production of a consumer good has a 0.8% chance of defective packaging that would lead the product to spoil before the expiration date printed on the packaging. The production process employs a binary test that can be considered as an imperfect indicator of the packaging being compromised. Suppose this binary test yields a 98% success rate for identifying a broken package when the packaging is indeed broken, and a 97% success rate for detecting that the package is not broken when indeed the package is not compromised.

a. During a single sample inspection a test indicates that the sample has broken packaging. Compute the a posteriori probability for this sample indeed has compromised packaging.
b. Suppose we conduct a second but independent test using the same testing procedure on the sample identified in part (a). Compute the a posteriori probability of the package being compromised given that the second test conducted yielded also an indication that the package is broken.

9.8 Research what the principal component analysis (PCA) method is and what it is useful for by consulting MATLAB's documentation page [20]. Use the Regression Learner app to perform a PCA analysis on the cities.mat data set.

9.9 Using the data set given by cities.mat, create a regression tree model.

a. Use the following categories as predictors: climate, housing, health, crime, transportation, arts, recreation, and economics, while using the education vector as the response.
b. Using the results from Problem 9.8, tailor your predictor selection to include only three categories for a tree model and compare the prediction to the results obtained in part (a).

9.10 Using the mass–spring–damper system from Section 9.7, alter the reward function to account for the actuation signal energy usage, with an objective to reduce the exerted energy while controlling the system to remain at a certain setpoint. Use the reinforcement learning environment provided in Appendix D.15.

9.11 Open the Simulink model rlwatertank.slx using the command

```
open_system('rlwatertank')
```

in MATLAB's root directory (from the Simulink Control Design toolbox)

a. Study the reward function and map out its value generation over a few iterations.
b. Train the AC model (agent) using the script provided in the Appendix (see Section 9.7 and Appendix D.15).
c. Alter the model to solely feature a PID controller (without reward, agent, and stopping criteria) and tune the PID controller with the built-in tuning function.
d. Compare a step and an impulse response of the trained system from part (b) and the tuned system in part (c).

9.12 Derive a mathematical model of the pendulum system from the project section in Appendix G. Program the model in Simulink and generate a reward function, an observation block, and a termination block similar to Problem 9.10. Use the code in Appendix D.15 and train your reinforcement learning control system to achieve a step response in minimal time.

REFERENCES

[1] Pohjalainen J, Räsänen O, Kadioglu S (2015). Feature selection methods and their combinations in high-dimensional classification of speaker likability, intelligibility and personality traits. *Computer Speech & Language*, 29(1): 145–171.

[2] Peng H, Long F, Ding C (2005). Feature selection based on mutual information criteria of max-dependency, max-relevance, and min-redundancy. *IEEE Transactions on Pattern Analysis and Machine Intelligence*, 27(8): 1226–1238.

[3] Widodo A, Yang BS (2007). Support vector machine in machine condition monitoring and fault diagnosis. *Mechanical Systems and Signal Processing*, 21(6): 2560–2574.

[4] Ma Y, Guo G (eds.). (2014). *Support Vector Machines Applications*. Springer, New York.

[5] Jakkula V (2006). Tutorial on support vector machine (SVM). School of EECS, Washington State University. https://course.ccs.neu.edu/cs5100f11/resources/jakkula.pdf.

[6] Refaeilzadeh P, Tang L, Liu H (2009). Cross-validation. *Encyclopedia of Database Systems*, 5: 532–538.

[7] Taunk K, De S, Verma S, Swetapadma A (2019). A brief review of nearest neighbor algorithm for learning and classification. *International Conference on Intelligent Computing and Control Systems (ICCS)*, pp. 1255–1260.

[8] Hart P (1968). The condensed nearest neighbor rule (corresp.). *IEEE Transactions on Information Theory*, 14(3), 515–516.

[9] Green PE, Rao VR (1969). A note on proximity measures and cluster analysis. *Journal of Marketing Research*, 6: 359–364.

[10] Weisberg S (2005). *Applied Linear Regression*. Wiley, Chichester.

[11] Tibshirani R (1996). Regression shrinkage and selection via the lasso. *Journal of the Royal Statistical Society: Series B (Methodological)*, 58(1): 267–288.

[12] Langley P, Sage S (1994). Induction of selective Bayesian classifiers. In *Uncertainty Proceedings*, pp. 399–406.

[13] Cestnik B, Bratko, I (1991). On estimating probabilities in tree pruning. *European Working Session on Learning*, pp. 138–150.

[14] Quinlan JR (1986). Induction of decision trees. *Machine Learning*, 1(1): 81–106.

[15] Sutton RS, Barto AG (2018). *Reinforcement Learning, and Introduction*, 2nd edn. MIT Press, Cambridge, MA.

[16] Sutton RS, Barto AG (1998). *Reinforcement Learning, and Introduction*. MIT Press, Cambridge, MA.

[17] Watkins CJ, Dayan P (1992). Q-learning. *Machine Learning*, 8(3–4): 279–292.

[18] MathWorks (2023). Create Custom MATLAB environment from template. www.mathworks.com/help/reinforcement-learning/ug/create-custom-matlab-environment-from-template.html.

[19] Boyer R, Savageau D (1985). *Places Rated Almanac: Your Guide to Finding the Best Places to Live in America*. Rand McNally, Chicago, IL.

[20] MathWorks (2023). Principal Component Analysis (PCA). www.mathworks.com/help/stats/principal-component-analysis-pca.html.

APPENDIX A
Modern Control System Tutorial

A.1　Introduction

In this appendix we review the essentials of basic modern control theory. The reader of the book is assumed to have seen this material in a prior class/book. However, in case such a background is missing or outdated, the following tutorial covers the main concepts of modern control theory in a brief and concise fashion.

Control in engineering applications is understood to include a feedback loop to compare the output of a system with a desired output (user defined) and take action based on any discrepancy between these two quantities. Any time such feedback is employed, we talk about closed-loop control. Another control structure exists where one controls the input to the system with the hope the output conforms to these aspirations. The latter system lacks continued measurement of the output and is referred to as an open-loop control system. This book primarily deals with the former structure, where feedback is utilized to take corrective measures in order to control the output.

Control systems has its origin with the invention of the steam engine. Though prior control systems existed, the advent of the steam engine – and in particular the governor controller designed by James Watt – resulted in efforts that produced many advancements and theories, especially during World War II, which constitute the topic of modern control systems. Much theory and specialized control topics have grown out of this topic. For example, the control of jet airliners required the development of adaptive control systems. Other engineering applications and mathematical advancements allowed for the development of control systems that are robust to uncertain models, control systems that are optimal in terms of energy consumption of the controller and actuator, or control systems that are tailored to nonlinear system dynamics, and many other specialized fields. This tutorial does not treat these special topics, and is not a substitute for a full treatment of basic control system theory. For textbooks covering modern control systems in detail, please see [1] or [2].

Figure A.1 Mass–spring–damper system.

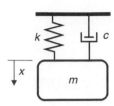

A.2   System Descriptions

Consider a simple dynamic system composed of three components: an inertia, an energy-storing component, and an energy-dissipation part. Such a mechanical system can be constructed using a mass, a spring, and a damper. For an electrical system the same system can be created using an inductor, a capacitor, and a resistor. The mechanical system is used here to showcase how such dynamics are represented in controls. We can imagine the system configuration as depicted in Figure A.1.

Assuming the motion is confined in one direction (i.e., a one degree of freedom [1DOF] system), the following differential equation can be developed using Newton's second law:

$$m\ddot{x} = -c\dot{x} - kx. \tag{A.1}$$

We introduce here the natural frequency, $\omega_n = \sqrt{k/m}$ and the damping ratio (also sometimes referred to as the modal damping factor) $\xi = c/2m\omega_n$. Using these two parameters, the three-parameter Equation (A.1) can be restated as a two-parameter differential equation:

$$\ddot{x} + 2\xi\omega_n\dot{x} + \omega_n^2 = 0. \tag{A.2}$$

To solve this equation we assume some initial conditions, such as $x = x_o, \dot{x} = v_o$, @$t = 0$ and a solution that contains the possible description of $x(t)$, such as:

$$x(t) = e^{pt}. \tag{A.3}$$

Here, p is a yet-to-be-determined complex constant. Utilizing the first and second derivative of the assumed solution in Equation (A.3), we obtain

$$\dot{x} = pe^{pt} \text{ and } \ddot{x} = p^2e^{pt}. \tag{A.4}$$

Substituting Equation (A.4) into (A.2) yields

$$\left(p^2 + 2\xi\omega_np + \omega_n^2\right)e^{pt} = 0$$

This allows us to formulate the characteristic equation of the system as

$$p^2 + 2\xi\omega_n p + \omega_n^2 = 0. \tag{A.5}$$

Hence, $x(t) = e^{pt}$ is a solution of Equation (A.1), conditioned that p is a root and given by

$$p_{1,2} = \left[\xi \pm \sqrt{\xi^2 - 1}\right]\omega_n. \tag{A.6}$$

The roots yield two solutions for Equation (A.1):

$$x_1(t) = e^{p_1 t} \text{ and } x_2(t) = e^{p_2 t},$$

where $p_1 = [-\xi + \sqrt{\xi^2 - 1}]\omega_n$ and $p_2 = [-\xi - \sqrt{\xi^2 - 1}]\omega_n$. Utilizing the form of the general solution as

$$x(t) = C_1 e^{p_1 t} + C_2 e^{p_2 t}, \tag{A.7}$$

where C_1 and C_2 are constants that have to be determined using the initial conditions:

$$x(t = 0) = C_1 + C_2 = x_o,$$
$$\dot{x}(t = 0) = p_1 C_1 + p_2 C_2 = v_o.$$

This can be used to infer C_1 and C_2 as:

$$C_1 = \frac{x_o p_2 - v_o}{p_2 - p_1} \text{ and } C_2 = \frac{v_o - x_o p_1}{p_2 - p_1} \tag{A.8}$$

Suppose the damping coefficient is restricted to be $0 < \xi < 1$, the roots defined in Equation (A.6) become:

$$p_1 = \left[-\xi + i\sqrt{1 - \xi^2}\right]\omega_n \text{ and } p_2 = \left[-\xi - i\sqrt{1 - \xi^2}\right]\omega_n. \tag{A.9}$$

Hence, Equation (A.7) can be expressed as

$$x(t) = e^{-\xi\omega_n t}[(C_1 + C_2)\cos\omega_d t + i(C_1 - C_2)\sin\omega_d t]. \tag{A.10}$$

Equation (A.10) can be restated as:

$$x(t) = e^{-\xi\omega_n t}\left(x_o \cos\omega_d t + \frac{x_o\xi\omega_n + v_o}{\omega_d}\sin\omega_d t\right), \tag{A.11}$$

where $\omega_d = \omega_n\sqrt{1 - \xi^2}$. Equation (A.11) is the response of a 1DOF system to a set of initial conditions. From a mechanical point of view, Equation (A.11) represents the response of a mass–spring–damper system due to an initial velocity and an initial displacement. Due to the conditions we imposed on the damping ratio, the response and hence the system is qualified as the response of an underdamped system. The oscillatory nature of this response is showcased in Figure A.2.

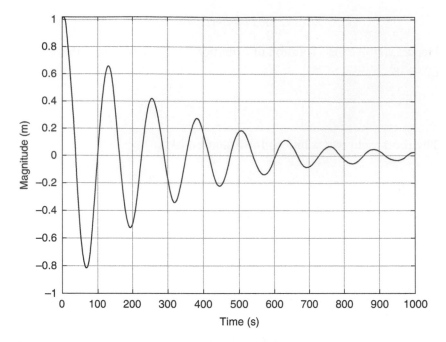

Figure A.2 Response of an underdamped system according to Equation (A.11).

Suppose we apply a force onto the mass – or inertia – to excite the system. For the purpose of our discussion, let's assume that this force acts only for a very short period of time (i.e., for Δt). From basic dynamics, we know that such a force can be classified as an impulse to the system. An impulse is equivalent to the difference in momentum before and just after applying the force, which can be expressed as

$$F\Delta t = mv_2 - mv_1.$$

Let's also represent the impulse by $\hat{F}$ as

$$\hat{F} = \int_{t}^{t+\Delta t} Fdt.$$

Using the impulse-momentum theorem, at time $t = 0$ the impulse is

$$\hat{F} = m\dot{x}\big|_{+0} - m\dot{x}\big|_{-0} = mv_o. \tag{A.12}$$

Equation (A.12) implies that the impulse causes the mass m to have an initial velocity equal to $v_o = \hat{F}/m$. Using Equation (A.11) and setting $x_o = 0$, we can write the response of a dynamic system due to an initial velocity v_o as

$$x = \frac{v_o e^{-\zeta \omega_n t}}{\omega_d} \sin \omega_d t.$$

Since $v_o = \hat{F}/m$, the response of the system due to an impulsive force $\hat{F}$ is

$$x = \hat{F} \frac{e^{-\zeta \omega_f t}}{m \omega_d} \sin \omega_d t. \tag{A.13}$$

Defining the unit impulse response function $g(t)$ as

$$g(t) = \frac{1}{m \omega_d} e^{-\zeta \omega_n t} \sin \omega_d t, \tag{A.14}$$

then the response of a system due to an impulse is given as

$$x(t) = \hat{F} g(t). \tag{A.15}$$

The response of the system due to an impulse as given by Equation (A.15) can be expanded to any input force $F(t)$ by constructing $F(t)$ with an infinite number of impulses $\hat{F}$, as indicated in Figure A.3, where $\Delta t \rightarrow 0$.

Using one of the impulses that is making out $F(t)$ at time τ (i.e., $F(\tau)$), and considering the duration of the applied impulse as $\Delta \tau$, the response of the system at t seconds after the impulse has been applied (i.e., $t - \tau$) is

$$\Delta x(t) = F(\tau) \Delta \tau g(t - \tau).$$

As $F(t)$ is composed of an infinite number of $\hat{F}$, the response of the system due to the force $F(t)$ is

$$x(t) = \int_0^t F(\tau) g(t - \tau) d\tau. \tag{A.16}$$

Equation (A.16) is referred to as Duhamel's integral or a convolution integral. For the 1DOF system introduced above, the solution or response due to a force

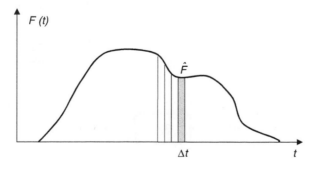

Figure A.3 Generic input force $F(t)$ composed by an infinite number of impulses $\hat{F}$.

$F(t)$ to the system is given by combining the particular solution given by Equation (A.16) and the homogenous solution given by Equation (A.11):

$$x(t) = \int_0^t \frac{F(\tau)}{m\omega_d} e^{-\xi\omega_n(t-\tau)} \sin(t-\tau)d\tau$$

$$+e^{-\xi\omega_n t}\left[x_o \cos\{\omega_d t\} + \frac{x_o\frac{c}{m} + 2v_o}{2\omega_d} \sin\{\omega_d t\}\right]. \tag{A.17}$$

It is easy to imagine that the convolution integrals in Equations (A.16) and (A.17) can be tedious to solve. A more convenient way to solve differential equations, integrals, etc., is to use the Laplace transform:

$$X(s) = L\{x(t)\} = \int_0^\infty x(t)e^{-st}dt, \tag{A.18}$$

where s is the complex variable given by $s = \sigma + i\omega$. Applying the Laplace transform to Equation (A.12) yields the transfer function $G(s)$:

$$G(s) = L\{g(t)\}. \tag{A.19}$$

The beauty of using the Laplace transform is that differential equations in the time domain become algebraic equations in the s-domain, convolution integrals become multiplications, and hence are easily solvable. Modern control systems theory makes ample use of this property and hence is almost exclusively dealt with in the s-domain. For example, applying the Laplace transform to Equation (A.16), the response of the system in the s-domain is given by:

$$X(s) = L\{x(t)\} = L\left\{\int_0^t F(\tau)g(t-\tau)d\tau\right\} = F(s)G(s).$$

This operation can be illustrated using block diagrams. For example, a system with a transfer function $G(s)$ and an input of $F(s)$ has a response of $X(s)$ (Figure A.4).

Block diagrams are used abundantly in modern control system as they provide an easy way to describe the control system with individual blocks, each

Figure A.4 Block diagram representation of transfer function operation.

representing a particular part/dynamic of the overall system. For modeling purposes, block diagrams are also very useful, as each subsystem can be modeled independently and interconnected using block diagram logic. To find the overall dynamic representation of the system, one can employ block diagram reduction methods, where individual components of the block diagram are combined until only one block is left. This block, most often composed of a numerator polynomial in s and a denominator polynomial in s, captures the entire dynamics of the system.

Some of the block diagram reduction rules are shown in Table A.1.

The procedure to reduce a complicated block diagram that contains a number of feedback loops, forward loops, multiple inputs, etc., into a single block can be given as follows:

Step 1: Combine all cascade blocks.

Table A.1 Block diagram reduction rules

Transformation	Equivalent equation	Block diagram	Equivalent block diagram
Combine blocks in cascade	$Y = (G_1 G_2)R$		
Combine blocks in parallel	$Y = G_1 R \pm G_2 R$		
Eliminate a feedback loop	$Y = \dfrac{G_1}{1 \pm G_1 G_2} R$		
Rearrange two summing points	$Y = R_1 \pm R_2 \pm R_3$		
Move a summing point before a block	$Y = G_1 R_1 \pm R_2$		
Move a summing point beyond a block	$Y = G_1 \{R_1 \pm R_2\}$		

Figure A.5 Block diagram for example using MATLAB and the `feedback()` command.

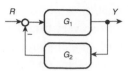

Step 2: Combine all parallel blocks.

Step 3: Reduce all minor feedback loops.

Step 4: Shift summing points to the left and takeoff points to the right of the major loop.

Step 5: Repeat Steps 1–4 until a cascade or parallel or feedback loop has been achieved for a specific input.

Step 6: Repeat Steps 1–5 for each input.

Alternatively, one can use MATLAB for block diagram reduction. The primary command in MATLAB is `feedback()`. For example, consider two transfer functions, G_1 and G_2, given by

$$G_1 = \frac{5s + 10}{s^2 + 17s + 12} \text{ and } G_2 = \frac{s + 120}{s^3 + 4s^2 + 7s + 2}.$$

We will construct a system according to Figure A.5.

The objective is to reduce it to one block. From Table A.1 we see the solution is

$$Y = \frac{G_1}{1 + G_1 G_2} R.$$

However, since G_1 and G_2 are rather complex, the evaluation of the above formula is tedious. Using MATLAB, we can write the following script:

```
num1=[5 10];
den1=[1 17 12];
sys1=tf(num1,den1);
num2=[1 120];
den2=[1 4 7 2];
sys2=tf(num2,den2);
sys = feedback(sys1,sys2)
```

Note, we defined the numerator and denumerator polynomials in s by using a vector notation and recording only the coefficient to each term of the polynomials in descending order. If one power in the polynomial does not exist, we record a zero. For example, the polynomial $s^2 + 5$ would be written as [1 0 5] as there is no s^1 term in the polynomial. Upon running the program, we receive the following result, representing the overall transfer function of the combined system:

$$\text{sys} = \frac{5\ s^4 + 30\ s^3 + \ 75\ s^2 + 80\ s + 20}{s^5 + 21\ s^4 + 87\ s^3 + 174\ s^2 + 728\ s + 1224}$$

Not all control analysis and design methods reside in the s-domain. Advanced control systems are usually treated in the time domain. When the system can be considered linear and time invariant, an alternative representation of the dynamic system to be controlled is the state-space description. The name is derived from the properties such a designation implies. In particular, a state has two properties: (1) it represents the complete system information; and (2) if a state is known at time t, it is possible to compute the state for all future times given the input. States are derived by using exclusively first-order differential equations of the system characteristics. Hence, a nth-order system will be converted into a system with n first-order differential equations. For example, the mass–spring–damper system described by Equation (A.1) is a second-order system, which can be converted into a set of first-order differential equations by the following substitution: $w_1 = x$ and $w_2 = \dot{x}$, which leads to the following sets of equations:

$$\dot{w}_1 = w_2, \tag{A.20}$$

$$\dot{w}_2 = \frac{1}{m}(-kw_1 - cw_2). \tag{A.21}$$

If we add an input, for example a force applied to the mass in the direction of the positive coordinate, Equation (A.21) will become

$$\dot{w}_2 = \frac{1}{m}(-kw_1 - cw_2 + F). \tag{A.22}$$

A convenient representation of state-space models is the use of matrices, allowing for the rich set of matrix analysis tools to be utilized. For example, Equations (A.20) and (A.21) can be expressed by the following matrix equation:

$$\dot{\mathbf{x}} = \mathbf{Ax} + \mathbf{Bu}, \tag{A.23}$$

where $\mathbf{A} \in \mathbb{R}^{n \times n}$ is the system matrix and $\mathbf{B} \in \mathbb{R}^{n \times n_i}$ is the input matrix, $\dot{\mathbf{x}} = [\dot{w}_1 \quad \dot{w}_2]^T = \mathbb{R}^{n \times 1}$ is the state vector, and $\mathbf{u} = \mathbb{R}^{n_i \times n}$ is the input vector, were we have a system with n_o outputs and n_i inputs. The system and input matrices are given by

$$\mathbf{A} = \begin{bmatrix} 0 & 1 \\ -\dfrac{k}{m} & -\dfrac{c}{m} \end{bmatrix}, \mathbf{B} = \begin{bmatrix} 0 \\ \dfrac{1}{m} \end{bmatrix}.$$

Suppose some of the states are being measured, then an output equation can be formed as

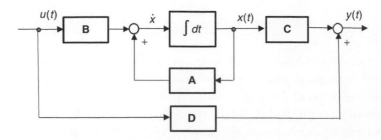

Figure A.6 State-space representation using blocks.

$$y = \mathbf{C}\mathbf{x} + \mathbf{D}\mathbf{u}, \qquad (A.24)$$

where $\mathbf{y} \in \mathbb{R}^{n_o \times 1}$ is the output vector, $\mathbf{C} \in \mathbb{R}^{n_o \times n}$ the measurement matrix, and $\mathbf{D} \in \mathbb{R}^{n_i \times n}$ the direct transmission matrix. The system can also be illustrated using block diagrams in the time domain. Such a schematic is given in Figure A.6.

A state-space system can be converted into a transfer function and vice versa. To do so, we take the Laplace transform of each term of Equations (A.20) and (A.22), assuming zero initial conditions and solve for $X(s)$, which yields $X(s) = (s\mathbf{I} - \mathbf{A})^{-1}\mathbf{B}U(s)$. Noting that the output in the s-domain can be found as $Y(s) = [\mathbf{C}(s\mathbf{I} - \mathbf{A})^{-1}\mathbf{B} + \mathbf{D}]U(s)$, the ratio output over input is found as

$$G(s) = \frac{Y(s)}{U(s)} = \mathbf{C}(s\mathbf{I} - \mathbf{A})^{-1}\mathbf{B} + \mathbf{D}, \qquad (A.25)$$

which is the transfer function of the system in the s-domain.

Any of these conversions is done easily in MATLAB, including the representation of the system in state-space as well as transfer function form. The latter we already reviewed in the above MATLAB example. For state-space, we can define each matrix separately and assign a variable representing the system. For example,

```
num1=[5 10];
den1=[1 17 12];
sys1=tf(num1,den1);
ST=ss(sys1)
```

which yields upon running the code in MATLAB the following output:

```
ST =

  A =
     x1 x2
  x1 -17 -3
  x2  4   0

  B =
     u1
  x1 2
  x2 0

  C =
     x1 x2
  y1 2.5 1.25
```

```
D =
   u1
   y1 0
```

To convert the state-space back to the transfer function form we can use the `tf()` command:

```
sys_tf = tf(sys_ss);
```

Alternatively, to convert a transfer function model to a state-space model we can use the MATLAB command `ss()` as follows:

```
sys_ss = ss(sys_tf);
```

To access each matrix separately, we can use the dot operator; for example, the A matrix is extracted by:

```
ST.A
```

```
      ans=
           -17 -3
             4   0
```

A.3 Transient and Steady-State Response

The response of a dynamic system can be separated into two parts: the initial transient response and the steady-state response. The transient response is primarily due to energy stored in the system imposed by its initial conditions, while the steady-state response is due to external energy flowing into the system. The transient response is in essence the homogeneous part of the solution to the corresponding differential equation, while the steady-state response matches the particular solution of the differential equation describing the dynamic system. In controls, often we seek to modify both parts of the response and hence we use key characteristics that describe these two responses quantitatively.

Figure A.7 shows the step response of an underdamped system with the key characteristics included. The horizontal dashed lines indicate a magnitude band around the steady-state response of $\pm$ 1%. One way to distinguish the steady-state response from the transient response is by using such a magnitude band and declaring the transient response to be over when the step response magnitude of the system stays within that magnitude band. The associated time is called the settling time. Other measures can be read out from such a response, as indicated in Figure A.7.

For example, the maximum percentage overshoot M_p is found as follows:

$$M_p = \frac{y(t_p) - y(\infty)}{y(\infty)} \times 100\% = e^{-\frac{\pi \xi}{\sqrt{1-\xi^2}}} \times 100\%, \tag{A.26}$$

where t_p is the peak time (i.e., the maximum magnitude of the response for underdamped systems). Here, ξ is the damping ratio (as defined in Equation (A.2)). The peak time can be computed by

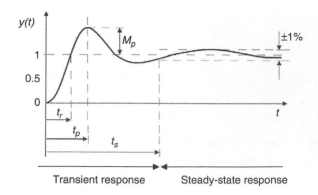

Figure A.7 Step response of an underdamped system.

$$t_p = \frac{\pi}{\omega_n\sqrt{1-\varsigma^2}}, \tag{A.27}$$

where ω_n is the natural frequency of the system. There is no such thing as maximum percentage overshoot or peak time for systems that are overdamped ($\varsigma > 1$), critically damped ($\varsigma = 1$), or not damped ($\varsigma = 0$). The rise time t_r is defined by the time it takes until the system reaches the input level for the first time:

$$t_r = \frac{\pi - \cos^{-1}(\varsigma)}{\omega_n\sqrt{1-\varsigma^2}}. \tag{A.28}$$

For overdamped and critically damped systems where no overshoot occurs, the rise time is defined by the time period of the response going from 10% to 90% in magnitude. For a 2% error band the settling time t_s can be computed as

$$t_s = \frac{4}{\varsigma\omega_n}. \tag{A.29}$$

The steady-state response is usually judged by the steady-state error. Errors in the system may be a result of imperfections in the system components, such as drift, friction, aging, deterioration, etc. For a unit feedback system (i.e., a system that has no transfer function on the feedback path [see Figure A.5 with $G_2(s) = 1$]), the error is given by the difference between the reference input and the output of the system:

$$e(t) = r(t) - y(t) = E(s) = R(s) - Y(s) = \frac{R(s)}{1 + G(s)},$$

where we used the Laplace transform on the two signals and solved for the error using the input and the open-loop system transfer function $G(s)$. Applying the final value theorem, the steady-state error is found as:

$$e_{ss} = \lim_{t \to \infty} e(t) = \lim_{s \to 0} sE(s) = \lim_{s \to 0} sR(s)\frac{1}{1 + G(s)}. \tag{A.30}$$

Suppose we have a polynomial input such as

$$r(t) = t^{k-1}u(t) \Rightarrow R(s) = \frac{1}{s^k}.$$

Hence, the steady-state error for a unit feedback system, using the final value theorem, is:

$$e_{ss} = \lim_{s \to 0} s \; \frac{1}{s^k} \; \frac{1}{1 + G(s)} = \lim_{s \to 0} \frac{s}{s^k}\frac{1}{1 + G(s)}. \tag{A.31}$$

The order of the polynomial in s (i.e., k) defines the type of the system. As a root at the origin of the s-plane indicates integral action (see Laplace transform for the term $\frac{1}{s}$) the higher the type of the system is, the better it can handle complex inputs with regard to the steady-state error. For a unit feedback system of type k, the feedback system ensures the following:

$$e_{ss} = 0 \quad \text{for } R(s) = \frac{1}{s^k}$$

$$|e_{ss}| < \infty \; \text{ for } R(s) = \frac{1}{s^{k+1}}.$$

The above can be restated as: *A unit feedback system is of type* k *if the open-loop transfer function of the system has* k *poles at* s = 0.

The system type is not only used to define a system, but also the so-called error constants. For example, for a type 0 system that can be factored as follows:

$$D(s)G(s) = \frac{(s - z_1)(s - z_2)\ldots}{s^k(s - p_1)(s - p_2)\ldots} = \frac{D_0(s)G_0(s)}{s^k}.$$

The error constant is given by

$$K_k = \lim_{s \to 0} s^k D(s)G(s) = D_0(0)G_0(0).$$

As we deal with integrators and differentiators, we can use the analogy to position, velocity, and acceleration to give names to these constants. For example, the position constant for a type 0 system is given by

$$K_p = \lim_{s \to 0} D(s)G(s) \;\; [-]. \tag{A.32}$$

Table A.2 Steady-state error and type of a system

System type	Step input	Ramp input	Parabolic input
Type 0	$\dfrac{1}{1+K_p}$	∞	∞
Type 1	0	$\dfrac{1}{K_v}$	∞
Type 2	0	0	$\dfrac{1}{K_a}$

For a type 1 system, the error constant is named the velocity error constant and defined by

$$K_v = \lim_{s \to 0} sD(s)G(s) \quad [\mathrm{s^{-1}}]. \tag{A.33}$$

For a type 2 system, we define the acceleration constant as the error constant as

$$K_a = \lim_{s \to 0} s^2 D(s)G(s) \quad [\mathrm{s^{-2}}]. \tag{A.34}$$

Hence, we can associate the type of a system and the expected steady-state error using the error constants. This is shown in Table A.2.

```
num1=[5 10];
den1=[1 17 12];
sys1=tf(num1,den1);% sys1 = open-loop transfer function
K=1;%Gain of feedback system
sse=(1-evalfr(feedback(K*sys1,1),0))% ss-error
```

In the above MATLAB code we made use of two MATLAB built-in functions: evalfr() and feedback(). The first function allows for the computation of the feedthrough term of the transfer function sys1, while the second function (i.e., feedback) constructs the closed-loop transfer function with a forward transfer function of sys1 and a feedback or sensor transfer function of "1" as we do with unit feedback systems. If you have sensor dynamics, the "1" can be changed to the transfer function representing the sensor dynamics (i.e., no longer unit feedback).

The result of the above computation yields a value of 0.5455 for the steady-state error. This can be easily verified by plotting the step response of the closed-loop system as follows:

```
step(feedback(K*sys1,1))
```

The plot for the above step input is shown in Figure A.8, which verifies the error to be 0.5455.

A.4 Stability

An important concept in controls and system dynamics is the notion of stability. Usually, we strive to achieve a stable system and impose dynamic characteristics beyond stability on its behavior using controls. However, in some rare instances,

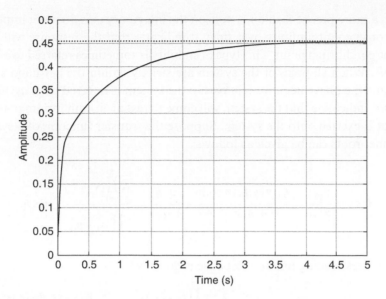

Figure A.8 Step response.

unstable systems are used and preferred. An example of a system that is chosen to be unstable is the flight dynamics of a fighter jet airplane, as it increases the maneuverability. Stability has been a subject of study for many decades, and hence many different ways to assess and determine stability have been proposed. In this brief review treatment of modern control, we limit our discussion to linear systems and the most popular stability tests available.

The bounded input–bounded output (BIBO) stability measure is one of the most generic and easy to understand stability criteria. It states that for every bounded input to the system, the system generates a bounded output and hence is assumed to be BIBO stable. Mathematically speaking, we can state that the system with an impulse response $g(t)$ is BIBO stable if and only if

$$\int_{-\infty}^{\infty} |g(\tau)| d\tau < \infty. \tag{A.35}$$

That is, the impulse response integrated over all times is finite. We can expand on this as we know from Equations (A.6) and (A.14) the roots of the transfer function have a direct effect on the exponent of the exponential of the impulse response function. Hence, if the roots of the transfer function are located in the left-hand side of the s-plane, the system is stable. For that matter, we can equally state that if the eigenvalues of the $\mathbf{A}$ matrix of the state-space description of the system are negative, the system is considered stable. Also, any pole on the imaginary axis or in the right half-plane implies lack of BIBO stability. For example, a pole at the origin of the s-plane indicates that the response is not finite and hence is not BIBO stable.

In a more precise fashion, a pole on the imaginary axis does not imply that the response goes to infinity over time, it indicates that the response will not go to some equilibrium point. This type of stability is sometimes referred to as *marginal stable*. When all poles of the system are strictly within the left-hand side of the s-plane (i.e., have negative real parts), the system is referred to as *asymptotically stable*, indicating that the system will come to rest at an equilibrium point after an input is exerted onto the system. Suppose the transfer function of a system with distinct roots can be given as follows:

$$T(s) = \frac{Y(s)}{R(s)} = \frac{b_0 s^m + b_1 s^{m-1} + \cdots + b_m}{s^n + a_1 s^{n-1} + \cdots + a_n} = \frac{K \prod_{i=1}^{m}(s - z_i)}{\prod_{i=1}^{n}(s - p_i)}, \quad m \le n. \tag{A.36}$$

The corresponding impulse response is given by

$$Y(s) = \frac{K \prod_{i=1}^{m}(s - z_i)}{\prod_{i=1}^{n}(s - p_i)}, \quad m \le n,$$

and converted into the time domain:

$$y(t) = \sum_{i=1}^{n} K_i e^{p_i t}, \tag{A.37}$$

which is analogous to Equation (A.14), where p_i are the roots of the system. Now we can see that asymptotic stability is given when $e^{p_i t} \to 0$ for roots p_i as $t \to \infty$, which was indicated above.

Stability criteria and its computation were a major research topic during World War II, when computers and fancy calculators were not available for determining all the roots of complex dynamic systems. Hence, alternative methods were developed that bypassed the explicit computation of the roots itself, but allowed determining how many roots are in the left-hand side of the s-plane, how many roots are in the right-hand side of the s-plane, and how many roots are on the imaginary axis. One of these methods is the Routh stability criterion. This method is shown in the following.

Consider the characteristic equation of a system in the s-domain:

$$a(s) = s^n + a_1 s^{n-1} + a_2 s^{n-2} + \cdots + a_{n-1} s + a_n. \tag{A.38}$$

One of the necessary conditions for Routh stability is that all of the roots of $a(s)$ in Equation (A.38) have negative real parts, which implies that all coefficients in the polynomial are positive and nonzero. We can construct the Routh–Hurwitz stability criterion as a tabular approach involving a few steps:

1. Form two rows and arrange the coefficients of the characteristic equation according to the following form:

$$\text{Row } n \qquad s^n: \qquad 1 \quad a_2 \quad a_4 \quad a_6 \quad \cdots$$
$$\text{Row } n-1 \quad s^{n-1}: \quad a_1 \quad a_3 \quad a_5 \quad a_7 \quad \cdots$$

2. Compute the coefficients of the next row:

$$\text{Row } n-2 \quad s^{n-2}: \quad b_1 \quad b_2 \quad b_3 \quad b_4 \quad \cdots$$

where $b_1 = -\dfrac{\det\begin{bmatrix} 1 & a_2 \\ a_1 & a_3 \end{bmatrix}}{a_1} = \dfrac{a_1 a_2 - a_3}{a_1}$, $b_2 = -\dfrac{\det\begin{bmatrix} 1 & a_4 \\ a_1 & a_5 \end{bmatrix}}{a_1} = \dfrac{a_1 a_4 - a_5}{a_1}$, etc.

3. Compute the subsequent rows:

$$\text{Row } n \qquad s^n: \qquad 1 \quad a_2 \quad a_4 \quad a_6 \quad \cdots$$
$$\text{Row } n-1 \quad s^{n-1}: \quad a_1 \quad a_3 \quad a_5 \quad a_7 \quad \cdots$$
$$\text{Row } n-2 \quad s^{n-2}: \quad b_1 \quad b_2 \quad b_3 \quad b_4 \quad \cdots$$
$$\text{Row } n-3 \quad s^{n-3}: \quad c_1 \quad c_2 \quad c_3 \quad \cdots$$

where $c_1 = -\dfrac{\det\begin{bmatrix} a_1 & a_3 \\ b_1 & b_2 \end{bmatrix}}{b_1} = \dfrac{b_1 a_3 - a_1 b_2}{b_1}$, $c_2 = -\dfrac{\det\begin{bmatrix} a_1 & a_5 \\ b_1 & b_3 \end{bmatrix}}{b_1} = \dfrac{a_5 b_1 - a_1 b_3}{b_1}$, etc.

4. Investigate the elements contained in the first column: If they are all positive, then the system has only roots in the left-hand plane of the s-plane. Also, the number of roots in the right-hand side of the s-plane equals the number of sign changes along the first column of the table.

There are three possible scenarios that one can encounter when constructing the Routh–Hurwitz table:

1. All elements in the first column are nonzero.
2. Some zeros in the first column with some nonzero values.
3. All zeros in one entire row, including the first column.

For the first case, no change in the analysis needs to be made and the condition spelled out above is applicable. For the second case, where the first entry in one row is zero but other entries in that row are nonzero, we simply replace the zero with a small positive number ε and after the array is completed we let ε go to zero and investigate the number of sign changes in the first column, just like for the first case. For the third case, where an entire row is zero, we simply form an

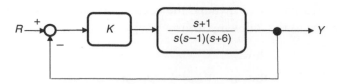

Figure A.9 Example system for Routh–Hurwitz tabulation.

auxiliary row and substitute it with the original one. This auxiliary row is constructed by utilizing the row above to form a polynomial in s (observing the skipping of power as was done in the first two rows) and taking the derivative of this new polynomial, which serves as the auxiliary row.

As an example of the third case, consider the closed-loop feedback system shown in Figure A.9.

The characteristic equation for the closed-loop system is given by

$$1 + K\frac{s+1}{s(s-1)(s+6)} = 0 \text{ or } s^3 + 5s^2 + (K-6)s + K = 0.$$

Suppose $K = 7.5$, constructing the Routh–Hurwitz table yields:

$$
\begin{array}{ccc}
s^3: & 1 & 1.5 \\
s^2: & 5 & 7.5 \\
(s^1: & 0 & 0) \\
s^1: & 10 & 0 \\
s^0: & 7.5 & 0
\end{array}
$$

where we formed the auxiliary equation $5s^2 + 7.5$ from the row above and took the derivative to arrive at the s^1 row with $10s + 0$. Note, the incident where we have one row with zeros implies the following:

$$5s^2 + 7.5 = 0 \text{ whose roots are: } s = \pm 1.22i.$$

As the roots are on the imaginary axis, we have a marginally stable system.

Most of modern control theory deals with continuous systems. However, many implementations are nowadays in discrete time. As we do not treat discrete time in any fashion in this tutorial, but have implementations in the book using microcontrollers and hence discrete time realizations, a brief note on how stability in continuous time is related to stability in discrete time. A system can be easily converted in MATLAB from continuous time to discrete time using the c2d or the d2c functions. The mapping that is employed maps the system from one domain (s-domain in continuous time) into another domain (z-domain for discrete time systems) and vice versa. It also maps the eigenvalues or characteristic values we employed to determine stability. This mapping results in the imaginary axis becoming a unit circle around the origin of the complex plane, where all of

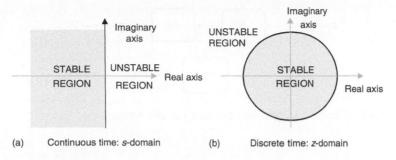

Figure A.10 Stability range for (a) continuous systems and (b) discrete systems.

the left-hand side of the *s*-plane got mapped into the interior of the unit circle. Hence, stability of discrete time systems is tested by evaluating the characteristic values against the boundary of the unit circle, as shown in Figure A.10.

A.5   Root Locus

In controls we adjust parameters that influence the outcome of the system's behavior. For that matter, great insight can be obtained by learning how the parameters affect the dynamics and in particular the stability of the system when these parameters are varied. If the variation of these parameters is mapped to the poles and zeros of the system, and the movements of the poles and zeros are tracked in the *s*-plane, we speak of the root locus of the system. As we learned, the location of the poles has a direct influence on the dynamics of the system (see Equation (A.6), where the system pole is linked to the system's natural frequency and damping ratio). In controls, we often are interested in finding the influence of the controller gain on the system dynamics and stability, hence root locus analysis is very useful to determine the range of such parameters. Consider Figure A.11, where we factored out the static gain and assigned it to act as the controller (in essence a proportional controller scheme). The system has a feedback transfer function $H(s)$ which represents the sensor dynamics.

The roots of the closed-loop feedback control system are determined by the following characteristic equation:

$$1 + KG(s)H(s) = 0. \tag{A.39}$$

Hence, the values of s that allow for the loop gain $KG(s)H(s) = -1$ are the closed-loop poles. $KG(s)H(s) = -1$ can be split into two parts, one where we equate the magnitudes and one where we equate the angles of both sides of the equation. Note, we do not need to compute the closed-loop transfer function for evaluating the root locus, all we need is the loop transfer function $KG(s)H(s)$. The

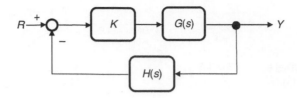

Figure A.11 Feedback system with proportional controller.

corresponding root locus is the locus of the closed-loop poles when the parameter K is varied from zero to infinity. To construct the root locus by hand, we refer to standard modern control textbooks as referenced in Section A.1. In this brief review, we rather make use of MATLAB's built-in function to draw the locus for such systems.

Suppose we want to draw the root locus for the system depicted in Figure A.11, with K serving as our proportional controller and

$$G(s) = \frac{s+2}{s^2 + 4s + 18} \text{ and } H(s) = \frac{1}{s(s+1)}.$$

The following MATLAB script plots the root locus for the system given above:

```
clear % clears work space
clc % clears screen
% Compute loop transfer function
den1=[1 1 0];den2=[1 4 18];
den=conv(den1,den2);% determine overall denumerator
num=[1 2];
r=rlocus(num,den);
plot(r,'-'); v=[-6 6-6 6]; axis(v); axis('square')
grid;
title('Root locus plot of G(s)=K(s+2)/[s(s+1)(s^2+4s+18)]');
xlabel('Real axis'); ylabel('Imag axis');
```

The corresponding root locus is shown in Figure A.12.

Note, the open-loop poles are at $s = 0, -1, -2 \pm 3.7417i$ and the open-loop zero is at $s = -2$. By varying K from zero to infinity, the poles move either to a system zero (-2) or to infinity. For example, the pole at 0 moves along a semi-circle (approximately) to -3 and then on the real axis to -2 when K reaches infinity. The two conjugate complex poles start off at $s = -2 \pm 3.7417i$ and then move in the complex plain across the imaginary axis into the right-hand plane of the s-plane, ending up at infinity when K reaches infinity (not shown in the graph for obvious reasons). It is interesting to know at what values of K these two poles cross the imaginary axis, as this point corresponds to the system becoming unstable. In order to find the gain value of K at specific locations of the root locus, use the following command in MATLAB:

```
[K,r]=rlocfind(num,den)
```

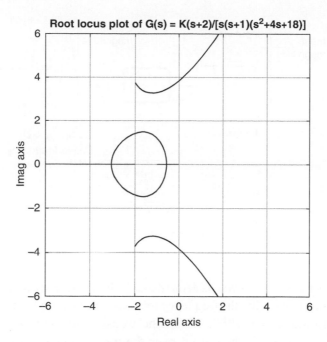

Figure A.12 Root locus of example system.

This allows the user to position the cursor at the location of interest in the root locus plot and, by initiating a click, MATLAB will provide the location (the variable *r*) and the gain value (*K*).

A.6   Frequency Response and Bode Plot

Consider a system setup as depicted in Figure A.4. Suppose the input is a sinusoidal wave with a constant amplitude and a constant frequency. The system represented by *G*(*s*) will respond by generating an output that mimics the input in form when arriving at a steady state (i.e., it will also have a sinusoidal wave form). From elementary differential equations we know that the magnitude of the response may be different; however, the frequency will be the same, though the output frequency will have some phase shift if the system has damping. This experiment can be repeated for different input frequencies and the resulting amplitude ratio between input and output as well as the phase shift can be recorded for each of the frequencies. A corresponding characterization of the described experiment is given in Figure A.13.

Hence, the magnitude of the response is a function of the input frequency and the system dynamics. The same can be said about the phase shift – that is, with different frequencies we obtain different phase shifts. Suppose the magnitude response is denoted as $M(\omega)$ and the phase response is given as $\phi(\omega)$, then the so-called

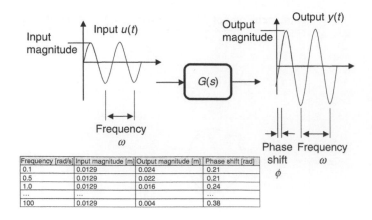

Frequency [rad/s]	Input magnitude [m]	Output magnitude [m]	Phase shift [rad]
0.1	0.0129	0.024	0.21
0.5	0.0129	0.022	0.21
1.0	0.0129	0.016	0.24
...	...		...
100	0.0129	0.004	0.38

Figure A.13 Frequency response experiment.

frequency response is given as $M(\omega)\angle\varphi(\omega)$. Note, there are a number of alternative options to describe the frequency response. For example, we could have used Euler's formula: $M(w)e^{i\phi(\omega)}$. Recall the transfer function $G(s)$ is given by the Laplace transform of the output over the input, where $s = \sigma + i\omega$. If we only consider the imaginary part of s (i.e., $s = i\omega$) then we can assert the following relationship:

$$M(\omega) = |G(s)| \quad \text{and} \quad \phi(\omega) = \angle G(s). \tag{A.40}$$

To compute the output based on the input values, we can use the following formulas:

$$M_o(\omega) = M(\omega)M_i(\omega) \text{ and } \phi_o(\omega) = \phi(\omega) + \phi_i(\omega).$$

Hence, we have

$$M_o(\omega)\angle\phi_o(\omega) = M_i(\omega)M(\omega)\angle[\phi_i(\omega) + \phi(\omega)]. \tag{A.41}$$

There are a number of ways we can utilize Equation (A.41) to illustrate its implication graphically. Some examples are the polar plot (which is commonly referred to as the Nyquist plot), where real and imaginary parts of the open-loop system are used as coordinates to plot the response as a function of frequency; the Nichols chart; and the Bode plot. In this brief modern control review tutorial we only focus on the Bode plot. The Bode plot consists of two diagrams: one for the magnitude $M(\omega)$ and one for the phase $\phi(\omega)$. A similar plot exists in vibration analysis, where two plots are used to depict the response of the system, and is termed the frequency response plot. However, the Bode plot differs in that logarithmic scales are used. For example, the magnitude in the Bode plot is given in dB:

$$1.0 \text{ dB} = 20 \log_{10}\left(\frac{M_o(\omega)}{M_i(\omega)}\right). \tag{A.42}$$

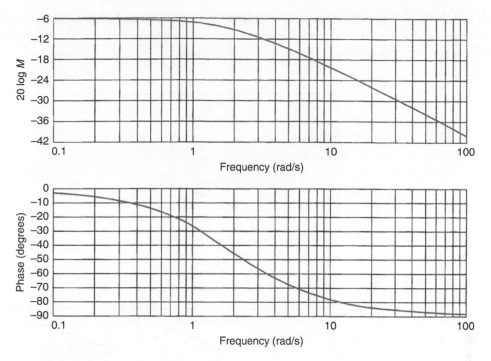

Figure A.14 Example of a Bode plot.

An example of a Bode plot is shown in Figure A.14. Note that the phase plot is usually given in degrees.

The reason for using log scale for the frequency and the magnitude is found in Equation (A.41): When cascading multiple transfer functions, the product of the magnitudes for each transfer function is computed simply as an addition when using the log scale. This property is already available for the phase without using the log scale on the phase itself.

This property is quite convenient, and utilized when constructing Bode plots, designing controllers, or when inferring transfer functions from Bode plots. We will review the design of controllers using this property in the next section. For now, let's consider constructing a Bode plot for a given transfer function.

To construct a Bode plot using Table A.3, one needs to first convert the transfer function into a form where the overall gain is found and then add each pole and zero graphically by adding graphically the magnitudes and phases for each individual component of the transfer function. Consider the following generic transfer function:

$$G(s) = \frac{K(s + z_1)(s + z_2)}{s^m(s + p_1)(s + p_2)}.$$

Table A.3 Bode plot straight line approximations for magnitude and phase plots

Transfer function element	Description	Magnitude	Phase
Gain element	$R \rightarrow K \rightarrow Y$	$20 \log_{10} K$	$0°$ ω
Zero at origin	$R \rightarrow S \rightarrow Y$	20 dB/decade, 1 ω	$90°$ ω
Pole at origin	$R \rightarrow \dfrac{1}{S} \rightarrow Y$	1 ω, -20 dB/decade	ω $-90°$
Simple zero	$R \rightarrow 1 + \dfrac{i\omega}{b} \rightarrow Y$	20 dB/decade, b ω	$90°$ $0°$ $b/10$ b $10b$ ω
Simple pole	$R \rightarrow \dfrac{1}{1 + i\omega/a} \rightarrow Y$	a ω, -20 dB/decade	$a/10$ a $10a$ ω $0°$ $-90°$
Nth-order zero ($N > 1$)	$R \rightarrow \left[1 + \dfrac{2i\omega\xi}{\omega_n} + \left(\dfrac{i\omega}{\omega_n}\right)^2\right]^N$	40N dB/decade, ω_n ω	$180N°$ $0°$ $\omega_n/10$ $10\omega_n$ ω
Nth-order pole ($N > 1$)	$R \rightarrow \dfrac{1}{\left[1 + \dfrac{2i\omega\xi}{\omega_n} + \left(\dfrac{i\omega}{\omega_n}\right)^2\right]^N} \rightarrow Y$	ω_n ω, $-40N$ dB/decade	$\omega_n/10$ $10\omega_n$ ω $0°$ $-180N°$

Its magnitude is computed by

$$|G(s)| = \frac{K|(s + z_1)||(s + z_2)|}{|s^m||(s + p_1)||(s + p_2)|},$$

which is given by

$$20 \log|G(s)| = 20 \log K + 20 \log|(s + z_1)| + 20 \log|(s + z_2)|$$
$$- 20 \log|s^m| - 20 \log|(s + p_1)| - 20 \log|(s + p_2)|,$$

while the phase is computed as:

$$\angle G(s) = \angle K + \angle(s + z_1) + \angle(s + z_2) - \angle s^m - \angle(s + p_1) - \angle(s + p_2).$$

To construct a Bode plot of a transfer function using MATLAB, one can simply use the command bode (num, den), where *num* stands for the numerator polynomial and *den* stand for the denominator polynomial.

The opposite can be done as well (i.e., given a Bode plot, we can infer the transfer function by deconstructing each possible element from Table A.3 and assembling the corresponding mathematical representation). An important characteristic of the Bode plot is the determination of the gain margin (GM) and phase margin (PM). We will demonstrate these characteristics along with some performance specifications and stability criteria on an example system.

Consider the following transfer function:

$$G(s) = \frac{10(1 + 0.1s)}{s(1 + 0.5s)(1 + 0.012s + 0.0004s^2)}.$$

Note, the overall gain is computed by converting the transfer function into the form where each polynomial in s starts with the coefficient 1. In this example the resulting overall gain is 10, the only zero is located at $s = -10$, and the poles are located at $s = 0, \ -2, \ -15 \pm i47.697$. Using Table A.3 or MATLAB's bode () function one can easily construct the Bode plot. For brevity we will use MATLAB in this example:

```
clear
num=10*[0.1 1];
den=conv([1 0],conv([0.5 1],[1/2500 0.6/50 1]));
bode(num,den)
grid
```

The resulting plot is shown in Figure A.15 along with some additional markings to indicate the different performance characteristics. To determine PM and GM, considering the additional dashed lines drawn in Figure A.15 at the gain cross-over point (i.e., the point where the magnitude of the gain crosses the zero dB line) and the dashed line associated with the phase cross-over point (i.e., the point where the phase crosses the $-180°$ line. The PM is found by the difference in phase at the gain cross-over frequency (in this example 4.5 rad/s) between the system phase (in this example approximately $-135°$) and the $-180°$ line. In this example the PM is approximately $45°$. The GM is found at the phase cross-over frequency by computing the difference in dB between the magnitude of the gain and the 0 dB line – in this example the GM is approximately 13.5 dB. Stability is guaranteed when both PM and GM are positive (see arrows in Figure A.15). A convenient way to find both PM and GM is by using MATLAB's function margin ():

```
[GM,PM,Wcg,Wcp] = margin(num,den)
```

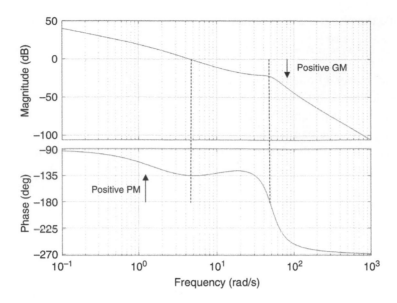

Figure A.15 Bode plot of the example system.

where Wcg is the gain cross-over frequency and Wcp is the phase cross-over frequency. Note, Bode plots are not only very useful for control system design, but also for an array of other applications, including filter design. In filter design, we aim to enhance certain frequencies while attenuating others. Since the Bode plot is a log-scale plot, negative magnitudes imply a reduced output compared to the input, while positive magnitudes indicate an amplification of the input.

A.7   Compensator Design

One of the most popular controllers in the process industry is the PID controller. PID stands for "proportional integral derivative" control, three terms with three distinct functions. The popularity of the PID controller is partially due to the fact that a number of tuning and even self-tuning methods exist that do not require knowledge of the system transfer function. This seems to make the entire treatment of the prior sections of this chapter redundant. However, more specific control performances can only be achieved by tailoring a controller to the given system dynamics, and hence the transfer function representing it. We have seen already the influence of the proportional part of a controller and its influence to the system's stability in the root locus section. The purpose of the integral part of the PID controller is to increase the system type, which we saw in Section A.3 can have a great influence on the steady-state error. The derivative portion of the PID controller is used to anticipate the system changes. MATLAB, and in particular Simulink, has a PID block that tunes its PID parameters automatically.

However, if we want to achieve very specific performance criteria of the closed-loop system, we can develop controllers that have the same functionality embedded as a PID controller but are able to address the steady-state and transient performance explicitly. For this we can utilize design methods based on the root locus diagram or the bode diagram, among others. We will only review the design of controllers using the Bode diagram here for brevity.

Using the Bode diagram as our instrument for designing a controller, it is helpful to associate the shape of the plot with characteristics of the closed-loop system. For this purpose, we divide the frequency spectrum into three parts: low-, medium-, and high-frequency band. The partitioning of the frequency spectrum is not based on some accurate science, but rather a simple guide, and needs to be adjusted to each particular system. As a rough rule of thumb, the upper limit of the medium frequency is approximately 5–10 times the upper limit of the low frequency, and beyond the upper limit of the medium frequency is the high-frequency spectrum. In general, a controlled system operates in the low- and medium-frequency bands, while disturbances operate in the high-frequency band. In the low-frequency range much of the steady-state performance is determined. The medium-frequency band hosts partially the transient response characteristics and the relative stability. For example, to have a high value of the velocity error constant in addition to a suitable relative stability, the slope of the magnitude plot in the Bode diagram near the cross-over frequency should be around −20 dB/decade. For the low frequency, one would like to have a large enough gain, whereas for the high-frequency region the gain should be attenuated as quickly as possible to reduce the effects of noise.

One of the traditional controller designs covered in modern control classes is the lead compensator. A lead compensator allows for improvements of the transient response performance while having little effect to the steady-state accuracy. However, it may accentuate the noise effect in the high-frequency region. The mathematical representation of the lead controller is given by Equation (A.43):

$$G_c(s) = K_c \alpha \; \frac{Ts + 1}{\alpha Ts + 1} = K_c \; \frac{s + \dfrac{1}{T}}{s + \dfrac{1}{\alpha T}}, \; 0 < \alpha < 1. \tag{A.43}$$

The lead controller increases the bandwidth (i.e., the range of the low-frequency region), and with that improves the response speed while reducing the overshoot by incorporating more damping. The condition that $0 < \alpha < 1$ implies that the zero's corner frequency appears before the corner frequency of the pole. This can be seen in the Bode plot of the lead controller, as shown in Figure A.16. The first corner frequency occurs at $\dfrac{1}{T}$, which is the one for the zero. It induces a 20 dB/decade increase in the magnitude plot. The second corner frequency belongs to the pole, at $\dfrac{1}{\alpha T}$, and reduces the 20 dB/decade incline into a

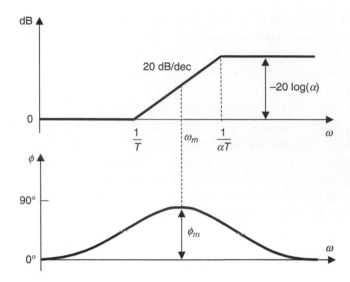

Figure A.16 Bode plot of lead controller.

flat line. The corresponding phase diagram is affected one decade before and after each of the corner frequencies and results in a "bump" of +90° phase at ω_m, which is in the middle of the two corner frequencies. Since we use logarithmic scales, the closed-loop magnitude plot of the Bode diagram is the addition of this controller magnitude and the magnitude of the transfer function. Similarly, the phase plot is obtained by adding the phase of the controller to the one of the transfer function. The effect of this lead controller becomes apparent by comparing the before and after pictures of the Bode plot: the gain cross-over frequency is increased. This is shown in Figure A.17.

Another interesting controller can be designed by also adding a zero and a pole; however, in this case we change the order when they appear first on the frequency band. The lag controller can be given as follows:

$$G_c(s) = K_c \beta \frac{Ts+1}{\beta Ts+1} = K_c \frac{s+\dfrac{1}{T}}{s+\dfrac{1}{\beta T}}, \quad \beta > 1. \tag{A.44}$$

The corresponding Bode plot is shown in Figure A.18.

As the higher frequencies are attenuated, the implementation of a lag controller results usually in an improved PM. A lag controller is equivalent to a low-pass filter, which results in allowing for a high gain at low frequencies. Also, the closed-loop pole near the origin of the s-plane imposes a slowly decaying transient response of the system, hence the settling time is affected adversely.

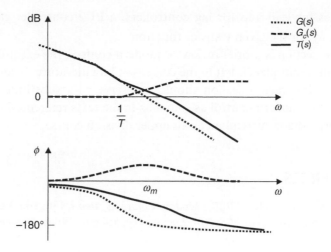

Figure A.17 Bode plot of compensated system using a lead controller.

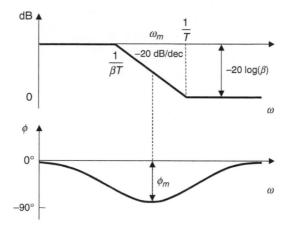

Figure A.18 Bode plot of a lag controller.

Another controller can be achieved by combining the lead controller with the lag controller, resulting in the lag–lead controller. The corresponding transfer function of the lag–lead controller is given in Equation (A.45):

$$G_c(s) = K_c\left(\frac{1 + T_1 s}{1 + \alpha T_1 s}\right)\left(\frac{1 + T_2 s}{1 + \beta T_2 s}\right), \tag{A.45}$$

where $\beta > 1, \alpha = \frac{1}{\beta}$. In Equation (A.45) the first part is associated with the lead controller, while the second part corresponds to the lag controller. Generally, the two corner frequencies of the lag portion are lower than the two corner frequencies of the lead portion. A lag–lead controller is analogous to the PID controller mentioned at the beginning of this section. However, using the design principles for lead and for lag controllers, a PID controller can be designed specifically for the given transfer function.

As this section is a brief review of modern control, the exact design steps for each of the controllers is left to the corresponding literature. It should be emphasized that this review section should not serve as a substitute for a rigorous treatment of modern control as found as in the texts referenced or instructed in an undergraduate curriculum as an upper division course.

REFERENCES

[1] Dorf RC, Bishop RH (2017). *Modern Control Systems*, 13th edn. Pearson, Harlow.
[2] Nise NS (2020). *Control System Engineering*, 8th edn. Wiley, Chichester.

APPENDIX B
MATLAB Tutorial

B.1 Introduction

MATLAB has become a universal programming tool to support many different disciplines. The diversity of different toolboxes and functions available to the user cannot be covered by a brief tutorial, as this section entails. Hence, we are presenting only the most basic information on some of the functionality and use of MATLAB related to the problems and tasks presented in this book. Additional material on the use of MATLAB and how to script and program specific computations is presented throughout each chapter. MATLAB provides for some excellent documentation that details not only the use but also provides example code. For new MATLAB users there is a wide range of resources available to get acquainted with and to learn the specific functionalities of MATLAB. One such source is MATLAB's free *onramp* online tool, which allows new users to learn the basics of MATLAB. In addition, MATLAB provides online tutorials in a more traditional form. These two resources can be found at:

www.mathworks.com/learn/tutorials/matlab-onramp.html
www.mathworks.com/help/matlab/getting-started-with-matlab.html

MATLAB also offers extensive courses on fundamentals as well as for intermediate users. Those courses include topics such as the ones found in this textbook. We hope the reader is familiar with MATLAB and uses the following tutorial as a refresher or look-up source and refers for more details to the resources listed above.

B.2 Basic MATLAB Operation

MATLAB provides for a number of different means of user interaction. The two primary venues utilized in this book are the command window and the option to use MATLAB scripts as *.m files. The command window operates in the folder you specify in the browser search bar and has access to any files stored at that location. Script files have the same access radius (i.e., you are able to read any files that are located in the same folder as the script file). There are ways to access other folders during the execution of scripted files, which we will show in a later section in this chapter.

When operating through the command window, in order to execute each command entered, it has to be followed by a carriage return (enter). Commands stored in a script file (file_name.m) are executed when the "Run" button is clicked. Alternatively, if the current folder is the location where the script file is stored, one can execute the script file by typing its file name into the command window followed by the enter key.

A running MATLAB routine (command in progress or script file is running) can be aborted by the keyboard combination of **Ctrl + c**. The command `clear` will erase any stored variables and values from the current memory. If selected variables need to be cleared, specify the variable name after the command `clear`. As an example, suppose you have two variables that you would like to clear, *speed* and *distance*. However, you would like to keep all the other variables in the current memory. The two variables can be cleared without affecting the other variables and stored quantities by the following command:

```
clear speed distance
```

Generally, predefined MATLAB commands are all lower case, and all commands and variables are case-sensitive. Variable names need to start with a letter, which can be followed by numbers, letters, and underscores. Variable names in MATLAB are restricted to consist of no more than 31 such letters, numbers, and underscores. To enter a line in MATLAB that is purely for commenting and not part of the program logic, use a percent sign (%) at the beginning of the comment. For example, if you want to denote that the next section of the program computes the average value for a given time series with the variable name measurement, you can use the following comment statement:

```
% Computing the average value of measurement
```

The command `clear` does not erase the command window's diary. For this, you need to use the command `clc`. However, unlike the clear command, `clc` will not erase the current memory in terms of values and variables, it simply clears the command window. When in doubt about an existing MATLAB command, one can easily find instruction on its use and explanation of the functionality using the MATLAB documentation. However, for a quick view of such information, type the command `help` followed by the command you are researching in the MATLAB command window. For example, if you want to find out how to use the function `svd`, type the following into the command window, followed by the enter key:

```
help svd
```

MATLAB will respond by providing you with the information on this particular command, as shown in Figure B.1.

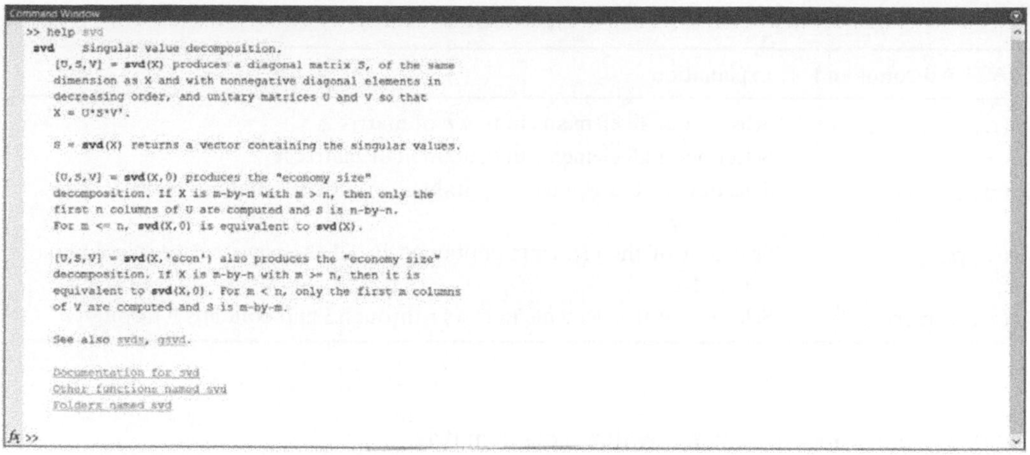

```
Command Window
>> help svd
svd    Singular value decomposition.
    [U,S,V] = svd(X) produces a diagonal matrix S, of the same
    dimension as X and with nonnegative diagonal elements in
    decreasing order, and unitary matrices U and V so that
    X = U*S*V'.

    S = svd(X) returns a vector containing the singular values.

    [U,S,V] = svd(X,0) produces the "economy size"
    decomposition. If X is m-by-n with m > n, then only the
    first n columns of U are computed and S is n-by-n.
    For m <= n, svd(X,0) is equivalent to svd(X).

    [U,S,V] = svd(X,'econ') also produces the "economy size"
    decomposition. If X is m-by-n with m >= n, then it is
    equivalent to svd(X,0). For m < n, only the first m columns
    of V are computed and S is m-by-m.

    See also svds, gsvd.

    Documentation for svd
    Other functions named svd
    Folders named svd

fx >>
```

Figure B.1 Command window response to a request for more information on the function svd using the `help` operator.

Variables do not need to be declared prior to use. However, it is good practice – in order to speed up the computation – to initialize variables to provide MATLAB with information about the amount of memory it needs to allocate. Assignments are always constructed so that the memory location – as given by the variable name – is on the left side and the value is on the right side. For example, if we want to assign the variable *distance* a value of 10 units, the MATLAB statement will look as follows:

```
distance = 10;
```

Notice that we used a semicolon at the end of the statement. Without the semicolon MATLAB would generate a response that essentially repeats what it has understood from your instruction given by the statement.

MATLAB has very powerful routines to work with matrices of all sizes. A matrix is constructed using square brackets. For example, to generate a matrix **A** which is given as

$$\mathbf{A} = \begin{bmatrix} 1 & 2 & 3 \\ 4 & 5 & 6 \\ 7 & 8 & 9 \end{bmatrix}$$

we can use the following MATLAB command:

```
A = [1 2 3;4 5 6;7 8 9];
```

Notice that each row of matrix **A** is given as a set of numbers separated by a single space and terminated by a semicolon within the square bracket of the MATLAB command. We can access certain elements and sections within a matrix. For this, one uses the column operator. Table B.1 provides for some of the different ways

Table B.1 Selection of elements within a given matrix

MATLAB command	Explanation
`A(n,:)`	Selection of all elements in row n of matrix **A**
`A(:,m)`	Selection of all elements in column m of matrix **A**
`A(p:q,:)`	Selection of the elements contained in all the columns between rows p and q
`A(:,p:q)`	Selection of the elements contained in all the rows between columns p and q
`A(r:s,p:q)`	Selection of the elements in rows r through s and columns p through q

Table B.2 Selection of elements within a given matrix

Arithmetic operation	Example
+	Addition
−	Subtraction
*	Multiplication
/	Right division
^	Exponentiation

to access matrix information. The above example and Table B.1 are shown for a two-dimensional matrix; however, MATLAB can handle easily higher-dimensional matrices.

Variables such as matrices but also scalar quantities can be manipulated using arithmetic operations. The primary operands used in MATLAB are listed in Table B.2.

Outputs can be formatted to display some or all of their digits when computing numerical results. The command `format long` changes the display of numbers so that maximal 15 decimal digits after the decimal point are used. The command `format short` changes the display of the value to four digits after the decimal point. For example:

```
≫ a = 29/7
  a =
        4.1429
≫ format long
≫ a
  a =
        4.142857142857143
```

MATLAB has some predefined variables whose name are not to be used for other purposes. Some of these variables are listed in Table B.3.

In addition to predefined variables, MATLAB provides for some specialty matrices. A partial list of such matrices is given in Table B.4.

Table B.3 Predefined MATLAB variables

MATLAB name	Explanation
ans	The response value to any MATLAB expression evaluation or input
i	The complex number given by $\sqrt{-1} = 0 + 1.0i$
inf	Infinity
pi	Mathematical constant π
NaN	Represents 'not a number'

Table B.4 Specialty matrices

MATLAB name	Explanation
eye(m,n)	Creation of an $m \times n$ matrix with 1 on the main diagonal and all other entries equal to 0
eye(n)	Creation of an $n \times n$ square matrix with diagonal entries equals to 1 and all other entries equal to 0
zeros(m,n)	Creation of an $m \times n$ matrix with all entries equal to 0
ones(m,n)	Creation of an $m \times n$ matrix with all entries equal to 1
rand(m,n)	Creation of an $m \times n$ matrix with each entry equal to a random number

B.3   Elementary MATLAB Functions and Structures

MATLAB comes with an abundance of built-in functions and options for additional specialized toolboxes. To see what your current installation entails, type in the command window the command help without any other specification, and MATLAB will list all the different toolboxes that are installed. If more details are needed – for example, what functions are entailed in a specific toolbox – you can type help followed by the toolbox name and MATLAB will provide you with a list of all the functions that are part of this particular toolbox. Table B.5 lists a few basic functions that every MATLAB installation possesses. Note, Table B.5 is only a sample; there are many more basic functions that your installation has access to.

When working with matrices it is often useful to find out the dimension of a matrix after some computation occurs or input to the program was provided that leads to a change of the matrix size. Hence, to determine the size of a matrix, the command size is used:

```
≫ M = [1 2;3 4;5 6; 7 8];
≫ size(M)
    ans =
          4     2
```

Another useful function for working with matrices is the transpose operator – defined by the single quotation mark – and the inverse function inv():

```
≫ Mt = M' % taking transpose of M by using '
  Mt =
    1   3   5   7
    2   4   6   8
```

Table B.5 Sample of basic MATLAB commands

MATLAB command	Explanation
abs(x)	Absolute mean value of x, value is in radians
atan(x)	Evaluates the arctangent of x, value is in radians and bounded as follows: $-\pi/2 \le \mathrm{atan}(x) \le \pi/2$
cos(x)	Evaluates the cosine of x, value is in radians
ceil(x)	Rounds the value of x to the nearest integer toward $+\infty$
exp(x)	Computation of the exponential of x, i.e., e^x
floor(x)	Rounds the value of x to the nearest integer toward $-\infty$
log(x)	Evaluates the natural logarithm of x
log10(x)	Computes the base 10 logarithm of x
round(x)	Rounds x to the nearest integer
sign(x)	Determines the sign of x by returning either a -1 for x is less than 0, a 0 when x is zero, and $+1$ otherwise
sin(x)	Evaluates the sine of x, value is in radians
sqrt(x)	Evaluates the square root of x
tan(x)	Evaluates the tangent of x, value is in radians
rand(p,q)	Computes a $p \times q$ matrix filled with pseudorandom numbers using a uniform distribution
randn(p,q)	Computes a $p \times q$ matrix filled with pseudorandom numbers using a normal distribution

Table B.6 Element-by-element arithmetic operations

MATLAB name	Explanation
.*	Element-by-element matrix multiplication
./	Element-by-element matrix division
.^	Element-by-element matrix exponentiation

```
≫ A = [1 5;0 2]
  A =
    1    5
    0    2
≫ inv(A)
  ans =
       1.0000   -2.5000
            0    0.5000
```

When dealing with matrices and arrays, computations often need to be performed in an element-by-element fashion. MATLAB uses the period character to indicate such array operations. As addition and subtraction operations are identical to element-by-element operations on arrays, the period operator is not used for those arithmetic operations. Table B.6 summarizes some of the different

Table B.7 Selected matrix functions

MATLAB name	Explanation
det	Computation of the determinant
diag	Diagonals of a matrix
eig	Computation of eigenvalues and eigenvectors of matrices
inv	Computation of matrix inversion
rank	Computation of the rank of a matrix

element-by-element arithmetic operations for two matrices **A** and **B**, which are of the same size (i.e., $\mathbf{A} \in \mathbb{R}^{m \times n}$).

For example, computing **A.*B** will result in having each element of matrix **A** multiplied by the corresponding element of matrix **B**. Matrix division in an element-by-element fashion therefore is given by **A./B**. Note that there also exists a matrix left division, which is given by the arithmetic backslash operator \, so we can also perform element-by-element left division using **A.\B**.

In addition to performing element-by-element arithmetic operations and the listed functions in Table B.4, MATLAB has some very useful matrix functions. A partial list is given in Table B.7.

Besides matrices/arrays and scalars, MATLAB offers other forms of variables, such as structures and cells. A structure in MATLAB is essentially a multidimensional array that can store data of different types. Structures are often compared to an object with a list of fields, with each field having a variable name or representing a subgroup of data. For example, creating the structure *food*, we can associate data with this structure such as type, calories, sugar content, and vitamin C level. In MATLAB such a structure would be created using the dot operator:

```
≫ food.apple = "Gala"
  food =
  struct with fields:
        apple: "Gala"
```

We can add other types of apples using new fields within the structure *food*:

```
≫ food(2).apple = "McIntosh"
  food =
  1×2 struct array with fields:
    apple
```

When investigating what the field *apple* contains, we can use the following access notation:

```
≫ food.apple
  ans =
        "Gala"
  ans =
        "McIntosh"
```

Notice that there are now two answers as there are two items stored in the *apple* field. One can also create an entire entry in one line. For example, if we were to record personal information about an employee in a structure, we could use the following syntax in MATLAB:

```
≫ employee = struct ('Name','John Smith','Salary',65000,
              'Position','Engineer','OfficePhone',229)
employee =

  struct with fields:

        Name: 'John Smith'
      Salary: 65000
    Position: 'Engineer'
 OfficePhone: 229
```

To add another employee to this structure we can use an index such as (2), (3), etc. As an example, if we add employee named Jane Doe, we use the following MATLAB statement:

```
≫ employee(2) = struct ('Name','Jane Doe','Salary',71000,
                  'Position','Scientist','OfficePhone',465)
```

Having different elements and categories within a structure allows us to apply suitable operations to these elements. For example, if the owner of the company where John and Jane work wants to know how much money she is expending yearly on their salaries, the information could be computed as follows:

```
≫ money=[employee.Salary];
≫ total=sum(money)
  total =
          136000
```

Another useful MATLAB data type for holding information is the cell array. A cell array is able to store indexed data in containers that are labeled cells. Each cell can store any MATLAB type of data. To create a cell array, we can utilize curly braces {}, which serve as array constructors. As an example, we create a cell with a 2×2 cell array and holding different data types in each of the cells as follows:

```
≫ M = [1 2 3;4 5 6;7 8 9];
≫ name = "Jane Doe";
≫ poles = -2.5+4i;
≫ newCell{1,1} = M;
≫ newCell{1,2} = name;
≫ newCell{2,1} = poles;
≫ newCell{2,2} = pi;
≫ newCell
newCell =

  2×2 cell array

    {3×3  double          }    {["Jane Doe"]}
    {[-2.5000 + 4.0000i]}    {[    3.1416]}
```

Figure B.2 Depiction of a cell array using the `cellplot()` function.

To find the entries of a cell array one can use the function `celldisp(newCell)` or use a plotting function to graphically depict the cell array:

```
cellplot(newCell)
```

which will result in a plot of each array element with its content graphically depicted. For our example, the resulting plot is shown in Figure B.2:

To create an empty cell array we can utilize the `cell` function directly:

```
Q = cell(3,4);
```

which creates a cell array with three rows and four columns. Access to any of the elements of the cell array is accomplished using curly braces. For example, if we want to find out what is in the (1,1) cell, we use the index within the curly braces preceded by the name of the cell structure:

```
≫ newCell{1,1}

  ans =
          1    2    3
          4    5    6
          7    8    9
```

B.4   Flow Control, Loops, and Vectorization

To impose a specific flow of executing a set of computations or evaluations, MATLAB provides for different ways to control the sequencing. The primary tools for this control are the for-loop, the while-loop, the if–else statement, and the switch statement.

Table B.8 Logic and rational operators

Operator	Function form	Explanation
<	C = lt(A,B)	Less than
>	C = gt(A,B)	Greater than
==	C = eq(A,B)	Equal
~=	C = ne(A,B)	Not equal
<=	C = le(A,B)	Less than or equal
>=	C = ge(A,B)	Greater than or equal
&	C = and(A,B)	Logic AND
|	C = or(A,B)	Logic OR
~	B = not(A)	Logic NOT

The if–else structure is defined as follows:

```
if expression
    statement;
else
    statement;
end
```

The expression is either a Boolean value or an evaluation using relational or logical operators. A set of such operators is listed in Table B.8.

The statement in the if–else structure may contain many lines of code and does not need to end with a semicolon. The indentation for each of the statements is for readability and does not affect the functioning of the structure. An end statement is required for any if or if–else structure.

The switch operator allows for selective execution of different statements while leaving out others based on the evaluation of its associated expression. This selective feature of the switch structure often has the benefit of a more efficient computation compared to the utilization of an if–else structure. The syntax of a switch structure is:

```
switch condition
    case condition_1
        statement;
    case condition_2
        statement;

        .
        .
        .

    otherwise
        statement;
end
```

The evaluation and selective execution of a statement in this structure is dependent on matching the condition defined with the switch statement at the beginning of the structure. It also entails an *otherwise* case that allows for the situation when none of the case conditions have been met.

The for-loop allows for the execution of a set of statements a predefined number of times. The syntax of the for-loop is:

```
for variable = expression
    statement;
end
```

An example of a simple for-loop is shown below. Here we use a variable x that goes from 1 to 10 in steps of 2. The single statement in the loop contains a formatted print function that lists the current value of x during the for-loop execution:

```
for x = 1:2:10
    fprintf('value of x: %d\n', x);
end
```

Note that we can substitute the value of x into the string provided by the `fprintf` function using a placeholder. In addition, we utilize a newline command within the statement string – that is, \n causes a carriage return in the output.

The while-loop allows for an undefined number of iterations as the termination criterion is based on a logic expression. The syntax of a while-loop is:

```
while expression
    statements
end
```

The statements are executed until the expression becomes false. In some instances your program may be structured such that the expression in the while-loop never evaluates to be false and hence becomes an infinite loop. To break out of such a loop you can utilize the keyboard combination **Ctrl + c**.

Generally, for-loops and while-loops are computationally expensive and time-consuming. Hence, many MATLAB programmers avoid using such loops and prefer to make use of MATLAB's matrix-based language. Those routines are executed internally using compiled C++ and other compiled codes that allow for a much more efficient deployment. Hence, rather than using loops, operations are structured so that they instead amount to a matrix operation, which is called vectorization. In addition to the increased execution speed of the program or function, the resulting code will be in a more compact form. As an example, consider the following brief MATLAB script that computes the sum of squares of the elements in a vector:

```
k = 1e8;
x = rand(k,1);
% starting clock
tic, p = 0;
for n = 1:k
    p = p + x(n);
end
toc
```

In the above code we utilize the `tic-toc` function to track the amount of time the program is engaged in computing the given script. Here we used a simple addition of square terms contained in a large vector using a for-loop. The same functionality and computation can be achieved using the vectorized form and avoiding the for-loop:

```
tic,p = sum(x.^2);toc
```

Evaluating the elapsed time, we find that the vectorized code is substantially faster in evaluating the sum of the square terms compared to the code using the for-loop. Generally, vectorization is accomplished using array operations. By using array operations, as summarized in Table B.6, MATLAB is able to use compiled code rather than interpret each line of the for-loop in sequence. To provide another example, we illustrate the commonly found nested for-loop in matrix computation. Suppose we have two matrices, **A** and **B**, and want to compute their multiplication – that is, $\mathbf{C} = \mathbf{A} \times \mathbf{B}$. If we were not using vectorization (i.e., matrix multiplication) the MATLAB code would result in the following nested for-loop:

```
A = magic(4); % magic generates a special matrix - see docs
B = magic(4);
C = zeros(4,4);
for row = 1:4
    for column = 1:4
        C(row,column) = A(row,column)*B(row,column);
    end
end
```

In vectorized form, this computation is done in a compact form as follows:

```
C = A.*B;
```

B.5   Graphics

MATLAB provides for an abundant selection of different graphing tools and functions. This section only provides for the most basic graphing utility, using the 2D plot. However, the reader is encouraged to use MATLAB's documentation page to explore the many other functions MATLAB provides for representing data and structures.

The `plot()` function comes with an array of different configurations. A few of these are reviewed in the following. The general structure of the `plot` function and its input parameters are visualized in Figure B.3.

To add labels on the axis as well as provide for a title and legends, use the following commands:

```
figure % creates new figure window
t=linspace(0,50,1000);
plot(sin(t),'k')
hold % hold current plot and window
plot(cos(t),'k-.')
title('Sin function plot')
```

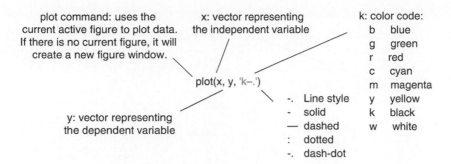

Figure B.3 MATLAB's plot function composition.

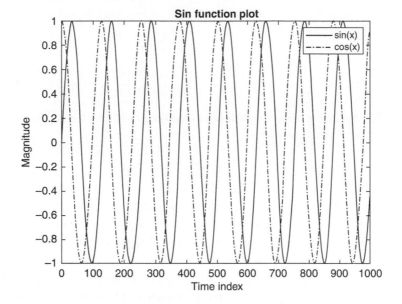

Figure B.4 Example of the plot command using two graphs in one window and the `legend` command to document each graph.

```
ylabel('Magnitude')
legend('sin(x)','cos(x)') % identify graphed lines
hold % release hold
xlabel('Time index')
```

The resulting graph is shown in Figure B.4. Note that we plotted two different functions onto one graph. This was possible by utilizing the `hold` command. We also identified the two resulting graphs using the legend function.

Newer versions of MATLAB provide for access to the figure properties using the Property Inspector, as shown in Figure B.5.

Some of this functionality embedded in the Property Inspector can be manipulated with simple code. For example, we can change the axis properties using a vector that defines the minimum and maximum values of each axis. An example of how to use such modification is:

```
axis([-10 1010 -1.2 1.2])
```

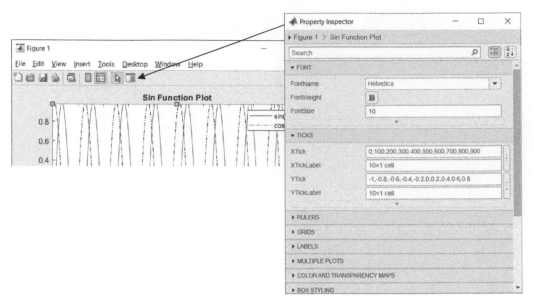

Figure B.5 Inspecting figure properties and modifying or editing labels, line characteristics etc. using the Property Inspector.

This will result in a view that is 10 units of the *x*-axis beyond what is depicted in Figure B.4 and 0.2 units beyond the *y*-axis. The axis command is applied automatically to the currently active figure window. If you need to specify which window, you can use the figure handle number – that is, `figure(1)`.

Another useful feature aligned with the `plot()` function is the ability to plot multiple graphs in one plot as an array or matrix. This is accomplished using the `subplot()` command. This command takes three integer values. The first two integers essentially define the dimension of a matrix, where each of the matrix entries represents a graph within the figure. The third integer identifies which element in the matrix is made active for plotting a graph. For example, to create a plot with two graphs stacked on top of each other we would create a 2×1 matrix and use the `subplot` function as follows:

```
figure
subplot(2,1,1)% creates a 2X1 matrix and activates the first one
plot(sin(t),'k')
title('sin(x)');grid
subplot(2,1,2)% activates the second window in the matrix
plot(cos(t),'k')
title('cos(x)');grid
```

The resulting plots are shown in Figure B.6.

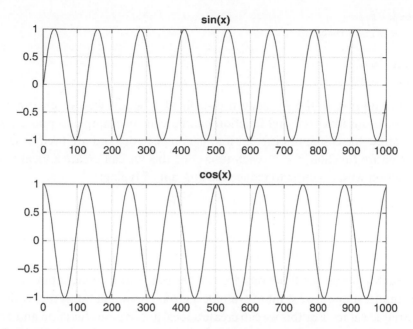

Figure B.6 Using the `subplot()` command to stack two graphs in one figure.

B.6   MATLAB Scripts and User-Defined Functions

Besides the pre-installed functions that come along with toolboxes, users can create their own functions. This is preferably done as a script file (i.e., a *.m file) but can also be executed in the command window. We will review the former implementation option in the following. The generic syntax for a function is such that the name function is utilized to indicate to MATLAB the type of code it is dealing with. The corresponding syntax for a function with *m* inputs and *n* outputs is:

```
function[output1, ..., outputn] = function_name(input1, ..., inputm)
```

Note, functions do not necessarily need to have input arguments or outputs. The input arguments are listed on the right-hand side of the function name in parentheses and represent local variables throughout the execution of the function itself. The outputs have to be assigned to the listed variable names within the function script. There are a few rules that functions need to follow: The name of a function cannot match an existing pre-installed MATLAB function name, and the naming convention follows the same rules as for variables in MATLAB.

Functions can be saved either as separate files (*.m scripts) or be included in the program that makes use of the function. For the latter case, usually the function scripts are placed at the end of the program. An example of a simple function detailing one input and two outputs is:

```
function [out1,out2] = examplefun(x)
% Example script for custom function with 1 input & 2 outputs
%_____
no = input('Please provide the number of measurements: ');
u = zeros(no,1);
```

```
for k = 1:no
    u(k,1) = randn(1)*x;
end
out1 = sum(u.^2)/no;
out2 = sqrt(out1);
```

In addition to the formally user-defined function we can also make use of anonymous functions and function handles. As an example, suppose we need to compute the cosine of different numbers repeatedly and are tired of writing out the entire function `cos()` each time. For this we can create a locally available function with a handle to reduce the amount of typing:

```
≫ f = @cos;
≫ f(pi)
  ans =
        -1
```

Such anonymous functions are not restricted to a single input variable. For example, suppose we want to evaluate a given polynomial structure with different variable values. For this we can create a local anonymous function and evaluate it in the workspace as follows:

```
≫ f = @(x,y)  5*x^3+2*y^2+7*x+9;
≫ f(4,5)
  ans =
        407
```

Note that the function handle identifies the two input arguments as variables it needs in order to compute the results of its definition.

B.7   File Management

As we deal with large data sets in machine learning and deep learning, file management and efficient access to such data are essential. The use of datastores and other means of handling large data sets is illustrated and explained in Chapter 8. In this section we will discuss some of the more basic tools for handling files using MATLAB.

Suppose your program created data that is contained in a number of variables currently loaded in the workspace. To save these variables, it is common to create a *.mat file and store the different variables in that file. For example, suppose we have three variables with assigned values such as:

```
name = 'rms_value';
voltage = 12;
A = [1 2 3;4 5 6;7 8 9];
```

To save these three variables in a new file named data.mat, we use the following command:

```
save data.mat name voltage A
```

The file data.mat will reside in the current active folder. To load this file, use the following command:

```
load data.mat
```

If the file you are seeking to load into the current workspace is located in a subfolder, you can add this subfolder to the current path that MATLAB keeps in mind when searching for files that have been requested by the user or a program. To do so, use the command `addpath()`. As the argument you will need to provide the entire string detailing the path to the directory. For example, if the data is located in a folder C5 that resides below the active folder Set1, you define the path as follows:

```
apath='F:\Investigation\Data\Set1\C5';
```

To add this path to MATLAB, use:

```
addpath(apath);
```

If this path needs to be removed from the current search path, use the command:

```
rmpath(apath);
```

Note, the above instructions do not lead MATLAB to change directories; rather, it will add a path to a folder in its memory when looking for files. In order to change to another folder using MATLAB script, use the `chdir` command:

```
chdir newFolderName;
```

If a new folder needs to be created prior to changing directories, use the `mkdir` command:

```
mkdir newFolderName;
```

The `mkdir` command creates the new folder named *newFolderName* under the default directory – that is, the current directory you are calling the script from.

B.8   Common Mistakes and Useful Tricks

In our brief review of MATLAB's basic functioning we have left out a few details that may become of importance when not considered. Hence, the following is a collection of common mistakes and useful tricks.

- When using loops, use indentations. Such form changes do not alter the functioning of the script and function of the program; rather, they allow for an enhanced reading experience when reviewing or learning the given code.
- Along with indentations, a useful – though often neglected – practice is to use commenting. Commenting is accomplished by the use of the percent sign (%). Any character appearing after the percent sign on the same line will be

disregarded by MATLAB's interpreter executing your code. Commenting allows for other users to easily become familiar with your program and programming style.

- Although often employed, loops should be avoided as they are mostly inefficient. Rather than loops, use vectorization as reviewed in Section B.4.
- Watch out for the different data types MATLAB uses and its implications for your computation. For example, using floating point variables in logic expression often can lead to misinterpretations. Instead, use integers.
- Although MATLAB keeps all digits of any number during the computations regardless of its formatting when displaying, these numbers are still limited to a finite accuracy. Hence, the more operations these numbers are involved with, the more error accumulation is possible due to the truncation of digits to a finite number.
- Avoid the use of predefined function names and variables. In many cases MATLAB will render a notice about such violations. However, there are exceptions to this as well: Consider the use of a variable i, which also serves as the MATLAB designation of a complex number. MATLAB will note your preference to use it as a variable. Problems arise once you follow such an assignment with the use of a complex number. For example, consider the following script:

```
i = 5;
m = 2+4i
m =
   2.0000 + 4.0000i
```

Note that MATLAB still uses i as the complex number generation and in your assignment for m, it did not recognize it as your variable i. However, if the second statement is changed to have i explicitly multiplied by 4 (i.e., m = 2+4*i), MATLAB recognizes that you are referring to the variable i and not the complex number.

- Wherever possible, pre-allocate variables using commands such as zeros() to speed up execution. Changing the size of an array or matrix during the execution of your script will cause MATLAB to redefine the space provided in the memory for this variable. A much more efficient implementation is to ensure that the size of the variable is found prior to any assignments. Use the size() command to find out about the current size of a variable.
- Generating a figure using MATLAB is reviewed in Section B.5. There are many other functions and tools to illustrate your results that are not detailed in this brief review, but are worth exploring. Once generated, such figures are easily copied to other text-processing programs such as MS Word using the pull-down menu Edit in the figure tool bar and selecting Copy Figure. Options to suppress or impose different backgrounds for the copied figure are available in the Edit menu.

APPENDIX C

Introduction to Simulink

C.1   Introduction

In this appendix we will provide a brief introduction to Simulink, its use, functioning, and some key features. We will review the basic structure of Simulink, introduce its main components, discuss how to create models, and briefly discuss simulation control and settings. The intention of this appendix is to provide a refresher on this graphical language. For readers who have not worked with Simulink before, this appendix will serve as a rather brief introduction. In addition to the material detailed here, there are a number of excellent resources online, including by MathWorks, with specific information and sample programs. MathWorks provides a freely accessible learning model for Simulink at: www.mathworks.com/learn/tutorials/simulink-onramp.html.

Simulink can be seen as a graphical version of MATLAB, where we use libraries of blocks and signal paths to construct a simulation. The programming is supported by a large set of toolboxes for specialized functions and applications. In addition, Simulink provides for ways to impose a simulation environment, offers tools to import and export data and simulation results, gives the ability to analyze systems, allows for documenting simulation programs, and has options for exporting simulation programs to microcontrollers. The latter feature will be explored in Appendix E. Some key features of Simulink are the ability to model nonlinear systems and entertain discrete time, continuous time, and hybrid models. Simulink is a graphical interface, where programs are constructed by connecting blocks that entail some kind of functionality, with lines that represent data transmission. In the following we will explore blocks and lines in more detail.

C.2   Basic Functioning of Simulink

There are two convenient ways to start Simulink: using the command window in MATLAB and typing `simulink` followed by the return button, or by utilizing the "Simulink" button under the Home tab of the menu bar in MATLAB. After calling Simulink, MATLAB will present you with a start window that serves for the selection of the file type. In this brief review we will start with a blank model, which is a blank canvas where we will draw our block diagram representing the

Figure C.1 Opening window for Simulink, serving for the selection of the type of work to be undertaken.

system we are attempting to model. Figure C.1 shows the opening window after calling Simulink from MATLAB, including a set of different options. Note, there is a wealth of other options for starting specific programs, such as working with a system from Git, a new project with folder, or any system detailed in the individual toolboxes that are sold as add-ons by MathWorks. You can also access previously developed and stored models. As many installations differ in terms of what toolboxes have been added, this list may look different in your window; however, you should have the Blank Model option available.

In the following we will introduce different elements and components that are utilized in the construction of these graphical simulation models. For this, a blank model or an existing model should be open in order to explore the block library Simulink provides.

When starting a new model we utilize the empty canvas of the blank model to drag blocks from the Library Browser and connect them with signal paths. The blank model provides the button to start the Library Browser. Both windows (i.e., the Blank Model and the Library Browser) are depicted in Figure C.2. The Library Browser provides access to a number of different tools and functions, and is dependent on your selection and installation of different toolboxes. However, it contains a base library containing most common functions. In the following we list a few key classes, noting that it only represents a partial list:

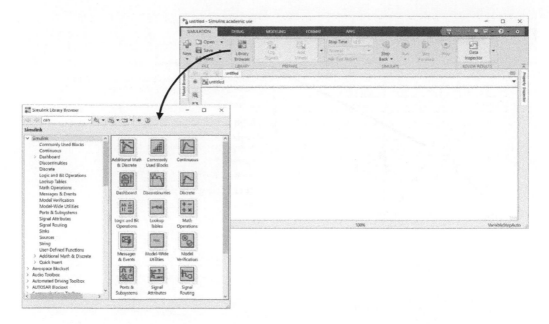

Figure C.2 Blank Model window and the button to open the Library Browser, as well as window of the open Library Browser.

- Sources: a repository of signal-generating function blocks;
- Sinks: a set of display and output containers;
- Math Operations: a collection of mathematical functions;
- Continuous: a set of continuous models and controllers;
- Discrete: a set of discrete time models and controllers;
- User-Defined Functions: a range of blocks whose functions can be modified.

Another useful category is the Commonly Used Blocks, which stores the blocks that have been frequently used by the user.

To program a specific functionality, browse for the block in the Library Browser, select the block of interest, and drag it onto the canvas. Blocks may have none, one, or more inputs as well as outputs. The block information and parameters are accessible by double-clicking the block once it is located on the program canvas. Figure C.3 depicts a Gain block and the Sine Wave block on the program canvas as well as the Block Parameter window that appears after double-clicking the Sine Wave block. Each block may have different parameters to tune and modify its functioning. For example, the Sine Wave block allows for changing the amplitude, the frequency, the bias, and the phase of the sine wave function. Many blocks, such as the Sine Wave block, also provide for brief explanations of the functioning and the parameters. MATLAB maintains excellent online documentation for each function.

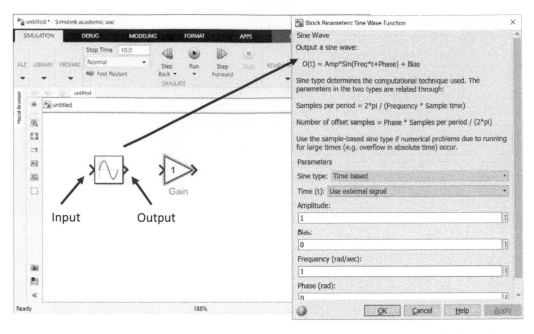

Figure C.3 Simulink program canvas with a `Sine Wave` block and a `Gain` block, indicating inputs and outputs, as well as the Block Property window.

Figure C.3 also identifies the input and output of the `Sine Wave` block. Those connections are defined by corner symbols. To connect two blocks, click and drag the output of one block to the input of the block you want to connect with. The functioning established by this connection is that the signal from the `Sine Wave` block is transferred to the `Gain` block, which has the property of multiplying the input by the selected gain – in our case the gain is 1.0. To change the gain, one can access the block parameters of the `Gain` block and change the corresponding value. The flow of information is given by the direction of the connection arrow. If a block needs to be turned 180 degrees to have its input at the right-hand side and its output on the left-hand side, right-click the block and select Rotate & Flip. A line that connects two blocks can also be branched off to feed other blocks. For that, use the input of the new block and drag a line to the existing line at the location of the branch point. Once your cursor is close, Simulink will suggest a branch point by depicting a small black circle. At that moment you can release the mouse button and you have created a signal branch. It is good practice to document programs, and since your canvas represents a simulation program in a graphical format you can add descriptions to signal paths as well as function blocks. To do so, double-click on the canvas at the location you want to add a label or text and select Create Annotation. When adding text, you will find that Simulink provides formatting options similar to any word processing application.

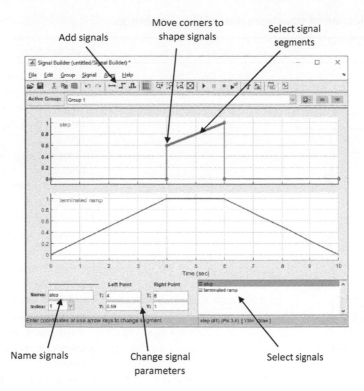

Figure C.4 Simulink's `Signal Builder` dialog box with functionality indicated.

C.3 Commonly Used Blocks

In the following we highlight a few important Simulink blocks that may be useful for some of the programming we encounter when addressing some of the control problems presented in this book. We categorize these blocks as signal-generating blocks, function-specific blocks, and output blocks.

Signal-Generating Blocks. Many of these blocks can be found under the title Source in the Simulink Browser menu. We have already used the `Sine Wave` generator. Very similar blocks such as the `Step` function, the `Chirp Signal` and the `Constant` are located here as well. However, in the following we shall explore the `Signal Builder` block in more detail. Figure C.4 depicts the `Signal Builder` block once dragged onto the canvas and opened by double-clicking the block. The `Signal Builder` block provides for a modular assembly of individual signals that are combined into a more complex signal. When opening the `Signal Builder` dialog box you will be presented by a group pane where individual signals are drawn using a time axis and a magnitude axis. Each signal is listed in the Signal List window and can be selected

Table C.1 Selection of signal-generating Simulink blocks

Block symbol	Block name	Block function
Group 1 step / terminated ramp /	`Signal Builder`	Allows for designing complex signals by superimposing simple signals, which are drawn in the Signal Builder dialog box.
	`Step`	Generates a step function with a final value and step time that can be defined in the Block properties window.
	`Sine Wave`	The `Sine Wave` block generates a continuous (or discrete if a sample time is selected) sine wave with choice of sine wave attributes.
	`Uniform Random Number`	This block generates a random number drawn from a uniform distribution. A seed value allows for making the sequence repeatable.
	`Pulse Generator`	Generates a pulse train with a user-specified amplitude, period, pulse width, and phase delay.

independently. To add signals, chose among three options in the top menu bar signal components that most closely fit your design. Alter the added signal by selecting components and toggle values in the parameter pan below the signal graph of the Signal Builder dialog box. You also can rename each signal using the parameter pan, as shown in Figure C.4. Some of the other signal-generating blocks are detailed in Table C.1.

MATLAB Function Block. To accommodate MATLAB script files and programs in a Simulink simulation, a MATLAB function block can be utilized. This block is found in the User-Defined Functions. Double-clicking on the block on the program canvas, we will find a MATLAB editor window for adding the functionality of the block using the MATLAB function format (see Appendix B.6 for details). Note that MATLAB has the general structure already embedded in the original MATLAB function block. It comes with a single input u and a single output y. However, this can be changed to suit the number of inputs and outputs the function requires. The connections on the graphical representation of the block in Simulink are automatically updated. The changes can be saved as a new file. To go back to the Simulink program canvas, use the Editor tab and select Go To Diagram.

Bus Creator Block. When dealing with many signals that assume the same path through your program, it is beneficial for the understanding and the presentation of the program to combine these signal lines into one line. For this, we use the `Bus Creator` block. It is a convenient function as this block allows bundling a group of signals that may be composed of different data types. The `Bus Creator` block is found in the Signal Routing menu section of the Simulink Library Browser.

Bus Selector Block. Having once bundled a set of different signals into a bus, you may have to access individual signals from the bus at certain stages of your Simulink program. For this, you can use the `Bus Selector` block. When accessing the `Bus Selector` block properties by double-clicking the `Bus Selector` block icon, we have the choice to select which elements of the bus is being selected.

Data Type Conversion Block. In some instances we may deal with different data types which are used to evaluate a functionality we are seeking to program. Some of the blocks in Simulink are data-type sensitive. A typical example of mismatch of data type for computation within a Simulink program is when we access external data through ports or microcontrollers. A convenient way to fix such compatibility issues is to use the `Data Type Conversion` block, which is found in Signal Attributes.

A select few signal routing and attribute blocks are summarized in Table C.2.

Table C.2 Selection of signal routing and attribute blocks

Block symbol	Block name	Block functioning
	Bus Creator	Creates a bus from input signals. The number of input signals can be modified using the Block Parameter window for the Bus Creator.
	Bus Selector	Allows for the extraction of individual signals from a bus signal.
convert	Data Type Conversion	Automatically changes the data type based on the required data type of the next block.
	Manual Switch	Connects individual signal routes using a switch that is activated using a mouse click onto the block symbol.

Table C.3 Selection of different data management blocks

Block symbol	Block name	Block functioning
untitled.mat	From File	Allows for access to data stored in *.mat files. Data are usually specified as time series or as a matrix.
untitled.xlsx Sheet1	From Spreadsheet	Reads data stored in an Excel spreadsheet. The first column of the spreadsheet is designated as the time data and following columns as signal data.
simin	From Workspace	Reads data from the MATLAB workspace.
out.simout	To Workspace	Allows sending data specified as time series, array, or structure to the workspace.
untitled.mat	To File	Writes in an incremental fashion data into a variable in a specified *.mat file.
	Display	Displays incoming data in the block screen.
	Scope	Generates a graph of incoming data. Changing Scope properties allows for multiple signal inputs, axis properties, etc. to be changed.

An important functionality is having access when programming using the Simulink environment to data from outside of the Simulink program environment, such as from the workspace or from files. The reverse flow of data is also often of interest: how to move data generated within Simulink to the outside, either to the workspace, a file, or to a connected peripheral. The latter functionality is covered in Appendix E, where we connect microcontrollers to the Simulink program environment. Table C.3 lists the most commonly used blocks for those types of data management functions along with a brief description.

C.4   Systems and Models

Simulink provides for a number of prebuilt system models for both continuous time and discrete time domains. This categorization is utilized to group these models in the Simulink Library Browser. We present a few models from each of these domains in Tables C.4 and C.5.

Table C.4 Selection of different continuous time system blocks

Block Symbol	Block Name	Block Functioning
$\dfrac{1}{s}$	Integrator	Computes the integral continuously of the input signal.
$\dfrac{1}{s+1}$	Transfer Function	Continuous time transfer function. Numerator and denominator are provided as coefficients of the s-domain polynomials.
$\dot{x} = Ax + Bu$ $y = Cx + Du$	State-Space Model	Continuous time state-space model. Matrices are defined in the Block Parameter window using MATLAB notation.
PID(s)	PID Controller	Continuous time PID controller with automatic tuning of controller parameters and noise filter for the differential portion.
	First Order Hold	Generates a continuous time, piece-wise linear approximation of the input signal.

Table C.5 Selection of different discrete time system blocks

Block symbol	Block name	Block functioning
$\dfrac{K\,Ts}{z-1}$	Discrete Time Integrator	Computes the accumulation of the input signal.
$\dfrac{1}{z+0.5}$	Discrete Time Transfer Function	Discrete time transfer function. Numerator and denominator are provided as coefficients of the z-polynomials.
$x_{n+1} = Ax_n + Bu_n$ $y_n = Cx_n + Du_n$	Discrete Time State-Space Model	Discrete time state-space model. Matrices are defined in the Block Parameter setting window using MATLAB notation.
PID(z)	Discrete Time PID Controller	Discrete time PID controller with automatic tuning of controller parameters and noise filter for differential portion.
	Zero Order Hold	Zero-order hold keeps the signal constant until the next data point of the signal is received.

When working with discrete time models it is sufficient to leave the sampling time specified in the block parameters as −1 (inherent) and have the first block of the overall model on the program canvas specify the sampling time. This specification of the sampling time will be copied to all other blocks in the model when each of them have *inherent* selected.

C.5   Simulation Parameters and Example

To run a Simulink program, Simulink employs a default simulation parameter set. Sometimes those parameters do not match the simulation goals and need to be altered. If the default parameters suffice, the simulation is executed by pressing the "Run" button in the Simulation tab of the Simulink program. Some of the default parameters are given as well in this menu bar, as shown in Figure C.5.

Along with the "Run" button you will find an option to change the default simulation duration, which is 10 seconds. However, for more access to the simulation parameters you will need to open the Modeling tab and select Model Settings. This window entails a menu list on the left side, and on the main part of the window the specific configuration settings based on the menu category. MATLAB includes a feature that allows you to inspect each parameter's meaning by right-clicking the name of the parameter and select the What's This? dialog.

The first category in the configuration menu list is the Solver category, which also entails the simulation time settings and the type of solver to be used. The solver is responsible for computing the model's state at each time instance. There are two types of solvers available to choose from: the fixed-step solver and the variable-step solver. The variable-step solver is the default solver and allows Simulink to adapt the step size during the computation of the next time instance of the model states. This is particularly beneficial when dealing with a model that entails discontinuities. The fixed-step-size solver does not adjust its time steps during the simulation and is required with the use of certain functions such as S-functions.

In Appendix F we use the configuration parameter window to allow for programming of microcontrollers. However, for a brief tutorial on Simulink we demonstrate some of the different components and settings using a

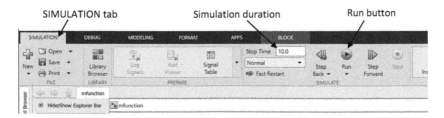

Figure C.5 Default simulation parameter settings in the Simulation tab.

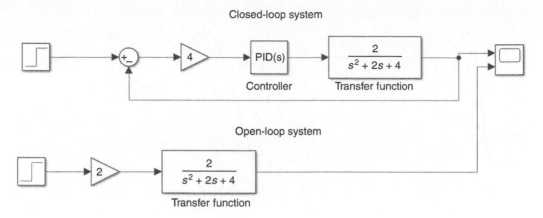

Figure C.6 Example Simulink program containing a closed-loop control system and an open-loop system.

simple example. Figure C.6 details two systems, one operating in a closed loop and the other in an open loop, employing the same transfer function and input command. To construct the depicted Simulink program we choose the components from the Library Browser and modify its parameters as shown. Note, the summing point in the closed-loop diagram entails an addition and a subtraction. The default summing point from the Math library entails two additions. In order to change the operation, open the parameter settings of this block and change the second operant from + to −. If more than two signals need to be operated on, add the operands to the list in the parameter setting window.

For the output we added an input channel to the scope block. This is done by opening the scope block and selecting under the File menu the option Number of Input Ports. When running the simulations with the default configuration parameters we notice that one of the responses does not show a smooth curve (the solid-line curve in Figure C.7(a)). This is not because the system is somehow piece-wise discontinuous; rather, the simulation parameters cause the solver to interact and compute at rather large time steps. To remedy this, go into the simulation parameter settings and change the simulation time to start at 1 second and end at 4. The new response is shown in Figure C.7(b). With this change we achieve smooth response curves, which are expected when simulating continuous functions.

In the above simulation program we employed a continuous time PID controller block. Simulink incorporated a tuning function with the PID controller block, which will lead to optimal PID parameter selection as well as the inclusion of a low-pass filter to support the D portion of the controller. To see the settings and perform the autotuning, open the PID controller block, which brings up a Block Parameter window that includes all the details of the settings.

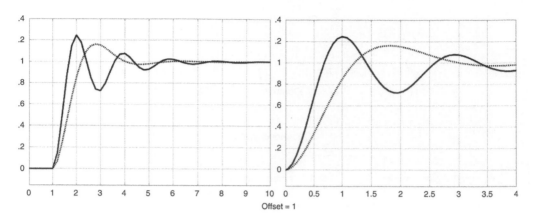

Figure C.7 Simulation results: (a) response of closed and open-loop system with default parameters; (b) response with modified simulation parameters.

C.6   Useful Features

Simulink offers a great number of options, functions, and programming aids. This brief tutorial is not designed to provide depth in exploring and explaining these features. However, a few simple useful items are briefly detailed in the following.

When constructing a simulation program using Simulink we assemble the functionality in a modular fashion. For complex systems we may end up with areas on the program canvas that contain a large number of such blocks that may represent a larger subsystem of the overall system. Along with the connection lines for data transmission, such large simulation models can easily become difficult to read. A remedy for this graphical complexity is the creation of hierarchical blocks that contain the assembly of all the functions and connections. Such a block – subsystem – is generated by selecting all the blocks and connections you want to place in the subsystem, right-clicking the selected group and selecting Create Subsystem from Selection. Once created, the individual components are still accessible by double-clicking the new block. In addition, such blocks or any other blocks can be modified graphically by accessing the Format tool in the Block Parameter settings.

When there are a number of blocks on the canvas, arranging them in some geometrical fashion is supported by utilizing any of the tools listed under the Format tab in newer versions and using Arrange, which is found in the same fashion as when creating a subsystem. Arrange aligns functions to a set of components and blocks, distributes them, or has their sizes match.

In Figure C.6 we utilized a Gain block with a hardcoded value of 4. However, we could have done this with a parameter whose value we obtain from MATLAB during the simulation run. This is not just for the Gain block, but for most blocks in Simulink. In order to have Simulink read the value from MATLAB, the parameter has to have the same name and an assigned value. For example, the Gain element could have recorded instead of the number 4 a parameter named K, which also exists in MATLAB's workspace and has an assigned value.

Finally, a convenient tool that MATLAB and Simulink provide is the automatic report generator. Newer Simulink and MATLAB installations utilize the Apps tab to initiate automatic report generation. The settings in this app allow for selection of the output format, such as Word, PDF, and HTML, among others.

APPENDIX D
Selected MATLAB Code

D.1   Min-Max Composition – Section 2.6

```
% MAX-MIN Composition
    clear
    R=[0.7 0.6;0.8 0.3];
    S=[0.8 0.5 0.4;0.1 0.6 0.6];
    for i=1:2
        for j=1:3
            T(i,j)=max(min(R(i,:),S(:,j)'));
        end
    end
    % Display results
    disp('The composition of R and S is:')
    disp(T)
```

D.2   Temperature Controller – Section 3.3

```
% Fuzzy control for room temperature
% Version 1.0 – Chapter 3 – Fuzzy Logic Control
% FuzzyRoomTemperature.m
%_____
clear;
% Create a FIS system
a=newfis('fuzz_temp');
% Adding membership functions for variable error e
a=addvar(a,'input','e',[-3,3]); % Error e
a=addmf(a,'input',1,'NB','trimf',[-3,-3,-1]);
a=addmf(a,'input',1,'NS','trimf',[-3,-1,1]);
a=addmf(a,'input',1,'Z','trimf',[-2,0,2]);
a=addmf(a,'input',1,'PS','trimf',[-1,1,3]);
a=addmf(a,'input',1,'PB','trimf',[1,3,3]);
% Adding membership functions for variable input r
a=addvar(a,'output','r',[-4,4]); % Input r
a=addmf(a,'output',1,'NB','trimf',[-4,-4,-1]);
a=addmf(a,'output',1,'NS','trimf',[-4,-2,1]);
a=addmf(a,'output',1,'Z','trimf',[-2,0,2]);
a=addmf(a,'output',1,'PS','trimf',[-1,2,4]);
a=addmf(a,'output',1,'PB','trimf',[1,4,4]);
% Fuzzy rules
```

```
rulelist=[1 1 1 1;2 2 1 1;3 3 1 1;4 4 1 1;5 5 1 1];
a=addrule(a,rulelist);
a1=setfis(a,'DefuzzMethod','mom'); % Defuzzying method specification
writefis(a1,'temproom'); %Save to fuzzy file "temproom.fis"
a2=readfis('temproom');
% Show fuzzy logic inference system in fuzzy logic MATLAB figures
figure(1);plotfis(a2);figure(2);plotmf(a,'input',1);
figure(3);plotmf(a,'output',1);
flag=1;
if flag==1
    disp('Fuzzy Rules Implemented');
    showrule(a,[1 2 3 4 5],'verbose','english')
    ruleview('temproom'); %Dynamic Simulation
end
for i=1:1:7
  e(i)=i-4;
  control_r(i)=evalfis([e(i)],a2);
end
control_r=round(control_r)
e=-3; % Example Error
r=evalfis([e],a2) % find control input
```

D.3   Gradient Descent Implementation – Section 4.4

```
clear
% plot contour of cost function
[x,y] = meshgrid(-3.5 : 0.5 : 3.5);
f=4*x.^2-4*x.*y+2*y.^2; % objective function
N = 20;        % number of contour lines
figure;
contour(x,y,f,N);
title('Gradient descent method with w_o=(2,3)')
xlabel('x');ylabel('y');
hold;
w = [2 3]'; % initial point
grad = zeros(2,1); % initialize gradient vector
grad(1) = 8*w(1)-4*w(2); % w(1) = x, w(2) = y
grad(2) = 4*w(2)-4*w(1);
etha = 0.1; % learning rate

for i=1:30
  plot(w(1), w(2),'k.')
  w = w – etha * grad
  grad(1) = 8*w(1)-4*w(2);
  grad(2) = 4*w(2)-4*w(1);
end
w
plot(w(1), w(2),'ro')
```

D.4   Newton's Method – Section 4.5

```
clear
% plot contour of cost function
[x,y] = meshgrid(-3.5 : 0.5 : 3.5);
f=4*x.^2-4*x.*y+2*y.^2; % objective function
N = 20;          % number of contour lines
figure;
contour(x,y,f,N);
title('Newton s Method with w_o=(2,3)')
xlabel('x');ylabel('y');
hold;

w = [2 3]'; % initial point
etha = 0.1; % learning rate

%Initialize
w = [2 3]';
G = [8*w(1)-4*w(2);4*w(2)-4*w(1)];
H = [8-4;-4 4]; % Hessian matrix

%Find the stationary point
for i=1:30
    plot(w(1), w(2),'k.')
    old = w;
    w = w - inv(H)*G*etha;
    G = [8*w(1)-4*w(2);4*w(2)-4*w(1)];
end
w
plot(w(1), w(2),'ro')
```

D.5   Conjugate Gradient Method – Section 4.6

```
function [w] = conjgrad(A,b,w)
% Solves Aw=b by conjugate gradient method with precision of 1e-10
% Using Fletcher Reeves gradient correction factor
% Supply A, b, and an initial guess w
dw=b-A*w;
d=dw;
rsold=dw'*dw;
for i=1:size(A(:,1))
    Ap=A*d;
    alpha=rsold/(d'*Ap);
    w=w+alpha*d;
    dw=dw-alpha*Ap;
    rsnew=dw'*dw;
    if sqrt(rsnew)<1e-10
        break;
    end
    d=dw+rsnew/rsold*d;
    rsold=rsnew;
end
end
```

D.6   Quasi-Newton Method – Section 4.7

```
function [min,i] = QuasiNewton(f,start,e,opt)
% f ... function to be optimized
% e ... stopping condition tolerance
% opt ... option: opt = 1 DFP and opt = 2 BFGS
% start ... initial guess
start=start';wk=start;
g=gradient(f);
B=eye(length(start)); % initial B as identity matrix
gk=double(g(wk(1),wk(2)));
i=1;
if opt == 1 % DFP
    while norm(gk,2)>e
        if gk==0
            break;
        else
            dk=-(B*gk);
        end
        wkn=wk+dk;pk=double(g(wkn(1),wkn(2)));
        qk=(pk-gk); kappa=qk'*B*qk;
        uk=dk/(dk'*qk)-(B*qk)/kappa;
        B=B+((dk-B*qk)*(dk-B*qk)'/(qk'*(dk-B*qk)))+0*kappa*uk*uk';
        wk=wkn;gk=pk;i=i+1;
        plot(wk(1),wk(2),'k.')
    end
    hold on
    plot3(wk(1),wk(2),double(f(wk(1),wk(2))),'ro');
    axis([-2 2-1 2])
    text(-1,0.5,'DFP Method')
    min=wk;
    print('Number of iterations using DFP method: ');i
else % opt ==2 - BFGS
    while norm(gk,2)>e
        if gk==0
            break;
        else
            dk=-(B*gk);
        end
        wkn=wk+dk;pk=double(g(wkn(1),wkn(2)));
        qk=(pk - gk); kappa=qk'*B*qk;
        uk=dk/(dk'*qk)-(B*qk)/kappa;
        B=B+((dk-B*qk)*(dk-B*qk)'/(qk'*(dk-B*qk)))+1*kappa*uk*uk';
        wk=wkn; gk=pk;i=i+1;
        plot(wk(1),wk(2),'k.')
    end
    hold on
    plot3(wk(1),wk(2),double(f(wk(1),wk(2))),'ro');
    axis([-2 2-1 2])
    text(-1,0.5,'BFGS Method')
    min=wk;
    print('Number of iterations using BFGS method: ');i
end
end
```

D.7   Levenberg–Marquardt Method – Section 4.8

```
% Solving Rosenbrock function
clear;clc
x=linspace(-2,2);y=linspace(-1,2);
[X,Y]=meshgrid(x,y);
fun=100*(Y-X.^2).^2+(1-X).^2;        % objective function
N = 60;                              % number of contour lines
contour(X,Y,fun,N);
title('Rosenbrock function using Levenberg-Marquardt')
xlabel('x');ylabel('y');
hold;
input('How many iterations: ');nd=ans;
count=1;
w(:,1)=[0 0]';
lambda=1e+5;                         % damping factor
Id=eye(2);epsilon=1e-5;
for i = 1:nd
    f_prev = 100*(w(2,i)-w(1,i).^2).^2+(1-w(1,i)).^2;
    j1=2*w(1,i)-400*w(1,i)*(-w(1,i).^2+w(2,i))-2;
    j2=-200*w(1,i).^2+200*w(2,i);
    J=[j1 j2];
    B=inv(J'*J+lambda*Id);
    w(:,i+1)=w(:,i)-B*(2.*J'.*f_prev);
    fnew = 100*(w(2,i+1)-w(1,i+1).^2).^2+(1-w(1,i+1)).^2;
    if fnew < f_prev*1.5
        lambda = lambda/1.1;
    else
        lambda = 1.1*lambda;
    end
    if abs(fnew-f_prev)<epsilon
        break;
    end
    count=count+1;
end
w(:,count)
plot(w(1,1:count),w(2,1:count),'k');
count
```

D.8   Tabu Search MATLAB Files – Section 5.3

The following MATLAB code corresponds to Section 5.3, detailing tabu search optimization.

```
% simpleTS.m
% Simple tabu search algorithm
clear;
% Define input parameters
input('Nominal 0, or specific 1: ');spec=ans;
if spec==0
    nx=2;;maxx=10;minx=0;tll=5;pll=5;r=2.5/100*(maxx-minx);r1=r/2;r3=r1/
2;n1=25;x=[rand*(maxx-minx)+minx,rand*(maxx-minx)+minx];nd=100;
ftheor=-18.5547;Sopt=[9.0389 8.6674];
    funchoice=12;
```

```
else
      input('Dimension of solution nx: ');nx=ans;
      input('Length of Tabu List tll: ');tll=ans;
      input('Length of Promising List pll: ');pll=ans;
      input('Tabu Ball Radius r: ');r=ans;
      input('Radius of Neighborhood r1: ');r1=ans;
      input('Radius of Promising Balls r3: ');r3=ans;
      input('Search space, maximum: ');maxx=ans;
      input('Search space, minimum: ');minx=ans;
      input('Number of Neighborhoods n1: ');n1=ans;
      input('Random point in Search space: ');x=ans;
      input('How many iterations to be carried out nd: ');nd=ans;

      input('Cost function: 1.Spher, 2.Quad, 3.Ackl, 4.Boha I, 5.Colv,
6.Eas, 7.Griew, 8.Hypel, 9.Rast, 10.Rosenb, 11.Schwl, 12.M: ');funchoice=ans;

      if funchoice==1 % Spherical parameters
          nx=2;maxx=100;minx=-100;tll=5;pll=5;r=2.5/100*(maxx-minx);r1=r/
2;r3=r1/2;n1=25;
          nd=100;ns=40;nd=100;ftheor=0;Sopt=[0 0];x=[rand*(maxx-minx)+minx,
rand*(maxx-minx)+minx];
      elseif funchoice==2 % Quadratic parameters
          nx=2;maxx=100;minx=-100;tll=5;pll=5;r=2.5/100*(maxx-minx);r1=r/
2;r3=r1/2;n1=25;
           nd=100;ftheor=0;Sopt=[0 0];x=[rand*(maxx-minx)+minx,rand*(maxx-
minx)+minx];
      elseif funchoice==3 % Ackley parameters
          nx=2;maxx=30;minx=-30;tll=5;pll=5;r=2.5/100*(maxx-minx);r1=r/2;
r3=r1/2;n1=25;
           nd=100;ftheor=0;Sopt=[0 0];x=[rand*(maxx-minx)+minx,rand*(maxx-
minx)+minx];
      elseif funchoice==4 % Bohachevsky I parameters
          nx=2;maxx=50;minx=-50;tll=5;pll=5;r=2.5/100*(maxx-minx);r1=r/2;
r3=r1/2;n1=25;
           nd=100;ftheor=0;Sopt=[0 0];x=[rand*(maxx-minx)+minx,rand*(maxx-
minx)+minx];
      elseif funchoice==5 % Colville parameters
nx=4;tll=5;pll=5;r=0.25;r1=0.125;r3=0.06;n1=25;x=[5,5,5,5];nd=100;
maxx=10;minx=-10;ftheor=0;Sopt=[0 0 0 0];
      elseif funchoice==6 % Easom parameters
          nx=2;maxx=100;minx=-100;tll=5;pll=5;r=2.5/100*(maxx-minx);r1=r/
2;r3=r1/2;n1=25;
           nd=100;ftheor=-1;Sopt=[3.1416 3.1416];x=[rand*(maxx-minx)+minx,
rand*(maxx-minx)+minx];
      elseif funchoice==7 % Griewank parameters
          nx=2;maxx=600;minx=-600;tll=5;pll=5;r=2.5/100*(maxx-minx);r1=r/
2;r3=r1/2;n1=25;
           nd=100;ftheor=0;Sopt=[0 0];x=[rand*(maxx-minx)+minx,rand*(maxx-
minx)+minx];
      elseif funchoice==8 % Hyperellipsoid parameters
                  nx=2;maxx=1;minx=-1;tll=5;pll=5;r=2.5/100*(maxx-minx);
r1=r/2;r3=r1/2;n1=25;
          nd=100;ftheor=0;Sopt=[0 0];x=[rand*(maxx-minx)+minx,rand*(maxx-
```

```
minx)+minx];
      elseif funchoice==9 % Rastigin parameters
          nx=2;maxx=5.120;minx=-5.120;tll=5;pll=5;r=2.5/100*(maxx-minx);
r1=r/2;r3=r1/2;n1=25;
          nd=100;ftheor=0;Sopt=[0 0];x=[rand*(maxx-minx)+minx,rand*(maxx-
minx)+minx];
      elseif funchoice==10 % Rosenbrock parameters
          nx=2;maxx=2.048;minx=-2.048;tll=5;pll=5;r=2.5/100*(maxx-minx);
r1=r/2;r3=r1/2;n1=25;
          nd=100;ftheor=0;Sopt=[1 1];x=[rand*(maxx-minx)+minx,rand*(maxx-
minx)+minx];
      elseif funchoice==11 % Schwefel parameters
          nx=2;maxx=500;minx=-500;tll=5;pll=5;r=2.5/100*(maxx-minx);r1=r/
2;r3=r1/2;n1=25;
          nd=100;ftheor=0;Sopt=[0 0];x=[rand*(maxx-minx)+minx,rand*(maxx-
minx)+minx];
      else % Haupt
           nx=2;maxx=10;minx=0;tll=5;pll=5;r=2.5/100*(maxx-minx);r1=r/2;
r3=r1/2;n1=25;
          nd=100;ftheor=-18.5547;Sopt=[9.0389 8.6674];x=[rand*(maxx-minx)
+minx,rand*(maxx-minx)+minx];
          end;
end;
TL=zeros(nx,tll); %TL(parameter 1, parameter 2, ..., tabu liste element 1, ...)

% main loop
xt(1,:)=x;clear
x;x=xt;globalbestcost=cost(x(1,:),funchoice);%globalbestcost evaluation
equals starting point cost
counter=1;globalbestx(counter,:)=xt(1,:);
for k=1:nd

    %create neighbhors
    xneighbor=creatneighbors(r1,n1,nx,x(k,:));

    % Select element from neighborhood
    for d=1:n1
        xtest(d,:)=xneighbor(d,:);%pick one candidate element from
        each neighborhood
        costtest(d)=cost(xtest(d,:),funchoice); % and evaluate its
        cost
    end;
    [minimumcost,bestelement]=min(costtest);%find minimum cost and
    element
    candidatex=xtest(bestelement,:);

    % test if it is in tabu balls
    flag=1;
    flag=CheckifinBall(candidatex,TL,r,tll,nx);
    if flag==0 % New position is in tabu ball
        if minimumcost>globalbestcost
            %'tabu'
            candidatex=xtest(n1,:);%use largest move
            minimumcost=cost(candidatex,funchoice);
        else
        end;
```

```
    else
    end;

      % use aspiration criteria
    if minimumcost<globalbestcost
      candidatex=xtest(bestelement,:);
      minimumcost=cost(candidatex,funchoice);
    else
              candidatex=x(k,:);
              minimumcost=cost(candidatex,funchoice);
              %change neigborhood
    end;

    % test if search limit reached
    for ik=1:nx
        if candidatex(1,ik)>maxx
            candidatex(1,ik)=maxx;%no wrap around
            minimumcost=cost(candidatex,funchoice);
        elseif candidatex(1,ik)<minx
            candidatex(1,ik)=minx; %no wrap around
            minimumcost=cost(candidatex,funchoice);
        end;
    end;
    x(k+1,:)=candidatex; mincost(k)=cost(x(k+1,:),funchoice);
    if mincost(k)<globalbestcost
        globalbestcost=mincost(k);
        counter=counter+1;
        globalbestx(counter,:)=x(k+1,:);
    else
    end;

    % update tabu list
    % move each element one down
    for jj=tll:-1:2
        TL(:,jj)=TL(:,jj-1);
    end;
    TL(:,1)=x(k+1,:);
end;
if nx<3
    figure(1)
    plot(x(:,1),x(:,2));axis([minx maxx minx maxx]);grid;title('Position of
agent during course of simulation');
    xlabel('nx(1)');ylabel('nx(2)');
    figure(2)
    plot(mincost);grid;title('Minimum cost');xlabel('Iteration index');
ylabel('Cost');
else
end
ftheor,Sopt
globalbestcost
bestx=globalbestx(counter,:)
```

MATLAB Code for Cost Function used in TS Optimization

```
function [y] = cost(x,funchoice)
% cost.m
% Computation of the cost function
[n,nx]=size(x);y=0;

if funchoice==12 %Haupt
    y=0;
    y = x(:,1)*sin(4*x(:,1))+1.1*x(:,2)*sin(2*x(:,2)); % fitness evaluation
(objective function)
elseif funchoice==1 %Spherical
    y=0;
    for i=1:nx
        y=y+((x(:,i))^2);
    end;
elseif funchoice==2 %Quadric
    y=0;
    for i=1:nx
        yy=0;
        for kj=1:i
            yy=yy+x(:,i);
        end;
        y=y+yy^2;
    end;
elseif funchoice==3 %Ackley
    y=0; %*
elseif funchoice==4 %Bohachevsky
    y=x(1)^2+2*x(:,2)^2-0.3*cos(3*pi*x(:,1))-0.4*cos(4*pi*x(:,2))+0.7;
elseif funchoice==5 %Colville
    y=100*(x(2)-x(:,1)^2)^2+(1-x(:,1))^2+90*(x(:,4)-x(:,3)^2)^2+(1-x
(:,3))^2+10.1*((x(:,2)-1)^2+(x(:,4)-1)^2)+19.8*(x(:,2)-1)*(x(:,4)-1);
elseif funchoice==6 %Easom
    y=-cos(x(:,1))*cos(x(:,2))*exp(-1*(x(:,1)-pi)^2-(x(:,2)-pi)^2);
elseif funchoice==7 %Griewank
    y=0; %*
elseif funchoice==8 %Hyperellipsoid
    y=0;
    for i=1:nx
        y=y+(i^2)*(x(:,i))^2;
    end;
elseif funchoice==9 %Rastrigin
    y=0;
    for i=1:nx
        y=y+((x(:,i))^2)-10*cos(2*pi*x(:,i))+10;
    end;
elseif funchoice==10 %Rosenbrock
    y=0;
    %y=y+100*(x(2)-x(1)^2+(1-x(1)^2));
    for i=1:nx/2
        y=y+100*((x(:,2)-(x(:,1)^2))^2+(1-x(:,1))^2);
    %y=y+100*(x(i*2)-(x(i)^2))^2+(1-x(i*2-1)^2);%
```

```
    end
elseif funchoice==11 %Schwefel
    y=0;
    for i=1:nx
        y=y+x(:,i)*sin((abs(x(:,i)))^(0.5))+418.9829*nx;
    end;
else
end;
```

MATLAB Code for Creating Neighbors used in TS Optimization

```
function xneighbor=creatneighbors(r1,n1,nx,x)
% creatneighbors.m
% Computation of neighbors by creating nx-dimensional spheres
% around current position x
    for d=1:n1 %for each neighborhood
        for j=1:nx %for each dimension
            ri(d,j)=(d-1)*r1; %inner diameter of neighborhood ball
            ro(d,j)=d*r1; % outer diameter of neighborhood ball
            dif=ro(d,j)-ri(d,j);difr=dif*rand;
            rneighbor(d,j)=(ri(d,j)+difr)*sign(randn);
            theta(d,j)=rand(1)*2*pi;
            xcomp(d,j)=rneighbor(d,j)*cos(theta(d,j))+x(:,j);
        end;
        xneighbor(d,:)=xcomp(d,:);
    end;
```

MATLAB Code for Testing Location against Neighborhood used in TS Optimization

```
function flag = CheckifinBall(Sx,TL,r,tll,nx);
flag=0;NTS=zeros(tll);
for i=1:tll
    for t=1:nx
        NTS(i)=NTS(i)+((TL(t,i)-Sx(1,nx))^2);
    end;
    NTS(i)=(NTS(i))^0.5;
    if NTS(i)>=r %tabu if NTS < r
        flag=1; %not tabu
    else
    end;
end;
```

D.9   Particle Swarm Optimization MATLAB Files – Section 5.4

The following MATLAB code corresponds to Section 5.4, detailing particle swarm optimization.

```
%   simplePSO.m
%
% Basic PSO algorithm using global best methodology
%
```

```
clear;
% Initialization of a ns x nx – dimensional swarm S
c1=.1;c2=.1;
input('Nominal 0, or specific 1: ');spec=ans;
funchoice=12;
if spec==0
     nx=2;maxx=10;minx=0;ns=40;nd=100;
else
     input('Which test function : ');funchoice=ans;
     input('Dimension of particle nx: ');nx=ans;
     input('Search space, maximum: ');maxx=ans;
     input('Search space, minimum: ');minx=ans;
     input('Number of particles ns: ');ns=ans;
     input('How many iterations to be carried out nd: ');nd=ans;
end
S.x=(maxx-minx)*rand(ns,nx)+minx; % uniform distribution of initial
particles
V=ones(ns,nx,nd);

% Initial personal and global best computations
for i=1:ns
     y(i,:)=S.x(i,:); % Personal best
     S.cost(i)=cost(S.x(i,:),funchoice); % Fitness evaluation
     costp(i,1)=S.cost(i);
end;
[gbestk,element]=min(S.cost); % Minimization problem
yhat=S.x(element,:); % Global best
costg(1)=S.cost(element);

% Main loop
   for k=1:nd
       for i=1:ns
           S.cost(i)=cost(S.x(i,:,k),funchoice);
           if S.cost(i)<costp(i,k)%y(i,:)
               y(i,:)=S.x(i,:,k);
               costp(i,k+1)=S.cost(i);
           else
               costp(i,k+1)=costp(i,k);
       end;
       if S.cost(i)<costg(k)%yhat
           yhat=S.x(i,:,k);
           costg(k+1)=S.cost(i);
       else
           costg(k+1)=costg(k);
       end;
     end;
     for i=1:ns
```

```
      % Update velocity vector
      V(i,:,k+1)=V(i,:,k)+c1*rand(1,nx).*(y(i,:)-
S.x(i,:,k))+c2*rand(1,nx).*(yhat-S.x(i,:,k));
      % Update position vector
      S.x(i,:,k+1)=S.x(i,:,k)+V(i,:,k+1);
      % Check border of search area
      if S.x(i,1,k+1)>maxx
          S.x(i,1,k+1)=maxx;
      elseif S.x(i,1,k+1)<minx
          S.x(i,1,k+1)=minx;
      else
      end
      if S.x(i,2,k+1)>maxx
          S.x(i,2,k+1)=maxx;
      elseif S.x(i,2,k+1)<minx
          S.x(i,2,k+1)=minx;
      else
      end
    end;
end;
for i=1:ns
    S.cost(i)=cost(S.x(i,:,k+1),funchoice);
    if S.cost(i)<costp(i,k+1)% y(i,:)
        y(i,:)=S.x(i,:,k);
    else
    end;
    if S.cost(i)<costg(k)% yhat
        yhat=S.x(i,:,k);
    else
    end;
end;
```

MATLAB Code for Cost Function used in Particle Swarm Optimization

```
function [y] = cost(x,funchoice)
% cost.m
% Computation of the cost function
```

```
[n,nx]=size(x);y=0;

if funchoice==12 % Haupt and Haupt
    y=0;
    y = x(:,1)*sin(4*x(:,1))+1.1*x(:,2)*sin(2*x(:,2)); % fitness evaluation
(objective function)
elseif funchoice==1 % Spherical
    y=0;
    for i=1:nx
        y=y+((x(:,i))^2);
    end;
elseif funchoice==2 %Quadric
    y=0;
    for i=1:nx
        yy=0;
        for kj=1:i
            yy=yy+x(:,i);
```

```
            end;
            y=y+yy^2;
        end;
    elseif funchoice==3 % Ackley
        y=0; %
    elseif funchoice==4 % Bohachevsky
        y=x(1)^2+2*x(:,2)^2-0.3*cos(3*pi*x(:,1))-0.4*cos(4*pi*x(:,2))+0.7;
    elseif funchoice==5 % Colville
        y=100*(x(2)-x(:,1)^2)^2+(1-x(:,1))^2+90*(x(:,4)-x(:,3)^2)^2+(1-x
        (:,3))^2+10.1*((x(:,2)-1)^2+(x(:,4)-1)^2)+19.8*(x(:,2)-1)*(x(:,4)-1);
    elseif funchoice==6 % Easom
        y=-cos(x(:,1))*cos(x(:,2))*exp(-1*(x(:,1)-pi)^2-(x(:,2)-pi)^2);
    elseif funchoice==7 % Griewank
        y=0; %*
    elseif funchoice==8 % Hyperellipsoid
        y=0;
        for i=1:nx
            y=y+(i^2)*(x(:,i))^2;
    end;
    elseif funchoice==9 % Rastrigin
        y=0;
        for i=1:nx
            y=y+((x(:,i))^2)-10*cos(2*pi*x(:,i))+10;
    end;
    elseif funchoice==10 % Rosenbrock
        y=0;
        for i=1:nx/2
            y=y+100*((x(:,2)-(x(:,1)^2))^2+(1-x(:,1))^2);
    end
    elseif funchoice==11 % Schwefel
        y=0;
        for i=1:nx
            y=y+x(:,i)*sin((abs(x(:,i)))^(0.5))+418.9829*nx;
        end;
    else
    end;
```

D.10 Genetic Algorithm MATLAB Files – Section 5.7

The following MATLAB code corresponds to Section 5.7, detailing genetic algorithm optimization. The code is based on the generic code provided by Haupt and Haupt [1].

```
% simpleGA.m
% Toolbox for genetic algorithm. Code for continuous
% genetic algorithm.

% Define variables:
maxiterations=input('Maximum number of iterations: ');
ipopsize=input('Population size of generation 0: ');
popsize=input('Population size for generations 1 - end: ');
%popsize=popsize*ipopsize;
keep=input('Number of chromosomes kept for mating: ');
```

```
%keep=keep*popsize;
pars=input('Total number of parameters in a chromosome: ');
mutaterate=input('Mutation rate: ');
hi=input('High end of parameter value: '); %for all parameters
lo=input('Low end of parameter value: '); %for all parameters
op=1; % pairing option;

% Create the initial population, evaluate costs, and sort
CHROMOSOMES=(hi-lo)*(rand(ipopsize,pars));
% CHROMOSOMES will be a matrix of random numbers within hi - lo

% Loop:
gen=0;quit=0;
h = waitbar(0,'Please wait ... ');
while (gen<maxiterations & (~quit))
    gen=gen+1;
    cost=costfunction(CHROMOSOMES);
    New=[cost,CHROMOSOMES];
    New2=sortrows(New,[1]);cost=New2(:,1);CHROMOSOMES=New2(:,2:3);
    mincost(gen)=min(cost);
    meancost(gen)=mean(cost);
    stdcost(gen)=std(cost);
     % Pairing,Mating, and Mutation
     [Mom,Dad]=pairing(CHROMOSOMES,cost,keep,popsize,op);
     CHROMOSOMES=matecon(Mom,Dad,CHROMOSOMES,keep,popsize,pars);
     CHROMOSOMES=mutatecon(CHROMOSOMES,mutaterate,popsize,pars,hi,lo);
     % Check for conversions
     % if mincost(gen)< ... and/or meancost(gen) < ... and or stdcost(gen)< ... quit=1
         waitbar(gen/maxiterations);
    end;
    close(h);
    plotcostfuncsurf;
    TopChrom=CHROMOSOMES(1,:)
    min(cost)
    figure;plot(mincost)
    title('Minimum cost')
    xlabel('Iteration number k')
    ylabel('Cost magnitude')
    grid
```

MATLAB Code for Pairing Function used in Genetic Algorithm

```
function [Mom,Dad]=pairing(CHROMOSOMES,cost,keep,popsize,op)
% Weighted random paring for the selection of the parents
% of the next generation. Other options can be included
% using the "op" variable

replacements=(popsize-keep)/2;denum=0;probn=zeros(replacements);
for r=1:replacements
    denum=denum+r;
end;
for n=1:replacements
    probn(n)=n/denum;
end;
```

```
cum=0;odds=zeros(1,replacements);
for i=1:replacements
    cum=probn(i)+cum;
    odds(1,i)=cum;
end;
pick1=rand(1,replacements);
pick2=rand(1,replacements);

Mom=zeros(1,replacements);Dad=Mom;
for i=1:replacements
   for j=2:replacements
      if (pick1(i)<odds(j) & pick1(i)>odds(j-1))
        Mom(i)=j;
      end;
   if Mom(i)==0
      Mom(i)=1;
   end;
   if (pick2(i)<odds(j) & pick2(i)>odds(j-1))
      Dad(i)=j;
   end;
   if Dad(i)==0
      Dad(i)=1;
   end;
 end;
end;
```

MATLAB Code for Mating Function used in Genetic Algorithm

```
function
CHROMOSOMES=matecon(Mom,Dad,CHROMOSOMES,keep,popsize,pars)
% Row index contains first offspring, row index +1 contains second offspring
% Mom-vector containing row numbers of first parent
% Dad-vector containing row numbers of second parent

replace=(popsize-keep)/2;
for ic=1:replace
     alpha=ceil(rand*pars);i=2*(ic-1)+1;
     beta=rand(1);
     CHROMOSOMES(keep+i,alpha)=CHROMOSOMES(Mom(ic),alpha)-beta*
(CHROMOSOMES(Mom(ic),alpha)-CHROMOSOMES(Dad(ic),alpha));

CHROMOSOMES(keep+i+1,alpha)=CHROMOSOMES(Dad(ic),alpha)+beta*(CHROMOSOMES
(Mom(ic),alpha)-CHROMOSOMES(Dad(ic),alpha));

end;
```

MATLAB Code for Mutation Function used in Genetic Algorithm

```
function
CHROMOSOMES=mutatecon(CHROMOSOMES,mutaterate,popsize,pars,hi,lo)
nmu=ceil(popsize*pars*mutaterate);
for i=1:nmu
```

```matlab
    row=ceil(popsize*rand)+1;
    col=ceil(pars*rand);
    CHROMOSOMES(row,col)=(hi-lo)*rand+lo;
end;
```

MATLAB Code for Cost Function used in Genetic Algorithm

```matlab
function cost=costfunction(CHROMOSOMES)
% Using Haupt and Haupt as objective function
[row,col]=size(CHROMOSOMES);
cost=zeros(row,1);
for i=1:row
    x=CHROMOSOMES(i,1);
    y=CHROMOSOMES(i,2);
    cost(i,1)=x*sin(4*x)+1.1*y*sin(2*y);
end;
```

D.11   Single-Layer NN – Section 6.2

```matlab
% SingleLayerMain.m
% Load training data X, desired output data D, and train single
% layer neural network using simple learning mechanism
%_____
clear;clc
[X,D,N]=example1data();
disp("Training data and desired output: ")
disp([X D])
input("Provide adaptation rate alpha: ");alpha=ans;
% Creating random weights for each connection:
W=2*rand(1,2)-1;
% Train NN:
for epoch = 1:50000
    [W,e(epoch)]=Update(W,X,D,N,alpha);
end
% Compute output of trained NN
for k = 1:N
    x=X(k,:)';
    v=W*x;
    y(k)=sigmoidActivation(v);
end
disp('Results: trained y and actual d: ')
[y' D]
figure
plot(e);title('Error vs. epochs');
grid;ylabel('Magnitude');xlabel('Epoch no.');

function [X,D,N] = example1data
% example1data.m
% Training data for example 1
%_____
X=[0 0;0 1;1 0;1 1]; % Input
```

```
D=[0 0 1 1]'; % Desired output
N = 4; % Number of training data sets

function [X,D,N] = example2data
% example1data.m
% Training data for example 2
%_____
X=[0 0;0 1;1 0;1 1]; % Input
D=[0 1 1 0]'; % Desired output
N = 4; % Number of training data sets

function [W,e] = Update(W, X, D, N, alpha)
% Function to compute the update on the weight of the NN element
%_____
[nw,~]=size(W);
for k = 1:N
    x=X(k,:)';d=D(k);
    v=W*x;
    y=Sigmoid(v); % sigmoid activation function
    e=d-y; % error evaluation
    grad=y*(1-y)*e; % use gradient information for update
    dW=alpha*grad*x; % change in weight
    for i=1:nw
        W(i)=W(i)+dW(i);
    end
end

function y = sigmoidActivation(x)
% Function to output of activation function
%_____

y=1/(1+exp(-x));
end
```

D.12 Multi-Layered NN – Section 6.3

```
% MainMultiLayer.m
% Load training data X, desired output data D, and train two
% layer neural network using backpropagation
%_____
clear;clc
[X,D,n]=example1data();
disp("Training data and desired output: ")
disp([X D])
input("Provide adaptation rate alpha: ");alpha=ans;
input("Provide number of epochs: ");epochs=ans;
% Creating random weights for each connection:
W1=2*rand(4,2)-1;
W2=2*rand(1,4)-1;
% Train NN:
for p = 1: epochs
    [W1 W2 err(p)] = Backpropagation(W1,W2,X,D,alpha,n);
end
```

```
for i = 1: n
    x=X(i,:)'; v1=W1*x;
    y1=Sigmoid(v1);
    v=W2*y1;
    y(i)=Sigmoid(v);
end
display('Desired output D and NN ouput: ');
[D y']
figure
plot(err)
title('Error vs. epochs');
grid;ylabel('Magnitude');xlabel('Epoch no.');

function [W1,W2,e] = Backpropagation(W1,W2,X,D,alpha,n)
% Training multi-layer NN using backpropagation
%_____
for i = 1:n
    x=X(i,:)';d=D(i);
    v1=W1*x;y1=Sigmoid(v1);
    v=W2*y1;y=Sigmoid(v);
    e=d-y;
  delt2 = y.*(1-y).*e;
  e1=W2'*delt2;
  delt1=y1.*(1-y1).*e1;
  dW1 = alpha*delt1*x';
  W1 = W1+dW1;
  dW2 = alpha*delt2*y1';
  W2 = W2+dW2;
end

function y = Sigmoid(v)
% Activation function for MLNN
%  _____
y=1./(1+exp(-v));
```

D.13 MATLAB Script for Example 8.17 – Section 8.6

```
% LSTMexample.m
% Forecasting wave heights from Pierson-Moskowitz wave
% spectrum data
%  _____
clear;
%% Generation of wave data
windvelocity = input('Enter wind velocity in [m/s]: ');
sigHeight = 0.209*windvelocity^2/9.81;
meanAngFreq = 11.1834/windvelocity;

% Frequency selection
freqmin = 0.025;
freqmax = 2.2;
```

```matlab
numbWaveComp = 50;
deltFreq = (freqmax-freqmin)/(numbWaveComp-1);
deltMeanAngFreq = 2*pi*deltFreq;

% Frequency
freq = freqmin:deltFreq:freqmax;
angFreq = 2*pi*freq;

% Time vector
kinit = 0;kend = 200;incr = 0.1;
tvector = kinit:incr:kend;

% Selection of random frequency components
angFreq = angFreq+deltMeanAngFreq*(rand(1,length(angFreq))-0.5);

% Compute wavespectrum
alpha = 0.11*(sigHeight^2)*meanAngFreq^4;
beta = 0.44*(meanAngFreq^4);
specMeanAngFreq = alpha./(angFreq.^5).*exp(-beta./(angFreq.^4));

% Generate random wave phases
p_shif = 2*pi*rand(1,length(freq));

data=zeros(1,kend*incr+1);
for i=1:length(tvector)
    data(i) = sum(sqrt(2*specMeanAngFreq*deltMeanAngFreq).*sin
(angFreq*tvector(i)+p_shif));
end;

plot(data)
figure; plot(data)
xlabel('Time index k')
ylabel('Wave height')
title('Irregular waves in ocean');

%% Partitioning of wave data
numWaveComp = floor(0.95*numel(data));
datTrain = data(1:numWaveComp+1);
datTest = data(numWaveComp+1:end);

%% Remove mean and standardize wavedata
mn = mean(datTrain);
sigma = std(datTrain);
dataTrainStandardized = (datTrain-mn)/sigma;

%% Define outputs
XvalTrain = dataTrainStandardized(1:end-1);
YvalTrain = dataTrainStandardized(2:end);

%% Define LSTM network architecture
n_Features = 1;
n_Responses =1;
n_HidUnits = 250;
```

```matlab
layers = [...
    sequenceInputLayer(n_Features)
    lstmLayer(n_HidUnits)
    fullyConnectedLayer(n_Responses)
    regressionLayer];
options = trainingOptions('adam', ...
    'MaxEpochs',150, ...
    'InitialLearnRate',0.0075, ...
    'LearnRateDropPeriod',125, ...
    'LearnRateDropFactor',0.1, ...
    'Verbose',0, ...
    'Plots','training-progress');
lstm_net = trainNetwork(XvalTrain,YvalTrain,layers,options);

%% Use LSTM to predict wave height data
datTestStand = (datTest-mn)/sigma;
X_Test = datTestStand(1:end-1);
lstm_net = predictAndUpdateState(lstm_net,XvalTrain);
[lstm_net,Y_Pred] = predictAndUpdateState(lstm_net,YvalTrain(end));
num_k = numel(X_Test);
for k = 2: num_k
    [lstm_net,Y_Pred(:,k)] =
predictAndUpdateState(lstm_net,Y_Pred(:,
k-1),'ExecutionEnvironment','cpu');
end
Y_Pred = sigma*Y_Pred +mn;
Y_Test = datTest(2:end);
rmse = sqrt(mean((Y_Pred-Y_Test).^2))
figure
plot(datTrain(1:end-1),'k')
hold
i = numWaveComp:(numWaveComp+num_k);
plot(i,[data(numWaveComp) Y_Pred],'k.-')
xlabel('Time index k')
ylabel('Wave height')
title('Forecasted irregular waves in ocean')
legend('Observed','Forecast')
```

D.14   Example 9.16 – Section 9.7

```matlab
classdef mountainPyCar < rl.env.MATLABEnvironment
  properties
    open_env = py.gym.make('MountainCar-v0');
  end
  methods
    % Constructor function: define action and observation spec.
    function this = mountainPyCar()
        % Initialize observation settings:
        ObsInfo = rlNumericSpec([2 1]);
        ObsInfo.Name = 'M_Car Disc';
        ObsInfo.Description = 'Pos, Vel';
```

```
        % Initialize Action Settings:
        ActInfo = rlFiniteSetSpec([0 1 2]);
        ActInfo.Name = 'accl dir';

        % Implement func. of environment int RL env.
        this = this@rl.env.MATLABEnvironment(ObsInfo,ActInfo);
    end

        % Reset environment
        function InitialObservation = reset(this)
            Result = this.open_env.reset();
            InitialObservation = double(Result)';
    end

    % Step function: take action, evaluate, check if done
    function [Observation,Reward,IsDone,LoggedSignals] - step(this,Action)
       Result = cell(this.open_env.step(int16(Action)));
       Observation = double(Result{1})';
       Reward = double(Result{2});
       IsDone = double(Result{3});
       LoggedSignals = [];
       if Observation(1) >= 0.495
         Reward = 0;IsDone =1;% Assign reward, goal = 0.5
             end
           end
         end
       end
```

D.15   Example 9.17 – Section 9.7

```
% MSDrl.m
% Mass-spring-damper environment
%_____
% Define observation and action
obsInfo = rlNumericSpec([3 1],'LowerLimit',[-inf -inf -4]','UpperLimit',[inf
inf 4]');
obsInfo.Name = 'observations';
obsInfo.Description = 'integrated error, error, and max displacement';
numObservations = obsInfo.Dimension(1);

actInfo = rlNumericSpec([1 1]);
actInfo.Name = 'Force';
numActions = actInfo.Dimension(1);

% Define Environment interface object
env = rlSimulinkEnv('MassSpringDamperRL','MassSpringDamperRL/RL Agent',
obsInfo,actInfo);

% Reset
env.ResetFcn = @(in)localResetFcn(in);

% Defining simulation times
Ts = 1.0; Tf = 100;
```

```
% Random Number Generator seed
rng(0)

% Create DDPG Agent
statePath = [

featureInputLayer(numObservations,'Normalization','none','Name','State')
    fullyConnectedLayer(50,'Name','CriticStateFC1')
    reluLayer('Name','CriticRelu1')
    fullyConnectedLayer(25,'Name','CriticStateFC2')];
actionPath = [

featureInputLayer(numActions,'Normalization','none','Name','Action')
fullyConnectedLayer(25,'Name','CriticActionFC1')];
commonPath = [
    additionLayer(2,'Name','add')
    reluLayer('Name','CriticCommonRelu')
    fullyConnectedLayer(1,'Name','CriticOutput')];

criticNetwork = layerGraph();
criticNetwork = addLayers(criticNetwork,statePath);
criticNetwork = addLayers(criticNetwork,actionPath);
criticNetwork = addLayers(criticNetwork,commonPath);
criticNetwork = connectLayers(criticNetwork,'CriticStateFC2','add/in1');
criticNetwork = connectLayers(criticNetwork,'CriticActionFC1','add/in2');

% Define options for critic
criticOpts = rlRepresentationOptions('LearnRate',1e-
03,'GradientThreshold',1);
critic = rlQValueRepresentation(criticNetwork,obsInfo,
actInfo,'Observation',{'State'},'Action',{'Action'},criticOpts);

% Create Actor
actorNetwork = [

featureInputLayer(numObservations,'Normalization','none','Name','State')
    fullyConnectedLayer(3, 'Name','actorFC')
    tanhLayer('Name','actorTanh')
    fullyConnectedLayer(numActions,'Name','Action')
    ];

actorOptions =
rlRepresentationOptions('LearnRate',1e-04,'GradientThreshold',1);
actor = rlDeterministicActorRepresentation(actorNetwork,obsInfo,
actInfo,'Observation',{'State'},'Action',{'Action'},actorOptions);

agentOpts =
rlDDPGAgentOptions('SampleTime',Ts,'TargetSmoothFactor',1e-
3,'DiscountFactor',1.0,'MiniBatchSize',64,'ExperienceBufferLength',1e6);
agentOpts.NoiseOptions.Variance = 0.3;
agentOpts.NoiseOptions.VarianceDecayRate = 1e-5;

agent = rlDDPGAgent(actor,critic,agentOpts);
```

```
% Define training options
maxepisodes = 5000;
maxsteps = ceil(Tf/Ts);
trainOpts =
rlTrainingOptions('MaxEpisodes',maxepisodes,'MaxSteps
PerEpisode',maxsteps,'ScoreAveragingWindowLength',20,'Verbose',
false,'Plots','training-progress','StopTrainingCriteria','AverageReward
','StopTrainingValue',800);

% Training
trainingStats = train(agent,env,trainOpts);
```

References

[1] Haupt RL, Haupt SE (1998). *Practical Genetic Algorithms*. Wiley, Chichester.

APPENDIX E

Introduction to Arduino Microcontrollers

E.1 Arduino Board Overview

Arduino boards have become very popular in the maker community due to their simplicity to program and ease of interfacing sensors and actuators. These devices can be programmed with open-source software (www.arduino.cc/en/Main/Soft ware). The software entails an integrated development environment (IDE) and sample programs. However, the projects in the problem sections of each chapter of this book utilize Simulink for programming these devices, so this brief introduction to Arduino microcontrollers will not elaborate on the use of the IDE to write programs for Arduino boards.

There are a multitude of Arduino boards available, both as originals and as compatible clones, as the board itself is open source. Table E.1 highlights a few Arduino boards and some of their characteristics that may be useful for the projects discussed in this book.

In Table E.1 the operating voltage refers to the voltage the board needs in order to be powered. Digital I/O refers to the number of digital input and output ports the device possesses, while the number of analog inputs details how many separate channels and pins are available to connect peripherals such as sensors that communicate with a continuous voltage signal. Most Arduino boards can accommodate a 0–10 V input signal for the analog channels. Each device that possesses analog input pins also has an associated A/D converter. The capability of the A/D is board-dependent and manifests itself at different resolutions. However, Arduino boards do not have analog output pins. For this, a pulse width modulation (PWM) is used to mimic analog signals. Those output signals have limited current supply and when driving motors or similar power-dependent peripheral devices, usually a current source such as an H-bridge is needed. The interface column in Table E.1 shows how to connect the board to a PC. The simplest version uses a USB connection that goes over the ATMEGA16U2 chip on the Arduino board.

Each board may have a different pin-out arrangement. However, Simulink provides for tables that detail each pin and their characteristics. To access such a table, go to the block parameters of an Arduino block, select *View Pin Map*, and you will be presented with a table that details all pin connections of the boards currently supported by your installation (see Figure E.1).

Table E.1 Common Arduino boards and their key parameters

Board name	Operating voltage (V)	Digital I/O	Analog inputs	PWM	Interface
Uno R3	5	14	6	6	USB
Leonardo	5	20	12	7	USB
LilyPad	3.3	14	6	6	USB
Mega 2560 R3	5	54	16	14	USB
Mega Pro	3.3	54	16	4	FTDI
Due	3.3	54	12	12	USB
Ethernet	5	14	6	6	FTDI

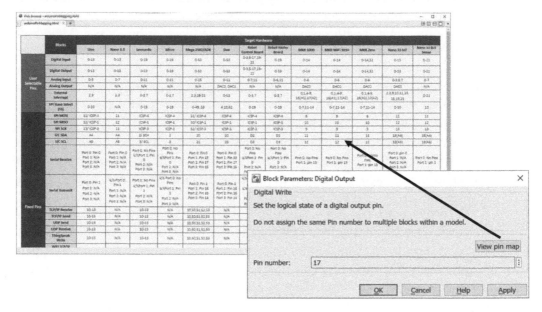

Figure E.1 Access to board information and characteristics including pin numbering for different Arduino boards using the Block Parameter window.

E.2   Arduino Board Basics

In the following we will briefly introduce the ports and connections of an Arduino Mega 2560 R3 board. Similar connectivity arrangements exist for the other boards listed, but are omitted here for brevity. Figure E.2 shows a top view of an Arduino Mega 2560 R3 board.

Power In and On Board Power. The operating voltage of the Arduino Mega 2560 R3 is 5 V. It is fed with an input voltage of 7–12 V, either through the USB connection or the power port (barrel jack). The DC current for each input and output pin is 20 mA, and the DC current for the 3.3 V pin is 50 mA. In addition to

Figure E.2 Arduino Mega 2560 R3 layout and general identification of components.

these pins, there are four ground pins, two located in the power section, one placed at the corner of analog in and digital I/O, and one at the end of the PWM section.

Memory. The board possess a 256 KB flash memory, 8 KB in SRAM and 4 KB in EEPROM.

Digital I/O. Pins 0–54 are configured as I/O pins, where pins 0–21 have dual configuration properties.

PWM. Pins 2–13 can be assigned to function as PWM output connections.

Analog Input. Pins A0–A15 serve as analog input connections. Each connection has access to an A/D converter with a 10-bit resolution.

Communication. Pins 0, 1, 14–19 can also serve as RX and TX communication ports. Essentially, the RX/TX 1–3 are located here. The Mega board also uses RX/TX 0 for communicating over the USB connection.

LED. There are several built-in LEDs on the Mega 2560 board; some are associated with the USB communication, plus there is a dedicated LED that is hardwired to Pin 13.

Reset. The board has a reset button and a reset pin located next to the 3.3 V pin in the power section.

AREF. This pin, located next to the ground pin in the PWM section, can be used to supply an analog reference voltage for the analog inputs. The range for the reference voltage must be 0–5 V.

E.3   Other Microcontroller Boards

MATLAB and Simulink support devices other than Arduino microcontrollers. Some of these support packages are similarly equipped, as detailed in Appendix F, while others may only have support through MATLAB or provide embedded coder support. There are also many other devices such as Android smart phones or Kinect cameras that are supported through these add-on apps. In this section, we present a few alternative microcontroller boards to the Arduino board, having similar access to functions in Simulink.

Raspberry Pi hardware support. There are support packages for both Simulink and MATLAB to program Raspberry Pi microcontrollers. Raspberry Pi hardware may be used for video and audio projects using Simulink and MATLAB.

Lego Mindstorms EV3 hardware. If working with the Lego Mindstorms EV3 hardware, you will find a Simulink support package available to support most functions of the Lego kit, including communication interfaces, interacting with the Lego sensors and actuators associated with the Mindstorms EV3 hardware.

BeagleBone Black hardware. Another popular microcontroller with similar capabilities to the Raspberry Pi is the BeagleBone Black device. You can interact with such a device using the MATLAB support package. As of the writing of this book, there is also a third-party device driver available for supporting Simulink programming of this device.

Programming Microcontrollers with Simulink and Support Package

F.1   Installation of Support Packages

MATLAB and Simulink support an array of third-party hardware items, including Arduino and Raspberry Pi microcontroller boards. To interact with and program these boards using Simulink you will need to install the appropriate support package. These support packages are free of charge; however, you need to have a MATLAB account in order to log in and to install them on your system.

To install a support package, navigate to the Home tab of MATLAB and select the "Add-Ons" button, as shown in Figure F.1. Select *Get Hardware Support Packages*, which will present a window with available programs that your MATLAB and Simulink version supports.

For Arduino boards you will find the Simulink Support Package for Arduino Hardware, as shown in Figure F.2. The installation process will inform you of any additional steps and software installations required (Figure F.3). Follow the instructions for the setup and configuration until the setup process is complete.

Once the support package is installed you will find the added toolbox for the hardware in the Simulink Library Browser, as shown in Figure F.4. The different blocks and functionalities depend on your MATLAB/Simulink version. Generally, the more recent the version of MATLAB, the more functionality these support packages contain.

F.2   Run Simulink Program on Arduino Hardware

To create a Simulink program that can interact with the Arduino hardware, you will need to open the corresponding template file, which you can find under the Simulink Start Page category of Arduino Support package, as shown in Figure F.5. Note, for older Simulink versions you will be able to use a regular blank Simulink model. However, you will have to manually adjust the configuration parameters. In this brief tutorial we restrict ourselves to the 2021 version of Simulink. The file has example code in the form of blocks that represent a source, as well as a digital output block to the Arduino board. You can change or delete any of these blocks to create

Figure F.1 Install a support package using the installer from MATLAB.

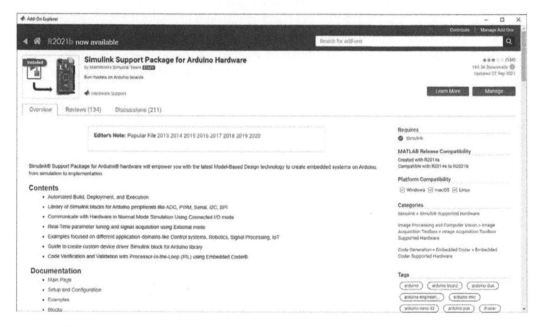

Figure F.2 Simulink support package for Arduino hardware information page during installation initiation.

your own program. To upload your Simulink program to the Arduino board you may want to change the *Stop Time* from 10 to *inf*, which means that your program will run on the Arduino board for as long as it is powered. Sometimes your computer fails to connect to the Arduino board and you will receive an error message. This communication error can be a result of your computer not finding the correct COM port, as the settings in your Simulink program are such that your computer will

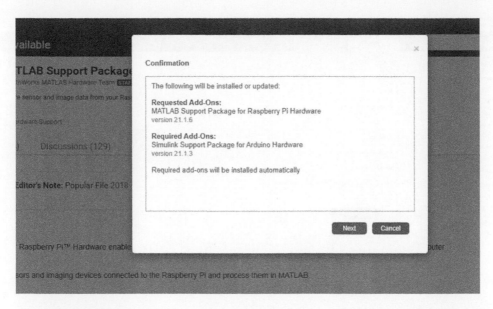

Figure F.3 Update notices during the installation process with information on the product and requirements.

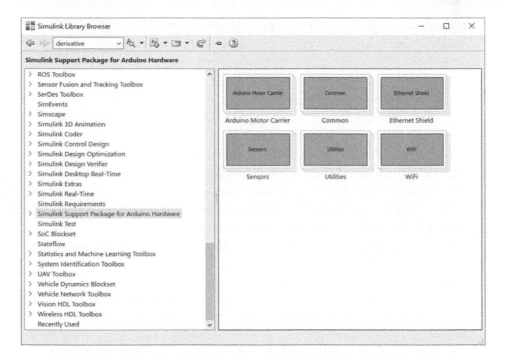

Figure F.4 Simulink Library Browser with Simulink support Package for Arduino hardware installed.

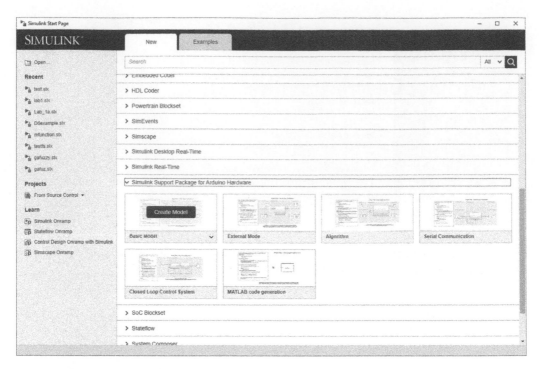

Figure F.5 New Simulink model for Arduino program: use the predefined Basic Model that includes all the hardware settings needed to interact with the Arduino board over a USB connection.

automatically detect the correct COM port. An alternative and sometimes helpful way to remedy such an error is to change the settings for the COM port to *Manual*. To do so, click on the "Hardware Settings" button of your Simulink program and select the Hardware Implementation tab on the Configuration Parameters window. This will provide you with a window that includes a category titled Target Hardware Resources. Expand this category and select under Host-Board Connection the Manually option for Set Host COM Port. You will need to manually identify the correct COM port your Arduino board is connected to.

When developing a Simulink program that is intended to run on an external board, your blocks in the program will need to function in discrete mode, as the implementation on the external board is usually only possible as a discrete time model. The block parameters of discrete time blocks in Simulink require the specification of a sample time. Rather than defining for each block a sample time, use one sample time of the first block in your Simulink program and set all other blocks to -1, which implies that they conform to the sample time defined by the first block.

To upload the Simulink program onto the board over USB, use the "Deploy" button and select the Build, Deploy & Start option. Simulink will compile and build your Simulink program and update you on the process in the bottom part of the Simulink program window.

APPENDIX G
Project Descriptions and Parts List

In this appendix we introduce a possible physical setup to support the learning objectives detailed in a number of chapters of this book. Although the following description is for a specific setup, there exists a large number of similar implementations that would serve the same purpose. Some setups are turnkey setups available for sale on a number of websites. The setup introduced here has been used in intelligent controls courses as well as modern controls and adaptive controls courses. The system detailed in this appendix possesses simple dynamics, can be constructed with a few relatively inexpensive components, and is portable. It also allows for modification and investigations to support class and laboratory instruction.

Figure G.1 depicts a simple pendulum that is actuated by a small propellor. The propellor is mounted at the end of the boom, while the boom is suspended and allowed to rotate by a potentiometer. The potentiometer not only serves as the mechanical component responsible for allowing rotation, but also as a sensor recording the angle of the boom. Different configurations can be achieved by adding a counter mass at the other end of the boom to define different equilibrium points and different inertias, or a second propellor. Depending on the inertia, a motor driver is needed to provide for adequate actuation.

For constructing the simple pendulum setup, the required electromechanical components are listed in Table G.1.

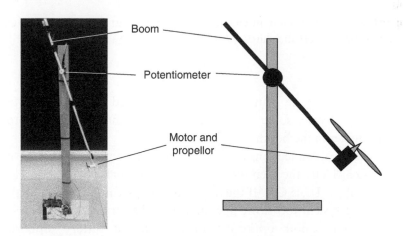

Figure G.1 Simple pendulum as physical system setup.

Table G.1 Electromechanical components for construction of a simple pendulum system

Name	Function	Specification
Microcontroller	Algorithm implementation and signal processing	Arduino Mega 2560 R3 or Arduino Uno R3
Motor and propellor	Actuator	DC 3.7 V 40,000 RPM 100 mA 7 × 16 mm motor with helicopter propellor
Potentiometer	Sensor/bearing	Rotational sensor V1
Motor driver	Power amplification	L298 N motor drive controller board DC dual H-bridge
Battery	Power source	18650 Li-ion
Breadboard	Circuitry assembly	Any

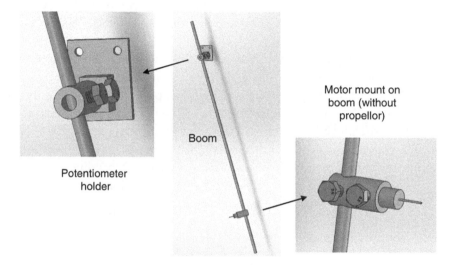

Figure G.2 CAD drawings of mechanical components to attach motor and potentiometer to the boom and support structure.

Connectors can be printed to mount the listed motor and potentiometer. Figure G.2 depicts some CAD drawings for connecting the motor and the potentiometer to the boom.

The electrical connections required to sense and actuate the system are listed in Figure G.3. Note, the identification of the terminals corresponds to the listed components in Table G.1. If the inertia of the system is too high or the response time of the system too slow, operation with battery support is recommended. Figure G.3 lists both options: with and without battery support, using an Arduino Mega 2560 and a L298 N motor driver.

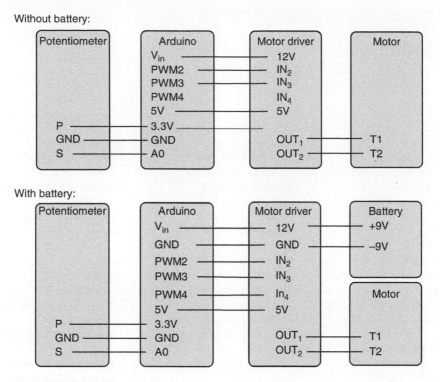

Figure G.3 Generic electrical connection between components of the physical system.

In the above wiring schematic, there are three PWMs used to control the motor motion. Note that this is not necessary; however, if you want to control the direction of the spinning movement of the motor, pins 2 and 3 are assigned opposing binary constants.

To utilize the sensor and actuator, some mapping has to be done where input and output are related based on the conversion count of the different Arduino Simulink blocks. For example, the potentiometer reading using the configuration depicted in Figure G.3 and utilizing the analog input channel A0 of the Arduino Mega 2560 controller will provide a number between 0 and 1023 (for other Arduino controllers this may differ). This number needs to be calibrated to the appropriate angle. Similarly, a correlation between the PWM4 input and the resulting motor voltage can be established.

Index